GARY K. GRIFFITH

The Quality Technician's Handbook

Fourth Edition

Prentice Hall
Upper Saddle River, New Jersey   Columbus, Ohio

Library of Congress Cataloging-in-Publication Data

Griffith, Gary K., 1950–

    The quality technician's handbook / Gary K. Griffith. --4th ed.

      p.   cm.

    ISBN 0–13–674250–5

    1. Engineering inspection.   2. Machine–shop practice.   I. Title.

    TS156.2.G75   2000   99–32127

    658.5′62--dc21   CIP

Editor: Stephen Helba
Production Editor: Alexandrina Benedicto Wolf
Production Supervision: Carlisle Publishers Services
Cover Design Coordinator: Karrie Converse-Jones
Cover Art and Design: Thomas Mack
Production Manager: Matthew Ottenweller
Marketing Manager: Ben Leonard

This book was set in Times Roman by Carlisle Communications, Ltd. and was printed and bound by R. R. Donnelley & Sons Company. The cover was printed by Phoenix Color Corp.

© 2000, 1996, 1992, 1986 by Prentice-Hall, Inc.
Pearson Education
Upper Saddle River, New Jersey 07458

Printed in the United States of America

10 9 8 7 6 5 4 3 2 1

ISBN: 0-13-674250-5

Prentice-Hall International (UK) Limited, *London*
Prentice-Hall of Australia Pty. Limited, *Sydney*
Prentice-Hall of Canada, Inc., *Toronto*
Prentice-Hall Hispanoamericana, S. A., *Mexico*
Prentice-Hall of India Private Limited, *New Delhi*
Prentice-Hall of Japan, Inc., *Tokyo*
Prentice-Hall (Singapore) Pte. Ltd., *Singapore*
Editora Prentice-Hall do Brasil, Ltda., *Rio de Janeiro*

Preface

This book acquaints readers with a variety of the basic skills that contribute to the quality of outgoing products, and covers various quality assurance, quality control, and inspection topics at the technician level. Operators, machinists, inspectors, quality technicians, and associate quality engineers will find the contents useful as a learning text and a quick reference.

This book can also be used by other manufacturing, quality, and purchasing personnel as a reference for quality-related subjects. Chapter discussions are arranged so that similar topics are covered with continuity, and the text is written at a level that is easy to understand. Numerous illustrations also help the reader to comprehend the material. In many instances, step-by-step instructions lead the reader through a particular method or technique.

This book has been very effective as a training aid and reference text for internal quality technician training and for courses designed to help the reader prepare for the American Society for Quality's Certified Quality Technician and Certified Mechanical Inspector examinations. This book can be kept in the machinist's or inspector's toolbox as a quick reference for some problems that come up in the shop. Several changes have been made in this fourth edition:

- All geometric tolerancing chapters and artwork have been updated to the latest revision of the standard (ASME Y14.5M–1994) from the previous version (ANSI Y14.5M–1982).
- Comparisons of the old and new geometric tolerancing standards are included in tables.
- Many figures have been improved for clarity, corrections, and up-to-date symbols.
- Various figures and tables (previously shown in the Appendix) have been integrated into the appropriate chapter where the subject is covered.
- Sampling plans have been renamed, such as (1) Mil-Std-105E changed to ANSI/ASQC Z1.4 and (2) Mil-Std-414 changed to ANSI/ASQC Z1.9.

For simplicity, many of the solutions to the mathematical problems in this book have been rounded off to one fewer place than the decimal places of the raw data using simple rounding techniques (e.g., 5 or higher, round up). In practice, one should not round intermediate calculations. Only the final answer should be rounded off to significant decimal places.

Product quality is everyone's job, and everyone should be aware of the methods for achieving quality. I sincerely hope that this book will continue to contribute toward that goal.

ACKNOWLEDGMENTS

I thank the following reviewers for their insightful recommendations: Edwin G. Landauer, Clackamas Community College; John Maxfeldt, Lakeshore Technical College; and J. Kent Snyder, Rose State College.

I also thank all of my friends, students, and coworkers for their assistance in the development of this book. Special thanks go to George Pruitt, Technical Documentation Consultants, Ridgecrest, California, for Chapter 8, "Graphical Inspection Analysis." A special thank-you also goes to Eugene Barker, Douglas Aircraft, Long Beach, California, for his input on Chapter 5, "Inspection Systems and Planning."

I would appreciate any comments regarding the improvement of this book as it is my concern to provide a worthwhile body of knowledge for people involved in quality.

Gary K. Griffith

Brief Contents

Contents

5   Inspection Systems and Planning   63

6   Reading Engineering Drawings   83

7   Geometric Dimensioning and Tolerancing   103

8 Graphical Inspection Analysis   159

9 Common Measuring Tools and Measurements   189

13 Statistical Process Control   399

14 Shop Mathematics   483

Introduction to Quality

Quality is a vital factor for the survival of any business. Quality (or the lack of quality) has a direct impact on other factors that are important to business such as costs, on-time delivery, inventory, reputation, market share, and so forth. Quality problems increase costs, delay delivery, increase or reduce inventories, damage a company's reputation, and can reduce market share. In the past, there have been several definitions of quality, including *fitness for use, meeting customer's expectations, doing it right the first time, degree of excellence,* and *conformance to requirements.* The following is the most recent definition of **quality:**

> Quality is the totality of features and characteristics of a product or service that bear on its ability to satisfy stated or implied needs now and in the future. (Ref. ANSI/ASQC A3-1987 "Quality Systems Terminology")

Quality is not necessarily a tight tolerance, a shiny surface, or a perfect fit. Quality is satisfying what the customer wants, needs, and is willing to pay for. Always keep in mind these broad steps toward achieving quality:

1. Determine the customer's requirements (wants and needs).
2. Set standards to define all requirements.
3. Prepare plans to meet those requirements.
4. Communicate requirements and plans effectively to all concerned.
5. Assess the ability to meet the requirements (or the changes required).
6. Select/use capable processes.
7. Control the processes.

Once the customer's requirements have been clearly defined, the product must be designed accordingly. The quality of design of a product, process, or service is vital. The phrase "make it like the specification" loses its meaning if the specification is wrong. When *quality of design* has been achieved, the next step is the *quality of conformance to requirements.*

Conformance to requirements bears heavily on the following factors:

1. The ability to meet design requirements (process capability)
2. A controlled process (to consistently produce within the process capability)
3. A well-trained and motivated workforce
4. Management commitment and support

THE QUALITY ORGANIZATION

The quality organization is actually the combination of all personnel and functions in the company. The "old school" of quality believed that the quality organization had to be a separate department of skilled personnel with direct responsibility for quality in the company. All functions in the company have responsibilities for quality, however, not just one department. The quality department has quality-related responsibilities but not responsibility for everyone's processes. Quality is not achieved simply by having skilled monitors or a policing function.

A quality organization should be a multidisciplined team effort that begins with the identification of the customer's wants and needs and extends throughout all phases of production or service beyond the delivery of the product or service to the customer. Every department in the company plays a key role in quality from the identification of the customer's wants and needs through field service.

Organizing for quality should recognize that this multidisciplined effort is best implemented using concurrent teams with focused objectives. Another vital aspect of quality is identifying and recognizing customers (both external and internal). Refer to Chapter 2, "Total Quality Management," for further information.

QUALITY SYSTEMS

A **quality system** is one in which all activities and responsibilities associated with quality have been clearly defined (typically through policies and procedures), or activities associated with quality have been integrated into the company's policies and procedures. The system is documented to provide assurance that personnel across different functions know their responsibilities and understand the responsibilities of others. There is a baseline for auditing and making improvements in the system, and there is a basis for training personnel in quality functions.

Military standards, international standards, and customer-specific standards are examples of "standard" quality systems. In all cases, these standards were developed with the express purpose of fostering the implementation of a quality system in the company. Various subsystems must be designed and implemented to form an integral part of a typical quality system.

A *system* is anything that has a specific objective and that has been planned, covered by appropriate procedures, and implemented to achieve that objective. The need for any of the following subsystems, of course, depends on the type of product being sold, service being performed, or processes being used. The following paragraphs give an overview of major quality subsystems and their application.

Inspection Systems

Inspection systems need to be established to ensure all of the activities associated with inspecting the product are covered by appropriate procedures, work instructions, personnel, planning, and equipment. At times, an inspection system is the main thrust of a quality system because often the manufacturer is only making product and is not involved with the end-product design, field support, or other aspects of the product. Refer to Chapter 5 for more information about inspection systems.

Calibration Systems

Calibration systems are necessary in any company that uses measuring and test equipment to provide assurance that the equipment is accurate or to determine the accuracy at the time of calibration. Calibration of measuring and test equipment should be based on the *stability, purpose,* and *usage* of the equipment.

Calibration systems should define frequencies for calibration of all measuring and test equipment, work instructions for methods of calibration, methods for identifying the status of calibration of the equipment (e.g., calibration stickers), and procedures for adjusting calibration frequencies based on the degree of usage of the equipment and the results of calibration. Calibration systems should also identify sources for calibration, environmental conditions for calibration, the action to be taken to correct equipment that is found to be out of calibration, and the impact on products that were measured by the equipment. If calibration stickers are used, the information on the sticker should specify when the equipment was calibrated, who performed the calibration, and when the next calibration should be performed.

Calibration systems are also concerned with the hierarchy of calibration standards and an objective audit trail (or traceability) to those standards. The hierarchy of standards is:

International standards: The highest order of standards. Examples are the master kilogram (maintained in France) and the ultimate standard for length, the wavelength of a certain radiation of light.

National (primary) standards: Standards that exist in all industrial countries that are calibrated and directly traceable to international standards. For the United States, these standards are maintained at the National Institute of Standards Technology (NIST).

Secondary (transfer) standards: Standards used by calibration laboratories, which are calibrated and directly traceable to primary, and ultimately international, standards.

Working standards: Used by companies to calibrate measurement and test equipment that will measure or test the product. These standards are calibrated and directly traceable to secondary standards.

Nonconforming Materials and Corrective Action Systems

Nonconforming materials must be identified and controlled to prevent the possibility of being mixed with conforming materials and to isolate and correct the cause(s). The elements of a nonconforming materials control system include provisions for *identifying* nonconforming materials, *segregating* those materials from conforming materials, *isolating the cause(s)* (for the purpose of taking corrective action), and *dispositioning* the materials. Identification of nonconforming materials includes some physical form of tagging or labeling them as nonconforming. Control of nonconforming materials includes *bonding* (locking up or effectively segregating them) from acceptable materials. Identifying cause(s) of nonconforming materials is the first step in corrective action.

When nonconforming materials are discovered, there should be a review for cause(s) (for the purpose of taking corrective action). Then the materials are dispositioned (e.g., complete missed operations, rework, sort, repair, scrap, or use as is). A Material Review Board (MRB) usually has the responsibility for material review and dispositioning. A preliminary review board may be assigned to review the nonconforming material and identify the cause, take corrective action, and make some of the aforementioned dispositions (except for use as is). Only the MRB function is authorized to make use-as-is dispositions.

Corrective action is taken to isolate and effectively correct (or eliminate) the root cause of nonconformance and to make changes in the process to prevent recurrence of the cause(s). Corrective action systems should be established to provide a systematic approach for taking corrective action on defective products or services. Corrective action systems include material review activities, quality reporting and analysis, problem-solving steps, bonding and segregating defective materials, corrective action, and follow-up procedures.

Quality Cost Reporting Systems

Quality cost reporting systems are an important aspect of quality and cost improvement. A good quality cost reporting system can be the source of information that provides improved products or services and reduced costs. Refer to Chapter 3 for more information about reporting and using quality costs.

Prevention-Oriented Quality Systems

Preventive quality systems are more effective than traditional detection systems. When considering the costs of quality, it becomes clear that investment in implementing and maintaining a preventive system is far superior to implementing any detection system. Consider the pie chart in Figure 1.1. This chart is similar to many that have been prepared when quality costs are measured for the first time in a company that has been practicing detection.

Figure 1.1 shows that costs associated with appraisal (inspection, auditing, and testing, for example) are somewhat equal to internal failure costs (such as scrap, rework, and repair). These costs often tend to trend together and increase the overall costs of quality. For example, as more defects are found, more inspection is performed. As more inspection is

FIGURE 1.1
Pie chart showing typical costs of quality the first time they are measured.

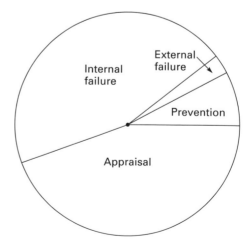

FIGURE 1.2
Difference in quality costs between
a detection system (left) and a
prevention system (right).

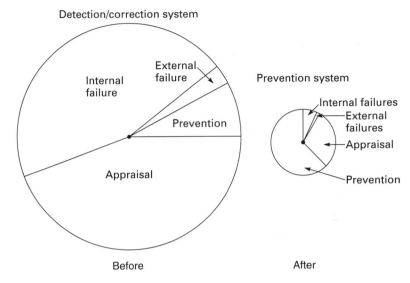

Prevention investment decreases all quality costs.

performed, more defects are found. Notice that prevention and external failure costs are
small in comparison.

A prevention system (see Figure 1.2) focuses on implementing preventive and regulatory actions that prevent failures (defects). An increased investment in prevention quality
costs will most often produce significant decreases in all of the other categories and the overall quality cost investment. Refer to Chapter 3 on quality costs for more detailed information.

QUALITY AND COMMUNICATION

One of the most important quality-related factors in any business is *communication.* Effective communication includes communicating quality policies, requirements, procedures,
work instructions, standards, and specifications. Many quality problems, large and small,
have been directly caused by a lack of communication of some kind. On the other hand,
many quality problems have been prevented by effectively communicating quality requirements to the workforce. Other forms of communication are teaming and training.
Cross-functional teaming is effective because it allows all of the people who are involved
in the process to communicate with each other and have the same objectives. Training, of
course, is communication with the expressed objective of teaching specific disciplines. Personnel should be trained in the requirements and technical aspects of achieving quality.

Breakdowns in communication can occur within or outside the company. Examples of
problems created by a lack of communication are:

1. Poor or no communication between producers and consumers (such as poor communications regarding customer requirements)

2. Poor communication between marketing and engineering (such as the lack of flowdown of customer requirements)
3. Poor communication between engineering and manufacturing (such as unclear, incomplete, or incorrect specifications or standards)
4. Poor communication between functional departments (leaving open gaps or causing problems) that leads to product or service quality problems
5. Poor communication between field services internal to the company's operations
6. Poorly written procedures (or lack of needed procedures), which can occur at any level of the organization
7. A lack of proper training that is necessary to perform the work

MOTIVATION (HUMAN FACTORS) AND QUALITY

Some processes are *people-dominant* and some are not. In a people-dominant process, the quality of the output of the process depends considerably on the training, skill, and/or attention of the person performing the work. Examples include any process performed manually. A process that is not people-dominant, then, is one in which the quality of the output of the process is not significantly influenced by the training, skill, and/or attention of a person. Examples are automated or computer-controlled processes.

Keep in mind that, regardless of dominance, people play a vital role in the success of any process. Understanding the dominance of a person and a process simply helps us to understand the requirements to successfully perform the work.

A *process* is a collection of six possible sources of variation (or problem causes): manpower, methods, materials, machines, measurements, and the environment. Manpower- (or people-) related problems can be related to the level of dominance that exists in the process.

People need certain things to provide assurance of quality output from their process:

Training or skill (especially in people-dominant processes) can be a major factor in assuring the quality of the output of a process. If people do not have the fundamental training, experience, or skills required, they are certain to make mistakes that will affect the output of the process. It is also probable that untrained personnel will not have the knowledge necessary to make any significant improvements in the process.

Proper tools are required in any process for people to get the job done and get it done correctly. In the case of production personnel, the proper tools can include machines, fixtures, gages, cutting or forming tools, and control devices. For office personnel, tools can include terminals and computers. Service process tools can include hand tools, analytical devices, testing devices, repair equipment, and so on. In all cases, tools can also include analytical methods for controlling the process (such as process control charts).

Clearly defined standards for quality must exist to define the quality requirements of the process and to provide a baseline for people to measure results. Without some basis for comparison, one rarely knows what to measure or how to measure the quality of the process.

Measurement (or the ability to accurately and repeatedly measure the output of the process) is essential for people to know that the output does (or does not) conform to requirements. No matter what the process or the output, the ability to measure the result against some applicable standard is crucial to the success of a process.

Control is an essential part of any process. The ability to monitor the process, and make timely and effective corrections is a vital aspect of quality, yield, and continuous improvement. There are several different types (or levels) of control in various processes. One type of control uses physical analytical devices (such as temperature recorders, in-line gages, and other devices). Another type of control uses computer-controlled devices to regulate the output of a process.

Graphical analytical tools such as statistical process control charts, histograms, and other methods are also used. Refer to Chapter 13 for more information about statistical process control methods. Note that these methods can be applied to any process for the purpose of controlling and reducing variation in the process.

Motivation for quality is an essential aspect of any quality system. Intrinsic causes of lack of motivation can be as simple as the absence of the needs previously expressed. For example, people can be unmotivated because they lack the proper training, skills, tools, standards, or ability to measure the output. Management style can be a motivator or unmotivator, according to Douglas McGregor's theories. Another important aspect of motivation is to consider the order of human needs according to A. H. Maslow's theory. The following is an overview of these two popular human motivation theories. Refer to the *Quality Control Handbook* (by Joseph Juran) for more detailed information about McGregor's, Maslow's, and other popular theories with respect to motivation for quality.

McGregor's theory separates managers into X and Y categories. A theory X manager, according to McGregor, is one who thinks people:
1. Dislike work.
2. Must be pushed to work.
3. Prefer to follow, not lead.
4. Have very little ambition.
5. Avoid responsibility.

A theory Y manager, on the other hand, thinks that:
1. Work is as natural to people as play.
2. Control or punishment are not always necessary to motivate people.
3. Most people seek responsibility.
4. People have imagination, ingenuity, and creativity.
5. The potential of most people has barely been tapped.

In many cases, it has been proven that the theory Y management style is more effective. People who work for theory Y managers are often highly motivated, successful, and happy about their work.

Maslow's theory says "people have different needs," and these needs must be satisfied in a specific order. If they are not satisfied, a person could be unmotivated to perform the job properly. The following is an overview of Maslow's list of needs:
1. Physiological needs (such as food and health)
2. Safety needs (people must feel that they are safe from harm)

3. Social needs (the need for companionship)
4. Esteem needs (the need to be noticed/recognized)
5. Self-actualization (all needs have been met)

REVIEW QUESTIONS

1. Which of the following is one of the largest single factors in quality improvement?
   a. knowledge of quality
   b. communication
   c. effective inspection
   d. end line testing
2. One of the first steps to be taken toward achieving quality is
   a. inspection planning
   b. flowchart the process
   c. determine the customer's wants and needs
   d. design review
3. Achieving quality of conformance must be preceded by achieving quality of _____ .
4. When all of the activities associated with quality have been planned and clearly defined in policies, procedures, and work instructions, we have a quality _____ .
5. Measurement and test equipment should be calibrated based on stability, purpose, and

   _____ .
6. The highest order of calibration standards is:
   a. primary standards
   b. transfer standards
   c. national standards
   d. international standards
7. Control of nonconforming materials includes identification, segregation, and
   a. disposition
   b. bonding
   c. tagging
   d. isolating cause(s)
8. A good source of information for improving quality and reducing costs is the _____ cost system.
9. In order for people to be successful in achieving quality, they must be given
   a. proper training
   b. proper tools
   c. clear standards
   d. ability to control the process
   e. all of the above
10. Which of the following is an example of McGregor's theory X manager?
    a. shows up to work early every day
    b. thinks that people must be pushed for results
    c. thinks that, to people, work is as natural as play
    d. feels that most people seek responsibility

11. Which of the following quality costs categories provides more return on the investment?
    a. appraisal
    b. internal failure
    c. prevention
    d. external failure

12. Which of the following is the first order of Maslow's needs hierarchy?
    a. social
    b. safety
    c. esteem
    d. physiological

13. The totality of features and characteristics of a product or service that bear on its ability to satisfy stated or implied needs now and in the future is the definition of _____ .

14. People should be able to measure results comparing them to clearly defined _____ .

15. Which of the following information does not belong on a calibration sticker?
    a. date calibrated
    b. who calibrated the equipment
    c. what method was used
    d. due date for next calibration

16. Secondary calibration standards must be directly traceable to
    a. transfer standards
    b. working standards
    c. international standards
    d. primary standards

17. Which of the following dispositions of nonconforming materials may not be made by a preliminary review board?
    a. rework
    b. complete missed operations
    c. scrap
    d. use as is

18. Any process contains six categories of possible sources of variation. Which of the following is not one of those main sources?
    a. manpower
    b. machine
    c. tooling
    d. measurement
    e. materials

19. Frequencies of calibration of measuring and test equipment should be based on the degree of usage of the equipment and the _____ of calibration.

20. All activities associated with inspecting the product are part of the _____ system.

Total Quality Management

The purpose of this chapter is to provide an overview of total quality management (TQM) and discuss the need for companies to monitor the quality of all operations. TQM helps remind us that all operations of the business are processes, and that three fundamental elements exist in every process:

Input → Processing → Output

The quality of the output of any process is adversely affected by unwanted variation in the input or processing stages. In every company there are processes that can, and should, be improved. Total quality management should not be applied just to manufacturing processes.

> TQM is the process of continuously improving performance at every level and in every area of the company. Improvements are directed at broad company goals such as costs, quality, and delivery. TQM combines fundamental management techniques, leadership, improvement efforts, and specialized skills under a rigorous and disciplined structure that is focused on continuously improving all company processes. TQM demands commitment and discipline, relies on people, and involves everyone.

TQM helps to remind us that quality is not just about the product we deliver to our customers but also about the quality of everything we do, including manufacturing, service, and business processes. In TQM, everyone has a responsibility for quality, and everyone has a customer and a supplier.

It has been said that the ultimate measure of quality levels is made by the customer. Note that the customer can be internal or external. An internal customer is defined as the next destination of the output of a process. (The term *external customer* is self-explanatory.) Many people have not identified their internal customer, nor do they view the next operation as their customer. The supplier is the previous operation or step that affects the process.

Manufacturing processes (those that produce a product) are more obvious than others. However, most companies have business processes and/or service processes that tend to be ignored in quality programs when the focus is on improving the quality of the product.

For many years *manufacturing processes* were the focus of quality programs because the key to success was assumed to be quality products alone. TQM does not change the importance of quality products. It enhances the quality program to include the quality of other company processes.

Business processes are internal company processes that can be improved in terms of their output quality. Business processes include, but are not limited to, the following examples (and each has many subprocesses): purchasing process, quotations process, contract review process, sales and marketing, order entry, accounting, engineering, and scheduling.

Each of these business processes can have an adverse affect on the company's quality, cost, or profit. Today's customer demands a quality product, on time, at a reasonable price.

Service processes involve providing a given service to the customer. The customer, once again, can be internal or external. The quality of service (often relating to time, cost, or the effectiveness of the service provided) can only be improved by treating the service as a process, then identifying and correcting the causes of poor process output.

TQM expands our view of quality from looking only at the quality of the product we make to looking at the quality of everything we do.

QUALITY FUNCTIONS

The following functions of quality assurance, quality control, and inspection are not intended to be interpreted just as quality department functions, but company-wide functions.

Quality assurance—A planned and systematic order of events that provides assurance regarding the quality of the product or service.

Quality assurance is primarily a planning and analysis function that was never intended to be the sole responsibility of the quality department. The assurance of quality begins in the early stages of identifying the customer's requirements and extends, in stages, throughout the life cycle of the product. The analysis functions of quality assurance are intended to provide facts and information that are fed back into product and process designs for improvement purposes. Analyses can include, but are not limited to, yield analysis, testing results, experimental models or results, chemical and metallurgical analyses, quality reports, and field failure analyses.

The quality assurance function should be a cross-functional activity in which all functional departments play a specific role in planning for and analyzing quality. It is important that the company's quality system be designed to provide assurance of quality from the identification of customer requirements through support functions in the field.

Quality control—The regulatory process through which we measure actual quality performance, compare it with the standard, and act on the difference.

The control function, once again, is a regulatory process. Each department measures the actual quality performance, compares it with the standard, and acts on the difference. All processes (manufacturing, quality, administrative, service, and many others) are involved in this activity.

Inspection—An appraisal activity that compares products to applicable standards.

The inspection function in a TQM approach involves more than just the traditional inspector. In a TQM approach, everyone is an inspector and is responsible for inspecting the work they perform. The best inspection is conducted at the source, at the time the output is made, and by the individual who makes it.

RESPONSIBILITY FOR QUALITY

Quality is not just the responsibility of the quality department. Quality is everyone's responsibility. Quality engineering is all departments working together to identify and plan for quality in the earliest possible phase of the product or service.

Marketing and Sales Responsibility

Marketing and/or sales departments have the responsibility to understand all customer requirements/needs, work with customers closely to help define their needs, and convey those needs early in the contractual phase to all concerned. Considerable effort should be made to understand the customer's application, special requirements, or particular performance criteria of the product. Then the effort should be focused on effectively interpreting and communicating those requirements to the rest of the organization.

The downward flow of effective communication of customer requirements is essential so that planning to meet those requirements can begin. The following case study shows one possible effect of communication gaps in the downward flow. Figure 2.1 shows a customer sketch of the application that is delivered to the marketing/sales department. It shows an antenna mount on which the company is to build a bracket that will be welded to a TV antenna. This figure, as we will soon see, does not give enough information.

Figure 2.2 shows the product as it was communicated to the engineering department. The part to be made and the application were assumed to be straightforward. Figure 2.2 still does not give enough information. The case study will continue under engineering responsibility.

Engineering Responsibility

The quality of design is a direct responsibility of engineering. The fit, form, function, and reliability of a product depend on product design quality. Poor design of a product (including applicable drawings, specifications, and standards) can be the root cause of quality problems with the product during manufacturing and inspection. Poor or inadequately communicated design

FIGURE 2.1

How the customer defined the application.

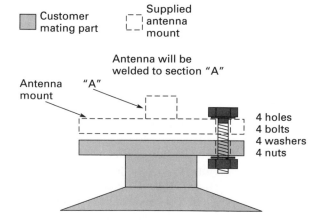

FIGURE 2.2
How marketing/sales interpreted the
application.

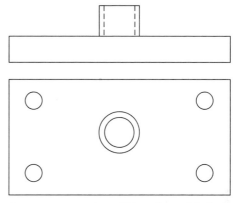

Antenna mount that will interface with the
customer's connection (removable); refer
to customer's drawing.

methods (such as improper tolerancing, nonfunctional interface descriptions, ambiguous notes, and many other examples) can be the direct cause of defects or false rejections of good product.

As early as possible in the design phases of the product, other functional departments such as quality assurance, manufacturing engineering, manufacturing, and process engineering should be involved in a concurrent approach toward identifying and resolving potential problems with the design, process, inspection, tooling, and many other areas of the product. This activity (often called *concurrent* or *simultaneous engineering*) is a prevention process that helps build quality into the product and process design. Engineering has the direct responsibility to convey customer and product requirements in the form of clear and complete specifications, drawings, and/or standards. In the case of products, engineering is responsible for providing drawings and specifications that have appropriate tolerances with respect to fit, form, function, reliability, and safety.

Improper tolerancing can be the root cause of poor products. Missing standards or specifications can be the cause of poor products or missed operations. Poorly developed standards or specifications can be the cause of false acceptance or rejections. In cases where quality requirements are cosmetic or could be subjective, it is engineering's responsibility to establish standards that are objective, not subjective.

It is also engineering's responsibility to develop tolerances for the part that are realistic, yet provide guaranteed fit, form, function, and reliability. Tolerances that are too tight for the product (e.g., tighter than necessary) add cost, not value, to the product. Tolerances that are too loose can impair fit, form, function, and/or reliability. *Fear tolerancing* is when engineers assign very tight tolerances to the part because (1) they have not been able to perform stack-up analysis or tests, (2) they assume that tolerances have to be tight for "better" quality, (3) there has been no time for functional testing and analysis, (4) the part has been toleranced based on an existing similar part that is working, (5) they don't know how to tolerance the part, or (6) "standard" tolerances have been set and communicated by design handbooks.

Part tolerances should be based on how the part works and how long the part will work. Early in the design phase, product designs should be analyzed and tested to properly tolerance

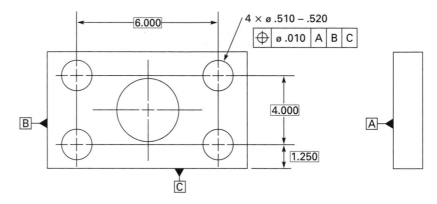

FIGURE 2.3
How engineering designed the product.

the part. A review of the actual application of the part should be performed to understand how the part works in assembly (functions).

Analysis of tolerance stack-up should be performed to develop fit, form, and functional tolerances for the component parts and to understand the envelope in which the assembly must fit. Prototype tests should be conducted and analyzed to a level sufficient to understand other variables that are involved and the effect of tolerances. Specific tolerancing experiments should also be conducted to tolerance certain characteristics.

The case of the antenna bracket is continued here as an example of the impact of overtolerancing. As demonstrated in Figure 2.3, this part is considerably overtoleranced with respect to its application in many different ways. The actual errors, in this case, are too numerous to mention. Problems were encountered late in the production phase or near the time of shipment. These problems are shown in Figure 2.4.

In this case, the function of the product was reviewed with the customer. The actual application is shown in Figure 2.5.

- 765 of 1,000 parts rejected for position of holes
- Owner requesting quotes for a Coordinate Measuring Machine (CMM) to measure parts
- Tool design prepared a $9,000 functional gage for operators
- Customer delivery schedule is 2 months past
- Customers are upset about delivery/quality problems
- The Material Review Board is trying to decide what parts (of the 765) can be "used as is"
- Company is consideing requesting a waiver
- Production costs are sky-high

FIGURE 2.4
Problems encountered with the original product design.

FIGURE 2.5

Manufacturer met with the customer and looked closely at the application.

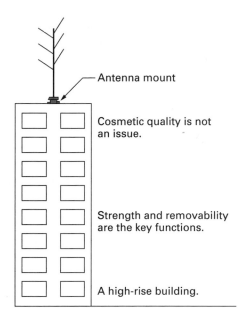

As a result of the functional review, the product design and tolerances were changed to depict the function of the part (as shown in Figure 2.6). Some of the changes were:

1. The outer datums were removed because they were not involved.
2. The regardless-of-feature-size (RFS) modifier was changed to maximum material condition (MMC) with bonus tolerances allowed.
3. The tolerance was increased because there was no reason for the previously tight tolerance.
4. The method of specifying the positional tolerance was changed from traditional to the "zero at MMC" method (refer to geometric tolerancing books for further information on this approach).

FIGURE 2.6

How the design was revised after understanding the application.

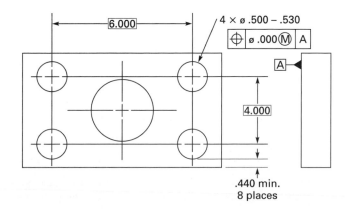

- Most of the rejected parts were acceptable
- No need for a CMM
- The functional gage was redesigned and the cost went down to $1,500
- No delivery problems
- Customers are now satisfied
- No need for MRB on quality products
- No need for waivers on quality products
- Production costs went down 65%

FIGURE 2.7
Results of the revised engineering design.

Figure 2.7 shows the results of the design changes and the improvements that were made. Because of the changes in the drawing, many costs were reduced, quality problems were eliminated, and the part function was better guaranteed.

In another example, the tolerance value was appropriate, but the drawing language did not properly communicate the tolerance. In this case, the drawing callout caused problems. Consider the drawing in Figure 2.8. The note specifies that the marked surfaces must be square (perpendicular) with the specified tolerance.

The machinist who made the part was satisfied that all of the parts made were good because the worst part found was .003″. The machinist checked the parts using the long side as the datum (as shown in Figure 2.9) either because there was a choice and the long side was more stable, or the long side was the side that was mounted for machining purposes.

FIGURE 2.8
Typical ("simple") engineering note about the characteristic.

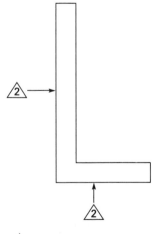

/2\ Surfaces to be square
within .005 max.

FIGURE 2.9
Machinist locates the parts on the
long side and accepts them.

The machinist produces 100 parts and inspects
them, referencing the long side on the surface plate.

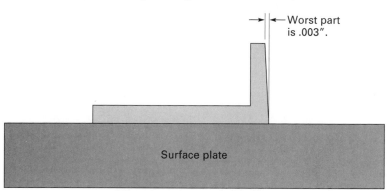

100 parts are sent to inspection.

Once the lot of parts were complete at all other operations, they were sent to the final
inspection operation. The inspector pulled a sample from the production lot and inspected
the sample using the short side as the datum (see Figure 2.10). The best part the inspector
found in the sample was out by .007″. Since the drawing stated .005″ maximum, the inspector rejected the lot for being out of squareness tolerance.

At this point (see Figure 2.11), the entire lot was on hold and the machinist and inspector argued about who was right. Both felt that they were using the best measurement
datum, but in fact, neither knew for sure. In this case, of course, either the machinist was

FIGURE 2.10
Inspector locates the parts on the
short side and rejects them.

Inspection samples the lot, referencing
the short side on the surface plate.

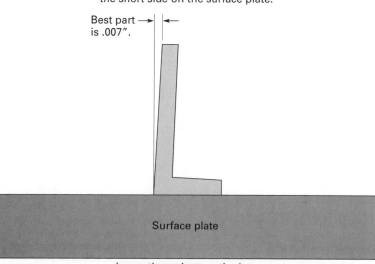

Inspection rejects entire lot.

FIGURE 2.11

Entire lot is on hold because of disagreements about quality.

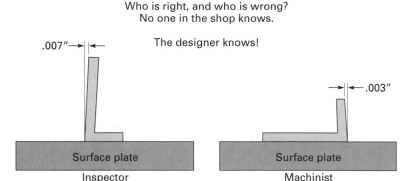

right (and the parts were good) or the inspector was right (and the parts were bad). Only the designer or product engineer knew for sure because the answer depends on how the part works in its assembly. If the primary interface is the long side, the machinist is right. If the primary interface is the short side, the inspector is right.

Once the designer was contacted (because parts were on hold and the required shipment date was near), the function of the part became known (see Figure 2.12). In this case, the inspector was correct because the primary interface of the part was the short side, and many parts were made out of tolerance. The machinist had to rework the parts to establish squareness to the short side as a datum.

In this case, the drawing was the root cause of defective parts and they all had to be reworked. It could, however, have gone the other way. If the long side had been the true datum, the inspector might have checked them from the short side and rejected good parts. The company might have reworked or scrapped good parts had that been the case.

FIGURE 2.12

Function of the part was later determined to be from the short side as a datum.

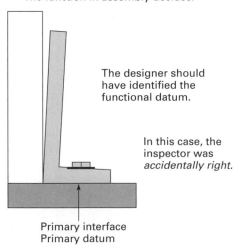

Process Engineering Responsibility

A process engineering function exists in some companies that have engineering specialists responsible for the design, standards, and specifications for special processes such as heat treatment, brazing, welding, coating, plating, and chemical processes. The specialist, depending on the process, could be a metallurgist or a chemist, for example.

In some cases, the specialists can improve process design and process control through the use of statistical experimental designs where the goal is to understand the variation of the process and eliminate unwanted source variables or make the process robust to these variables. Process engineering supports its customer (manufacturing) by designing production processes that produce output quality when they are properly operated.

Manufacturing Engineering Responsibility

Manufacturing engineering is responsible for supporting its customer (manufacturing) by identifying and selecting the most capable manufacturing processes, tooling and equipment, processing sequence, and applicable work instructions. It also provides ongoing shop floor support to maintain control and capability of processes. Although the manufacturing engineering function varies slightly from company to company, the manufacturing engineer is usually required to:

1. Plan the sequence of manufacturing operations.
2. Select capable processes for producing the part.
3. Prepare shop travelers and operational work instructions and/or sketches for production.
4. Identify and design/select proper tooling for the processes.
5. Take action to correct out-of-control or incapable processes that are caused by problems related to these responsibilities.

Manufacturing engineering supports the manufacturing department by providing these services. Many quality problems can result (or be prevented) depending upon the quality of the manufacturing engineer's work. Insufficient manufacturing planning, work instructions, sketches, tooling, sequence of operations, process capability, and so on can cause recurring manufacturing problems. Manufacturing engineering's quality depends a great deal on the work of its internal supplier, engineering.

Manufacturing Responsibility

Manufacturing (production) personnel have considerable responsibilities regarding quality, or producing parts to specifications *the first time:*

1. Follow work instructions, sketch sheets, and applicable specifications.
2. Properly operate the process and take proper care of tooling, fixtures, and the like.
3. Inspect the output of the process.
4. Use the results of inspecting output to reject defective parts and control the process.
5. Take local action to correct out-of-control problems, incapable processes, and root causes for defective parts.

Quality control is a regulatory activity that uses various methods during the production process to control and improve product quality. Therefore, quality control is more associated with manufacturing than with the quality department. In the past, the term *quality control* was presumed to be a function of the quality department (using inspectors). The control of quality is most effective, however, when it is controlled *at the source.* Production personnel should be directly responsible for monitoring the quality of the output of their process and taking action on the process to ensure that quality products are produced (which will be sent to the next operation, their customer). Manufactured quality depends a great deal on the work of their suppliers, engineering and manufacturing engineering.

Procurement Department Responsibility

The procurement (purchasing) department plays a vital role in quality. This role is not necessarily a new role in a TQM operation, but one that is better defined. For example, purchasing has always had the direct responsibility to select suppliers for raw materials, purchased parts, and/or processes based on quality, price, and delivery. In the traditional purchasing system, price and delivery were the prime directives since quality was the quality department's job.

Today, there is growing concern about the major difference between price and cost. Price is what one pays for goods and/or services to be performed. Cost is the ultimate amount that is paid (price plus the cost of poor quality, downtime, failures, and many other realities). Since quality affects the usability of the product, the actual cost can far exceed the price.

Procurement's job has always been to:

1. Purchase only quality raw materials, purchased parts, processes, and/or services.
2. Select qualified suppliers.
3. Rate suppliers based on price, delivery, and quality (PDQ).
4. Effectively communicate requirements to suppliers.
5. Monitor suppliers' results.
6. Give feedback on problems to suppliers.
7. Take action to improve supplier quality.
8. Maintain the importance of controlling quality at the source.

Procurement's direct customer is manufacturing: those who will use the raw materials, purchased parts, and processing to make or assemble the product.

Quality Assurance Department Responsibility

Quality assurance departments play a vital support role in the data collection, analysis, and improvement of quality. In cross-functional teams, quality assurance representatives use their experience to help identify inspection equipment problems, recurring defects, sampling requirements, quality data requirements, and customer special measurement, reporting, or record requirements.

Quality representatives also provide decisions regarding measurement or gaging of product characteristics, inspection planning, quality-related training, and/or reporting. Some of the typical responsibilities of the quality assurance department include:

1. Provide for the calibration of inspection equipment (no matter who inspects the product).
2. Monitor the results of in-house and supplier quality and report the status in a timely manner (including quality cost reports).
3. Provide internal and external quality audits for the purpose of improvement.
4. Provide needed inspection of the product or material at stages where it is impossible or impractical for other personnel to perform inspection.
5. Provide for training as necessary with regard to quality, inspection, and so on where the quality department has the expertise.
6. Prepare inspection plans.
7. Evaluate and implement sampling plans.
8. Be the central source of quality information and quality-related input for the company.

Storage Department Responsibility

Storage is responsible for storing only acceptable product; storing the product in a manner that will prevent problems such as damage, oxidation, and other defects; and having a system for properly locating and issuing the right product. When the acceptability of the product is in question, storage is responsible for finding all suspect products and submitting them to the appropriate area for evaluation. Storage is also directly responsible for issuing the proper materials or products requisitioned by manufacturing or assembly.

Packaging/Shipping Department Responsibility

The quality of packaging, packing, and shipping plays a vital role in the quality of the product, delivery, and other related factors. Packaging must be performed in a manner that meets internal transit and/or customer packaging requirements, prevents damage or other defects to the product, protects the product during shipment, and maintains proper identification of the product. Shipping must perform shipping functions correctly, e.g., ship the product to the proper destination, on time, free of shipping damage, and with the proper documentation.

Field Service Responsibility

Field service is an important customer support function that occurs "in the field" after the product has been shipped. Field service responsibilities can include analyzing problems, performing on-site repairs, being the focal point for customer complaints, and working with the customer toward needed improvements. Field service not only serves the customer directly, but also should serve as a feedback loop for communicating problems with the product in the field (such as field failures, maintainability, reliability, availability, and specific customer use-related problems). This feedback loop is one of the most important aspects of field service. It helps to close the gap between the customer's wants and needs and the quality of the product delivered.

TQM Systems Are Preventive, Not Detective

The traditional "detection" type of system doesn't work. Quality is a value that must be built into the product (hopefully the first time it's built). Quality cannot be inspected into a product. The only meaningful system is one that prevents defects from occurring and identifies/corrects defects if they occur. A prevention system improves quality, productivity, and profits. A detection system only helps to improve outgoing quality at increased costs.

Figure 2.13 shows a comparison of the detection system and prevention (e.g., process control) system. The detection system tends to build products, then inspect them to find and segregate defects, whereas the prevention system controls the process to prevent defects from occurring.

Summary

No single department or person can provide assurance of quality. Quality takes a dedicated effort coordinated among all functional departments and personnel in the company. Quality involves everyone in the company working together in a team approach. The words *total quality management* try to convey this vital message.

TQM TOOLS

A wide variety of tools can be used in TQM. This chapter introduces the tools that are most widely used. Some of these tools are used primarily for problem solving, some are for problem analysis, and some are for process control. Note, however, that most of the tools can be used for more than one function. Some of the tools covered in this chapter will be expanded

FIGURE 2.13
Numerous inspection operations versus process control.

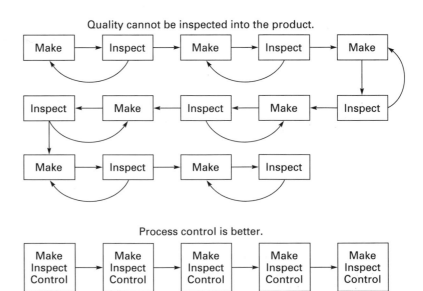

FIGURE 2.14
Checksheet for swing assembly defects.

Defect	Count	Total
Missing nameplate	### ### ###	15
Handling damage	### ### ### ### ###	25
Failed test	###	5
Poor torque	### ### //	12
Improper assembly	### ### ///	13
Wrong part	### ###	10
Missing components	### ///	8
	Grand total	88

upon and used in later chapters. The TQM tools covered in this chapter are checksheets, Pareto analysis, brainstorming, cause and effect diagrams, flowcharts, matrix charts, run charts, control charts, histograms, and scatter diagrams.

Checksheets

A checksheet (see Figure 2.14) is typically a list of inspected items that permits data to be collected quickly and easily in a format that lends itself to quantitative analysis. Checksheets facilitate data collection by providing a standardized format for recording information. In many cases, checksheets can be set up to automatically produce a histogram or Pareto chart.

Pareto Analysis

A Pareto chart (see Figure 2.15) is a chart arranged in descending order that shows the contribution of the vital few versus the trivial many problems, causes, sources, or defects. This chart assists in identifying the problems, causes, sources, or defects that should be addressed first in the problem-solving process.

Brainstorming

In the brainstorming approach (see Figure 2.16) team members are allowed to pool their knowledge about a given problem to reach overall agreement on the best course of action chosen from many possible options. Participants offer their ideas until all ideas have been exhausted. Ideas are not discussed until just before voting. All ideas are accepted, no matter how wild they may be. Then a discussion/voting process helps participants to decide on the best ideas.

Brainstorming can be structured or nonstructured. In nonstructured brainstorming, participants offer ideas randomly. Structured brainstorming involves going around the room in clockwise or counterclockwise order. Each participant either offers an idea or says "pass." Structured brainstorming is initially over when everyone "passes" in a row (ideas have been exhausted).

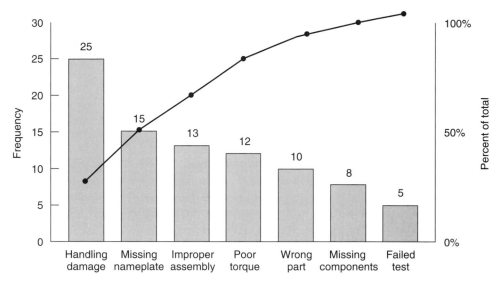

FIGURE 2.15
Pareto chart for defects, including cumulative line. See Figure 2.14 for data.

Brainstorming methods provide numerous, varied ideas, involve everyone, and promote team ownership. Brainstorming is used considerably by problem-solving teams. It is used to define problem areas, probable causes, probable solutions, and other factors. The technique is very simple, but the uses for brainstorming are many.

The brainstorming technique brings together a collection of minds and directs it toward one particular idea or solution. Everyone participates in the brainstorming, and everyone has his or her own thoughts on the subject.

Brainstorming is also a lot of fun. The team members are allowed (in one phase of the brainstorming) to come up with any idea, no matter how wild, and enter it into the analy-

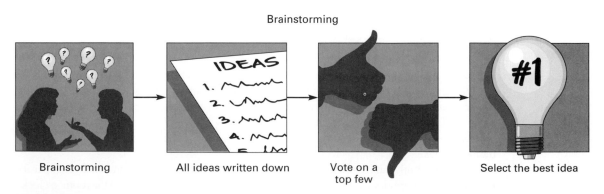

FIGURE 2.16
Brainstorming is a powerful method for generating ideas.

sis. Sometimes, the wildest of ideas actually has merit. If you do not believe it, consider the Pet Rocks that sold so successfully.

Collective thinking is a very powerful tool in problem solving. When one person might be assigned to solve a problem (with only his or her single outlook on that problem), the task may be overwhelming. However, when several people are studying the problem, there is a good chance they will more readily come up with a solution. This is especially true when the problem exists in the team's own area of experience.

Ground Rules for Brainstorming

1. Every member shall have a turn to voice his or her idea (only one idea per turn).
2. If no thought comes to mind, the member simply says "pass," and the turn goes to the next member.
3. Ideas can be any possible idea (no matter how wild) during the idea-gathering part of brainstorming.
4. No one is allowed to ridicule a person's idea. This can cause ideas to be kept inside and damage the brainstorming process.
5. During the initial brainstorming, there should be no critique made about a member's ideas. They are simply listed.
6. The brainstorming is finished when every member in the room has said "pass" in sequence.
7. The ideas are listed by the team leader or delegate as they are given. (Large poster paper is good for this.)
8. By team voting, all of the ideas are narrowed down to a few; then they are voted on again to determine the priorities of the remaining ideas.
9. The end result is that one or a few ideas (or probable causes) are selected. Brainstorming is used to define problem areas, probable causes, and solutions.

Cause and Effect Diagrams

Often called the "fishbone" diagram, the cause and effect diagram (Figure 2.17) is a method of representing the relationship between a problem (effect) and its potential causes. The power of a cause and effect diagram is that it forces participants in the problem-solving process (and the brainstorming) to structure the ideas in categories and helps foster new ideas, knowing the categories. It also helps the team to isolate the source of the problem.

Major Areas of Probable Causes.   Cause and effect diagrams typically have at least six major areas considered to be the primary causes for problems:

1. People
2. Methods
3. Machines
4. Materials
5. Measurements
6. Environment

FIGURE 2.17
Cause and effect ("fishbone")
diagram.

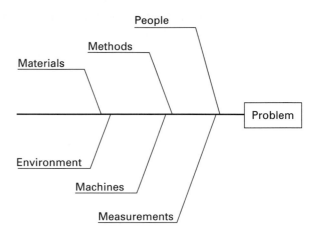

People.   This category states that the probable is related to the people involved. Some of the reasons why people create problems are:

1. Inadequate training
2. Not following work instructions (methods)
3. Health (eyesight, coordination, and similar problems)
4. Personal problems (inability to concentrate on job needs)
5. Personal tools (poor)
6. Boredom (may be caused by lack of motivation)
7. Fatigue
8. Lack of communication

Methods.   This category states that the probable cause for the problem is related to the methods by which something is done. The cause for poorly assembled units could be the method (or technique) used to assemble them, or that work instructions are inadequate, or that personnel skills are not suitably matched to those required for the job.

Materials.   This category states that the materials used could be the problem cause. The raw materials used to make the product could be:

1. Too hard
2. Too soft
3. Dimensionally poor
4. Internally defective
5. Externally defective

Machines.   This category states that the machines could be the cause of the problem. Some machine-related causes include:

1. Devices that are worn out or incapable of the tolerances needed
2. Poor tooling

3. Hydraulic problems
4. Coolant problems
5. The wrong machine for the job

Measurements.   This category states that measurements could be the cause of the problem. Some measurement-related causes when using variables data include:

1. Gage problems, such as the wrong tool, wrong gaging method, poor discrimination, and other errors identified in Chapter 9 that can cause defective products to appear to be acceptable and acceptable products to appear to be defective
2. Observer measuring problems, such as parallax, manipulation, flinching, rounding, feel, training, and several others that are also covered in Chapter 9

    Some measurement-related causes when using attributes (gaging and visual inspection) data include:

1. Gage problems, such as poor design, too much wear allowance, wrong datums contacted, and wrong-size gaging members
2. Observer problems, such as no standards, no gage or inspection instructions, fatigue, and boredom

Environment.   This category states that the environment could be the cause of the problem. Some environment-related causes include temperatures, dust, dirt, noise, distractions, locations, and other causes that have to do with the conditions in which work is performed.
    One of the best ways for the team to assess the possible causes of a problem is the use of cause and effect analysis. The cause and effect analysis is often referred to as the *fishbone technique* (or Ishikawa Diagram) because of the way it is structured.
    Each of the main causes has other branches from it where the detailed causes are listed. The fishbone technique displays a picture of the probable causes and a definition of the problem. It is very useful in problem solving and should be kept as a permanent record. Once the selection has been made, the primary cause (according to the voting of the team) is circled on the diagram.

Flowcharts

The flow (or order of events) of a process can, at times, be the problem that needs to be solved. A flowchart (Figure 2.18) is a powerful tool for looking at a process in terms of the sequence of operations (or steps), the type of operations, and the number of operations.
    In many cases, when operations are depicted in a procedure or traveller, problems in the sequence, types, or number of operations are not readily seen. Flowcharts help to show a picture of the operations in the order in which they occur. Flowcharts help to identify problems with the order of operations and nonvalue-added steps that could be removed. The user can see the relationship among steps and their relative timing and occurrence. Current processing steps and proposed steps for improvement can be documented. Flowcharts have been used in a wide variety of processes to improve the sequence, reduce overall cycle time, reduce defects, and improve effectiveness.

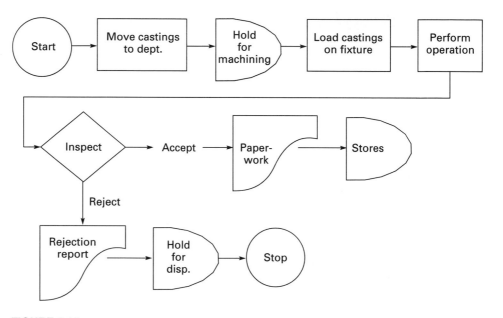

FIGURE 2.18
Process flowchart.

Matrix

A matrix is a chart that provides a simple tool for visually analyzing and comparing different alternatives to aid in the selection of the best alternative. This matrix can be used to view a variety of different effects for different chosen alternatives (see Figure 2.19), or it can be used to view a variety of different defects for different processes. The matrix is useful when there is more than one possible choice with more than one possible effect.

FIGURE 2.19
Decision matrix.

Alternatives → Criteria ↓	A	B	C
Defects	12	50	8
Cycle time	7	21	11
Cost	25	38	48
Delivery	4	18	8

Vote: (1) weak, (5) great

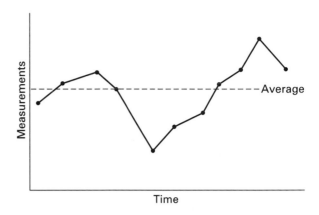

FIGURE 2.20
Run chart.

Run Chart

A run chart (see Figure 2.20) is a chart that displays measurements or observations over a specified period of time (or sequence). The run chart, not always as effective as a control chart, can show simple trends or shifts in a process. Unlike control charts, run charts have no statistically derived control limits.

Control Chart

A control chart (Figure 2.21) is a statistical chart that provides a tool for users to identify special (assignable) causes of variation in a process at the time they occur, monitor and control process variation, and control the process (or make it repeat its inherent variation). Control charts are tools that help us understand and regulate the variation in any process. Refer to Chapter 13, "Statistical Process Control," for more information.

Histograms (Frequency Distributions)

Both histograms and frequency distributions are graphs that display the frequency of data in column format. The histogram (see Figure 2.22) is a bar chart that shows the frequency

FIGURE 2.21
Control chart.

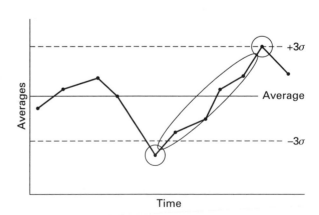

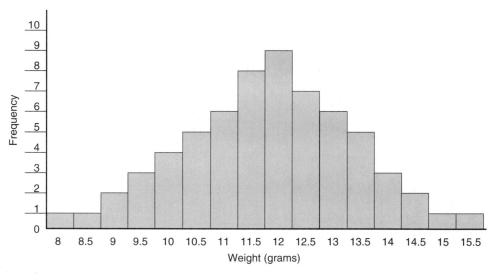

FIGURE 2.22
Histogram of measures of weight in grams.

(or the number of times of occurrence) of raw data from a process, and the frequency distribution is a chart of tally marks (or Xs) that shows the same thing. These tools help the user understand the location of the process center, the spread (or variation) of a process, and the shape of the distribution of a process (e.g., a bell curve). Histograms and frequency distributions do not show time (like a run chart or control chart).

Scatter Diagrams

Scatter diagrams (Figure 2.23) are used to study the relationship between two variables (X and Y). Where two variables are assumed to be related, scatter diagrams (and associated regression and correlation statistics) help us to see if there is a relationship between the two variables and to test the strength of the relationship. In scatter diagrams, there are two variables of concern: the independent variable (X) and the dependent variable (Y). Many types of problems involving two variables can be solved using scatter diagrams. For example, the Y variable may be the important characteristic but it cannot or should not be measured, whereas the X variable can be measured. If there is a relationship, the Y variable can be controlled measuring the X variable. In another example, the Y variable is found using an expensive (or destructive) test, but finding the X variable requires only a simple (or nondestructive) test. If they are correlated, the X variable could be measured instead.

A SYSTEMATIC APPROACH TO TQM

Quality policies must be defined and communicated. Policies have more to do with what is to be accomplished than how it will be accomplished and by whom. Procedures are in-

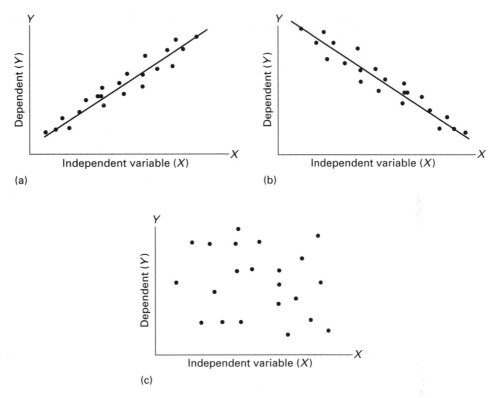

FIGURE 2.23
Scatter diagrams (positive correlation, negative correlation, and no correlation).

tended to cover the methods and responsibilities of carrying out the policy. Manuals of procedures provide a baseline that allows groups of different personnel to perform a function with consistency and a baseline for auditing the performance of a quality system. Written procedures allow for constant coordination of the methods that must be applied for successful output of a process. They also provide a baseline for change toward improvement of a method or system. Company policies should be documented and, usually in a separate manual, procedures should be written that will carry out each policy.

The typical company has a quality manual intended to establish procedures for carrying out quality policy. It should be noted that just because a manual exists doesn't mean that it is working. The typical quality manual of the past has been the source of problems such as:

1. It only covered quality department procedures.
2. It wasn't properly implemented.
3. It did not agree with other departmental manuals.

With regard to any manual, three questions are vital to answer. Answering no to any of these questions identifies a problem that needs to be solved.

1. Is there a procedure for the activity? (Existence)
2. If the procedure were followed to the letter, would it work? (Adequacy)
3. Is the procedure being followed? (Compliance)

If a procedure does not exist for covering operations that require one, it is a sure bet people will be doing things very differently (especially when a topic crosses department lines or involves groups of people with different experiences or training). If a procedure does exist, it should be reviewed to make sure following it will achieve the objective(s) successfully. If both existence and adequacy are acceptable, the only question left is whether it is being followed. Audits are a helpful tool to provide answers and action to correct any of these three questions.

With regard to procedures (manuals) the author has a preference. In many companies, different manuals in different departments share the same topic or objective. It is rare for these manuals to agree with each other, and in most cases, they create barriers between departments that should be working together. The author's preference is for *one* book, an operations manual that covers all procedures in a manner that shows the steps of each process in order and the responsibilities of each function at that step. This type of procedures manual promotes:

1. Teamwork
2. Concurrent changes
3. Removal of departmental procedural barriers
4. Easier upkeep
5. More effective implementation

The perfect procedure is never written the first time out. The rule should be, "Follow the procedure, or get it changed." Procedures should be adaptable when new ideas are born. Nothing in the company is set in concrete, and anything can be changed by the team. Ideas for improved operations are always welcome and evaluated. Procedures should constantly be put to the test.

MANAGEMENT INVOLVEMENT AND COMMITMENT

Without the commitment and involvement of management, a TQM process is doomed to failure. Senior management must ensure that the TQM system is communicated, supported, and working for the company. Here are some activities that management should follow to provide assurance that TQM is being implemented properly and the implementation is successful:

1. Identify, communicate, and enforce quality policies across all departments.
2. Establish quality and productivity objectives.

3. Take action on quality system deficiencies.
4. Ensure that quality and productivity responsibilities are well defined.
5. Remove traditional barriers between departments.
6. Foster teamwork throughout the organization.
7. Train the organization in using TQM tools.
8. Establish metrics to monitor improvement.
9. Help translate top-level metrics into lower-level measurements.

METRICS

A **metric** is a measurement of performance. Metrics should be established that are related to top-level objectives. Metrics and objectives have different terms (e.g., the original objective could be related to dollars, product quality, efficiency, schedule, or other factors). The following are some examples of metrics:

Quality—Some common quality metrics are number of defects, percent defective, parts per million (ppm), percent yield.

Delivery—Common delivery-related metrics are past due, missed schedule, cycle time.

Profit—Some common profit-related metrics are margin, throughput, cost of goods sold.

Inventory—Inventory-related metrics are items such as turnaround (turns), excess inventory, inventory costs.

Marketing—Some common marketing metrics are contract wins and losses, customer evaluation survey results.

Using Metrics for Improvement

Once metrics are established, the following general guidelines should be followed to make them a viable tracking and improvement tool:

1. Ensure that the data source for each metric is accurate and timely.
2. Establish the metric, or method of tracking (e.g., run chart or pie chart).
3. Establish a frequency for reporting the metric and the personnel responsible to prepare and maintain the metric.
4. Establish the reporting audience (who will get the metric).
5. Have management steering teams monitor metrics and help to apply them to processes, low-level associated variables, and teams of people.
6. Have action teams use TQM tools for making improvements in whatever processes or variables are associated with the metric. Establish a TQM training program for TQM tools and techniques. Conduct the training with predetermined teams and improvement projects.
7. Ensure that the management team supports well-developed improvement projects with the necessary resources to implement the best alternative.

PROBLEM SOLVING

To identify and solve problems effectively, there are three primary considerations: the problem-solving team, steps, and techniques used. Every problem has one or more true causes and usually several symptoms. The elimination of the true causes of problems will correct the problems themselves and often will prevent the problem from recurring. A fact of life is that symptoms of a problem are generally easier to find than are causes. Subsequently, symptoms are often solved, but the problem remains.

Problem-Solving Teams

The best time to solve a problem is when it occurs, and the best team to solve the problem is a *natural work team.* A natural work team is a group of people with different responsibilities who are closest to the process or the problem. Companies should not just gather a team of individuals to solve a variety of company problems. The best results come when the problem-solving team is selected by the problem statement.

The *direct members* of a problem-solving team are those who work the process where the problem originated on a daily basis. In the shop, the direct members of a team are the operator, the shop supervisor, process setup personnel, and the inspector (if applicable). This membership applies whether it is a process or a product problem. The *indirect members* of a problem-solving team are those personnel who have technical responsibility for given subjects related to the support of the product or process. Indirect problem-solving team members may be on more than one team at a time. For the shop, they include the manufacturing engineer, quality engineer, and often the industrial engineer. For problems in the machine shop, these direct and indirect members are naturally the ones closest to the problem.

There are times when *consultants* from other departments are also requested to participate in the solution of a problem. These consultants may be, for example, a tooling engineer, metallurgist, chemist, or a statistician. The outside consultant requested depends on the expertise required by the team to obtain the answers necessary in the problem-solving process.

Problem-Solving Steps

Effective problem solving entails five steps used by problem-solving teams:

1. *Identify the problem* and prepare a clear problem statement.
2. *Quantify the problem* (measure the impact and extent).
3. *Identify the cause(s)* (not symptoms).
4. *Take action* to correct the cause(s).
5. *Follow up* to ensure that the action taken was effective.

SUMMARY

Total quality management takes commitment, planning, setting objectives, training, and teamwork to be successful. It also involves the willingness to make changes in the organization, processes, and products for the purpose of improvement. By its very nature, TQM is not just a quality improvement approach, but rather a productivity improvement approach. Improving the quality in the total operation will automatically improve the efficiency of those operations and help make the company more profitable.

If a natural work team is assembled and supported and if it uses these problem-solving steps and tools, the problem will be solved more quickly and efficiently. Care should be taken during problem solving to avoid "symptom" solutions and to focus instead on eliminating the root cause of the problem. The problem-solving tools covered in this chapter help the team to use and analyze data at each problem-solving step and make effective decisions.

REVIEW QUESTIONS

1. The three fundamental elements that exist in every process are _____ , processing, and output.
2. Everyone in the company has a customer and a(n) _____ .
3. The order entry function is an example of a(n) _____ process.
4. Which of the following are examples of variables related to a service process?
   a. time it takes to repair something
   b. time it takes to arrive on the scene of needed repair
   c. cost of the repair
   d. effectiveness of the repair
   e. all of the above
5. A planned and systematic order of events that provides assurance with regard to quality is the definition of _____ _____ .
6. Which of the following is a regulatory process through which we measure process performance, compare it to a standard, and act on the difference?
   a. quality assurance
   b. inspection
   c. performance measurement
   d. quality control
7. Which of the following functions are an appraisal activity?
   a. quality planning
   b. quality control
   c. inspection
   d. prevention methods
8. Defining a customer's wants and needs is a primary responsibility of
   a. engineering
   b. quality assurance
   c. manufacturing engineering
   d. marketing and sales
9. Which of the following functions has the primary responsibility for the quality of design?
   a. engineering
   b. quality assurance
   c. manufacturing engineering
   d. marketing and sales

10. From a design point of view, which of the following can be the root cause of quality problems in the product?
    a. improper tolerancing
    b. missing standards
    c. unclear standards or specifications
    d. all of the above
11. The design of special processes in some companies is the responsibility of
    a. product engineering
    b. process engineering
    c. marketing and sales
    d. manufacturing
12. Manufacturing engineering's responsibilities includes
    a. planning the sequence of operations
    b. setting product design tolerances
    c. flowdown of customer requirements
    d. none of the above
13. Control of quality is most effective when it is controlled at the _____ .
14. Which of the following functions has always had the direct responsibility for the quality of products, parts, and services from a supplier?
    a. manufacturing engineering
    b. engineering
    c. quality assurance
    d. procurement
15. Suppliers should be rated with respect to required levels of price, quality, and _____ .
16. Calibration of measuring equipment, quality reporting, quality audits, and certain inspection operations are the direct responsibility of the _____ department.
    a. engineering
    b. procurement
    c. marketing
    d. quality
17. Which of the following is not a quality responsibility of the shipping department?
    a. proper packaging
    b. proper identification
    c. damage prevention
    d. field service
18. Who is responsible for providing information that helps close the gap between customers' wants and needs and actual delivered quality?
    a. marketing
    b. quality assurance
    c. field service
    d. shipping
19. An analysis tool that helps separate problems or causes with respect to their degree of importance is:
    a. Pareto analysis

    b. brainstorming

    c. checksheets

    d. run chart

20. A well-structured _____ can be used to collect data in a meaningful manner.

    a. Pareto analysis

    b. checksheet

    c. matrix

    d. flowchart

21. Which of the following tools helps to identify the central tendency and variability of a process variable?

    a. Pareto analysis

    b. run chart

    c. histogram

    d. control chart

22. A method, often used along with cause and effect diagrams, that helps to generate ideas about a problem is

    a. run chart

    b. flow chart

    c. brainstorming

    d. histogram

23. Trends in manufacturing process can be seen using a _____ chart.

    a. histogram

    b. Pareto

    c. cause and effect

    d. run

24. A tool that is used by inspection planners, problem-solving teams, and auditors for understanding the sequence of operations is called the _____ chart.

25. A tool that can help to compare different alternatives to solving a problem is the

    a. histogram

    b. matrix

    c. run chart

    d. control chart

26. Which of the following charts can be used in place of a histogram to obtain the same information?

    a. Pareto chart

    b. run chart

    c. frequency distribution

    d. matrix

27. A tool that helps us to understand if there is a relationship between X and Y variables is the _____ diagram.

    a. cause and effect

    b. fishbone

    c. matrix

    d. scatter

28. Which of the following tools helps us to identify the extent of variation in a process variable?

    a. histogram

    b. Pareto analysis

    c. variable matrix

    d. variable cause and effect

29. An approach early in the design phases of a product, which involves cross-functional teams in the engineering process, is called a(n) _____ engineering approach.

30. Which of the following are poor methods for engineers to establish part tolerances?

    a. "fear" tolerancing

    b. tolerance the part the same as a similar part

    c. standard tolerances that apply to all parts

    d. all of the above

3 *Quality Costs*

One of the methods used to improve quality and provide effective management is an accurate system of reporting quality costs. If a company can accurately measure the costs of quality, it can improve quality and productivity. A quality cost system defines the areas of high costs, which, in turn, define areas of concentration for corrective action to reduce those costs.

CATEGORIES OF QUALITY COSTS

There are four categories in which quality costs are reported:

1. Prevention
2. Appraisal
3. Internal failure
4. External failure

Prevention

Prevention costs are related to any function in a company that attempts to prevent poor quality from being produced. Some examples are:

- Quality engineering tasks (planning)
- Manufacturing engineering tasks (work instructions)
- Design engineering tasks (design reviews)

Appraisal

Appraisal costs are related to functions that appraise (or evaluate). Inspection and testing are two examples.

Internal Failure

Internal failure costs are related to failures (nonconformance) that occur in-house. Two examples are scrap and rework.

External Failure

External failure costs are related to failures (nonconformance) in the field (or at the customer's facility). Some examples are:

- Returned products from the customer that must be inspected, scrapped, or reworked
- Handling customer complaints
- Products that must be reinspected and/or reworked at the customers' facility

Other quality cost examples are discussed below. Most costs will fit one of the four categories. The important thing to remember is that these costs should be reported and corrective action taken to reduce them.

More Examples of Quality Costs

Appraisal costs include the costs of staffing full-time inspectors, inspection supervisors, inspection clerks, and people performing tests. They also include the costs of maintaining materials and equipment used for testing the product, preparing and handling test or inspection records, and performing product audits and field tests.

Internal failure costs include the direct and associated costs incurred due to scrap, rework, and repairs. The direct cost of scrap, rework, and repair is the cost of the product. The associated costs of scrap, rework, and repair (often called the "hidden factory") are the costs of handling, paperwork, replacing, material review board functions, and loss in profits due to reduced prices for substandard products.

External failure costs include the cost of repairing or replacing products returned by the customer, handling customer complaints, recalls for repair or replacement parts during warranty periods, and the costs of testing, legal services, settlements, and other costs associated with product liability problems. External failures also come with "hidden factory" costs, such as processing paperwork, corrective action teams, and other actions that were originated directly because of returned materials. Other costs are not directly measurable, such as customer dissatisfaction (or the impact on the company's business due to customer dissatisfaction).

Prevention costs include the costs of quality planning, process control, quality training, designing equipment and processes to measure and control quality, special studies, vendor relations, variability analyses, design reviews, manufacturing planning, and other activities that are performed for the purpose of preventing defects.

QUALITY COST RELATIONSHIPS

When it first establishes a quality cost monitoring system, a company may be startled by the following relationships and proportions of quality cost expenditures:

1. The total quality cost expenditures are usually a large proportion of sales, profits, or other selected cost denominators.
2. The amounts of internal failure costs (such as scrap, rework, and repair) and appraisal costs (such as inspection, test, and auditing) are the largest portion of the total (and sometimes equal to each other).
3. The amount of external failure costs (due to products delivered to the customer that are not to specifications) are fairly low because of the mass amount of inspection and testing being performed. In general, the more you inspect, the more defects you find. The more defects you find, the more you inspect.
4. The amount of prevention cost expenditures is also a small portion of the total.

Prevention versus Corrective Action

Prevention is a vital part of quality costs, mainly because it should be the major area of investment by a company to reduce overall quality costs. Prevention activities help stop poor quality *before* it occurs. Any action taken after poor quality occurs is corrective action. Corrective action is necessary to improve poor quality, but preventive action saves money by preventing poor quality. Prevention is the one category where increased investments will reduce the other categories and the total costs of quality.

Preventive Action.   Preventive action costs money because extra time and effort are needed to plan for good quality.

Corrective Action.   Corrective action costs more than prevention because of the costs of investigating and correcting in addition to the costs of the defective products (if they were scrap, they must be replaced, and if they can be reworked, it costs to rework them).
Remember: Quality costs are not only the costs of the quality assurance department; they are the costs of all departments that have an influence on product quality.

PARETO'S LAW AND QUALITY COSTS

One important subject to understand when attempting to solve problems is Pareto's law. The basic concept of Pareto's law is that in almost all cases, 80 percent of the problems are caused by 20 percent of the people, machines, and other factors. Pareto's law says that you should concentrate on the major problems immediately and work on the minor problems later. This is often referred to as "the vital few over the trivial many." When setting up a quality costs reporting system, one of the most important parts is applying Pareto's law to the reports so that people are concentrating on the vital few.

■ **Example 3–1**
You collect several scrap costs (internal failure). These costs are shown below.

Department	Quantity Scrapped	Cost per Unit
Welding	12	$3.00
Grinding	3	25.00
Stamping	103	.30
Forging	1	10.00
Heat treat	3	100.00

Since you are reporting quality costs, one of the first things you must do is find the total cost for each department (the quantity scrapped times the cost per unit). They are shown here:

Welding = $ 36.00 scrap cost

Grinding = $ 75.00 scrap cost

Stamping = $ 30.90 scrap cost

Forging = $ 10.00 scrap cost

Heat treat = $300.00 scrap cost

Next, you report these costs in Pareto's form (the highest cost first):

Heat treat = $300.00 scrap cost

Grinding = $ 75.00 scrap cost

Welding = $ 36.00 scrap cost

Stamping = $ 30.90 scrap cost

Forging = $ 10.00 scrap cost

  The Pareto list you just made shows that the highest quality cost area to the company is in the heat treat department. This scrap should be investigated first, and attempts made to correct the problem. Next, the grinding department should be investigated to correct the $75.00 scrap problem, and so on.
  It would be a great mistake for a company to investigate the problem in the stamping department (where the scrap quantity is high, but the cost is only 30 cents each) while letting the high-cost problems in the heat treat department continue to occur.

  If you are not on a quality cost reporting system, it would be a good idea to start one. Until then, you may use Pareto's law in other ways. One approach is to consider quantity instead of costs.

■ Example 3–2

A company makes one product where overall scrap costs remain approximately the same. Here, you can use Pareto's law by the quantities of scrap. This can be done companywide, departmentwide, or by operation.

Department	Quantity Scrapped	Quantity Produced
Grind	123	1,000
Heat treat	21	100,000
Stamping	1100	3,000

Pareto Analysis by Quantity

Stamping = 1100
Grind = 123
Heat treat = 21

Pareto Analysis by Percent of Product[a]

Stamping = 37%
Grind = 12.3%
Heat treat = 0.021%

---

[a]Quantity scrapped/quantity produced $\times$ 100.

In the above examples, no matter how you report it, the stamping department is at the top of the Pareto analysis and requires immediate attention to correct the problem(s).

Now, you see that understanding Pareto's law can be an asset to any company (or even at home to find out where most of your money is spent).

GRAPHS, CHARTS, AND QUALITY COSTS

There are many ways to report quality costs in meaningful relationships to each other or over time. Most of them have a lot to do with Pareto's law (the vital few versus the trivial many). Almost all reports include charts or graphs of some kind, used mainly to represent trends (the magnitude of the problem) pictorially. Several different types of charts and graphs are used, but we will only look at the three most widely used: line graphs, bar graphs, and pie charts.

Line Graphs

Line graphs are usually drawn on square grid paper. They are used for many things, but for our purposes, we will use them for scrap reports. The graph in Figure 3.1, for example, shows $1100 worth of scrap in week 1, $1200 in week 2, and so on.

Some line graphs serve a dual purpose. For example, Figure 3.2 shows a line graph of scrap cost and quantity of scrap (% of production) for each week. The graph in this figure shows that in week 2 there was $1200 worth of scrap, or 4 percent of production.

FIGURE 3.1
Line graph.

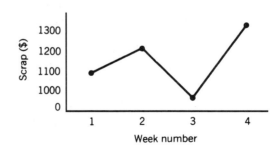

FIGURE 3.2
Dual-purpose line graph, including legend.

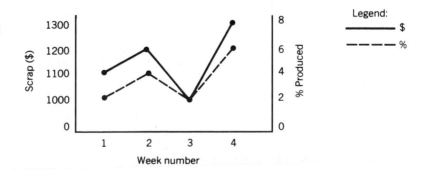

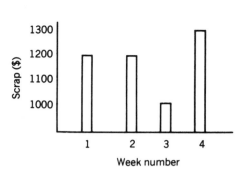

FIGURE 3.3
Bar graph.

FIGURE 3.4
Comparison bar graph.

Bar Graphs

Bar graphs are also used in reports of various kinds. Two main styles are used. One shows the magnitude of a single problem, and the other makes comparisons.

Bar Graph—Magnitude.   The graph in Figure 3.3 shows, for example, that there was $1300 worth of scrap in week 4.

Bar Graph—Comparison.   The graph in Figure 3.4 shows the total quantity of scrap for each week and the machine that contributed most to that scrap (machine 201). The length of the bar surrounding the machine number indicates the magnitude (in this chart, the quantity of scrap).

Pie Charts

Pie charts are very effective in showing Pareto problems. Two examples of the use of pie charts are shown in Figure 3.5 and 3.6.

Figure 3.5 is a pie chart that reflects total company scrap costs by department. This shows that departments 21 and 19 contribute most of the scrap. (*D* means department and *K* means thousands of dollars.) Figure 3.6 is a pie chart of the percentages of total scrap by machine. This chart shows that if you solve problems at machines 200 and 185, most of your scrap will be reduced.

Graphs and charts are very effective in showing trends and relationships in whatever you are reporting or comparing. They show you when things are getting worse or better, and they also help you recognize the most important thing—*the specific areas that need improvement.*

PREVENTION: THE RETURN ON QUALITY INVESTMENT (ROQI)

A primary consideration in any business has to do with the *return on investment* (ROI). Management must consider that if they invest in necessary quality programs, they will

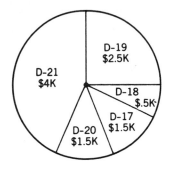

FIGURE 3.5
Pie chart.

FIGURE 3.6
Pie chart.

receive a return on their investment and more. This return is not always immediate, however.

One investment that management often discards is in the area of prevention. This is because returns are seldom immediate and because it is difficult to measure them. I call prevention techniques a *return on quality investment* (ROQI). Management makes a certain investment in the people, tools, systems, training, and other factors that will help prevent problems downstream. An ROQI is not directly measurable because you do not know the problems that might have occurred if you did not take preventive measures. You can measure the effects of prevention, providing you have measured quality costs (scrap, rework, warranty, etc.) prior to taking preventive action. All you need to do is continue measuring the progress you are making in these areas, and you can say that some of this progress is due to prevention and some would have happened anyway.

It is just good business to take preventive action. Common sense tells you that if you prevent the occurrence of one problem one time, then you may have prevented its occurrence several times. What are some areas of prevention that could use more investment?

1. *Training.* Training is one of the most important areas in which to invest. People who are highly trained will still make some mistakes because it is human nature; however, people who are not trained will make even more frequent and costly mistakes.
2. *Process control.* Statistical process control (SPC) is another prevention technique. It is costly to implement but will be a major factor in quality improvement. With SPC, costly bad parts (and scrap and rework) can be prevented.
3. *Quality engineering.* This group of people shares a sole purpose: to plan for quality and analyze the results. Planning always pays off because if quality is well planned, things are usually done right the first time.
4. *Calibration.* Equipment calibration (performed properly) is a major investment in people and tools that can be a turning point in quality improvement. Without proper calibration, inspection and test personnel cannot be certain their measurements are accurate.
5. *Effective reporting of quality costs.* Reporting the various quality costs incurred during given time periods can detect areas in need of immediate improvement. Correcting these problem areas can be a factor in preventing future costs in these areas.

6. *Communication.* The improvement of communications between customers and vendors can be a major factor in quality improvement. Many problems occur simply because people cannot or do not communicate.

Other prevention areas exceed the purpose of this chapter, but some of the major areas have been discussed. Management must be aware of their options:

1. Take no action and leave quality to chance.
2. Take sound preventive action and make an impact on quality and productivity levels.

Preventive action is costly, but corrective action costs more in the final analysis.

QUALITY COST REPORTING

The main purpose of collecting and reporting quality cost data is to identify areas of excessive costs and trends that reflect poor performance and to improve the profits of the company. It's important to note that reporting quality costs is not as simple as just providing data on the costs of quality. Quality costs reporting is most effective when accurate cost data are converted into *information* so that accurate decisions can be made. Many companies have huge amounts of data, but no information. In most cases, charts (such as those previously covered) will help to transform data into useful information. When implementing a quality cost reporting system, make sure that costs are collected, analyzed, and reported from all four cost categories.

Some critical factors in establishing a quality cost reporting system are:

1. Identify reliable sources for quality cost data. Note that, in most cases, one of the best sources for quality cost data is the financial department. Other sources include existing scrap, material, labor, and salary reports.
2. Identify reliable data collection methods and techniques.
3. Choose the most effective graphical methods to transform the data into information, and choose the most effective method of preparing the graphs for the report.
4. Identify how often the report will be distributed and any problems with the timing of the information. Some quality cost data may not be available for collection in a timely manner; therefore, the information may be late when reported.
5. Identify key personnel for the distribution of the report. Key personnel include any person at any management level who has direct interest in the results, is responsible for monitoring progress, and/or can instigate or take improvement action.
6. Establish a baseline for comparing overall quality costs (such as the ratio of quality costs to net profit). Refer to the following discussion on baselines.
7. Try out the system before it is fully implemented so that improvements can be made.

In the search for quality cost information, it is also critical that the information should be related only to costs of quality, not to other company cost areas such as capital equipment costs or inventory costs. A company that purchases a forklift is not incurring a quality cost, for example. Forklifts are capital equipment costs. Quality costs always have a direct relationship with preventing, finding, or making defects.

Baselines for Quality Cost Reporting

In every quality cost reporting system, some baseline must be established to compare quality costs to overall company results. Since quality costs are measured in dollars, it is only practical (and analytically correct) to compare quality cost dollars to other company-related dollar baselines. For instance, management may want to know and track the percentage of quality costs to overall company costs. In other cases, quality cost expenditures may be compared to profits. Whatever baseline is selected, the proportion of quality costs (in dollars) is compared to that baseline (in dollars).

Examples of baselines for quality cost comparisons include net profit, gross receipts, total operating costs, or total sales. Net profit is the baseline most widely used. As mentioned earlier in this chapter, once quality cost reporting begins for the first time, the company is often surprised at the proportion of quality costs with respect to their baseline.

QUALITY COST SYSTEM RESULTS

It is important to understand that collecting, analyzing, and acting on excessive quality costs is not done only to improve quality. The costs of quality incurred by a company also have a major effect on profits. A successful quality costs reporting system will improve the quality of the product and process but will also reduce costs, which is a factor in improving profits. Effective quality cost systems also play a vital role in managing the company. Any manager automatically has a direct interest in quality cost reporting because the information is reported in the language of management (dollars) and has a direct relationship to drains in profit, which are opportunities for improving profit.

Customer satisfaction improves because (1) the product improves, (2) the process improves, or (3) the cost of the product is reduced. Note that cost is not the same as price. Price is what someone pays for something. Cost is the price plus the cost of ownership. In the case of defective products, processing, or service, the cost is always higher than the price. Customers are often lost because of cost, not price. Most customers are aware that price and cost should at least approach equality. With quality products, processes, and services, price and cost tend to be nearly the same.

Quality Cost Practice Problem

The information shown in Figure 3.7 refers to activities that took place or expenditures that were made in a company during one month. These activities include all categories of quality costs. Review the information and complete the blanks in the quality cost data table shown in Figure 3.8.

FIGURE 3.7
Quality cost data report.

Activity/ expenditure	Actual cost	Activity/ expenditure	Actual cost
Final inspection	$24,500	Product audits	$ 1,200
Design review meetings	$ 3,360	Scrap	$21,500
Training	$ 7,000	Warranty claims	$ 1,200
Repair	$ 2,400	Final test	$ 3,300
Rework returned parts	$ 2,000	Rework in-house	$ 8,500
Handling customer complaint	$ 3,000	Prepared work instruction	$ 500

FIGURE 3.8
Quality costs data table.

Quality cost category	Total cost per category	Percent of total quality cost
Prevention	$_____	_____%
Appraisal	$_____	_____%
Internal failure	$_____	_____%
External failure	$_____	_____%
	Total of all quality costs for the month: $_____	

Solution. Upon completing the quality cost table in Figure 3.8, compare your results to these totals for the month:

Prevention totaled $10,860 (13.8% of all quality costs). The applicable prevention cost areas were design review meetings, training, and prepared work instruction.

Appraisal totaled $29,000 (37% of all quality costs). The applicable appraisal cost areas were final inspection, product audits, and final test.

Internal failure totaled $32,400 (41.3% of all quality costs). The applicable internal failure costs were repair, scrap, and rework in-house.

External failure totaled $6,200 (7.9% of all quality costs). The applicable external failure costs were rework returned parts, handling customer complaint, and warranty claims.

Without some baseline for comparison, one cannot tell whether the total quality costs for the month were excessive or not. In this problem, then, only cost categories can be compared to each other.

Notice, however, that prevention and external failures are lower than appraisal and internal failures. This could be an indicator that the quality system is more detective than preventive, although this is not always true. Further investigation would help prove or disprove this statement. Keep in mind that a detective system tends to have appraisal and internal failure costs that are high in comparison to others. In fact, internal failure costs and appraisal costs tend to trend together in a detection system because of their direct effect on each other.

REVIEW QUESTIONS

1. Which of the following quality cost categories, with investment, will decrease costs in the other three categories?
   a. appraisal
   b. prevention
   c. external failure
   d. internal failure

2. The cost of handling nonconforming materials found during the operation is reported under which category of quality costs?
   a. appraisal
   b. prevention
   c. external failure
   d. internal failure

3. Which of the following is not a quality cost?
   a. inspection
   b. process control
   c. inventory
   d. audits

4. (True or false) Quality costs are dependent only upon the costs incurred in the quality assurance department.

5. Quality costs are associated with all activities that find or prevent _____ .

6. The primary purpose of collecting, reporting, and acting on quality costs is to reduce
   a. failure costs
   b. defects
   c. overall quality costs
   d. profits

7. Which of the following examples is not an internal failure cost?
   a. scrap
   b. calibration
   c. rework
   d. authorized repair

8. Which of the following is not a prevention quality cost?
   a. quality planning
   b. process control
   c. design reviews
   d. calibration

9. Which of the following is not an appraisal quality cost?
   a. inspection
   b. audits
   c. inspection planning
   d. testing
10. Which of the following is not an external failure quality cost?
   a. defective products that have been shipped
   b. defective products found in shipping
   c. warranty claims
   d. analysis of returned products
11. The following costs have been reported for a month:

Capital equipment costs	$100,000
Design reviews	$ 30,000
Inspection and test	$ 60,000
Excess inventory	$ 20,000
In-house scrap and rework	$ 50,000
Process control	$ 10,000
Customer returns rework	$ 4,000
Reinspecting in-house rework	$ 1,000

What is the total amount spent on appraisal costs for the month?
12. In question 11, what is the total amount spent for prevention costs during the month?
13. In question 11, what is the total of all quality costs for the month?
14. In question 11, how much money was spent on internal failure costs?
15. In question 11, how much was spent on external failure costs?

4 *Quality Audits*

The purpose of this chapter is to provide an introduction to modern quality audits. A quality audit is a planned and documented assessment of a system, product, or process that is usually performed by an independent qualified auditor or audit team. Audits can be used to help determine the efficiency of the quality system, verify the quality of a product, and measure the results. The primary purpose of quality audits should be *improvement.* Quality auditing is intended to be a management tool for monitoring and improving the performance of a quality system. The benefits of quality audits are numerous and include the following:

1. Audits provide a baseline assessment for improvement purposes.
2. Audits help prevent customer complaints.
3. Audits serve to identify specific operations that require improvement (hopefully in advance of producing poor product).
4. Audits foster some immediate improvements during the conduct of an audit.
5. Audits provide an objective point of view that is often necessary to find problems.
6. Audits provide information on which management decisions can be based.

Quality audits can be conducted by a variety of people. Auditors can be internal company personnel (who are independent from the area being audited), customer representatives, or outside auditing agencies (such as regulatory agencies or a third-party audit firm).

WHY AUDIT?

Audits can be either a powerful tool to help achieve significant improvements or a complete waste of time if they are not planned and conducted properly. If the actual objective is not improvement, then the audit can cause more problems than it corrects. Auditing involves a considerable amount of time, effort, and expense that are worthwhile when improvements are made in quality and productivity.

Conducting quality audits for the purpose of improvement is simply good business. Quality audits help to determine the effectiveness of the quality system, identify areas for improvement, improve quality problems, satisfy customer requirements, and improve supplier quality. Quality audits help to determine whether we are meeting our own quality objectives and policies, and they help to provide assurance that we are meeting all of our customers' requirements.

KEY STEPS OF THE AUDITING PROCESS

Five critical steps should be observed in any auditing process:

1. *Initiation* of the audit is the responsibility of the client. The client should specify the requirements authorizing the audit, the specific objectives or purposes of the audit, and the scope of the audit.
2. *Planning* the audit is the responsibility of the audit team and lead auditor. They review requirements documents and/or specifications, prepare audit checklists (as applicable), select the audit team members, schedule the audit, and make initial contact with the organization being audited.
3. *Conducting* the audit includes responsibilities of the audit team and the auditee. This is the actual audit itself, including an entrance meeting with the auditee, performing the audit, daily auditor meetings and auditee debriefings, formulating the audit report, and the exit interview.
4. *Reporting* audit results is the direct responsibility of the lead auditor. Audit reports should be factual, concise, and written so that readers can understand and recreate all findings. Audit reports are sent to the client and to the organization that was audited.
5. *Closing* the audit includes responsibilities of the audit team and auditee. The auditee is required to correct all findings (by removing root causes) and give dates when these actions will be implemented. The audit team reviews the auditee's response, discusses any problems with the response with the auditee, and follows up on actions that are taken. Once follow-up results are acceptable, the audit can be closed. The report and backup documents are then filed for the required period of time. At times, reports are used for future follow-up audits.

AUDIT FINDINGS AND OBSERVATIONS

All quality audits will produce findings. In the old school of auditing, this was true because auditors were instructed to "find lots of problems." Even under today's auditing practices, there will be findings because the perfect quality system has yet to be made and implemented. Auditors also make observations. There is a big difference between findings and observations.

Audit findings are statements of fact regarding actual noncompliance with a given requirement. Audit findings always require corrective action. *Audit observations* are statements of potential problems that *may* result in noncompliance in the future. Observations deserve attention but do not require formal corrective action. An audit team that is considering the prime objective (improvement) will have a mixture of findings and observations to report.

TYPES OF QUALITY AUDITS

There are three main types of quality audits: system audits, product audits, and process audits. A **system audit** (typically used to see if the quality system meets contractual requirements) can encompass policies, procedures, documentation, specifications, work instructions, and so on. A **product audit** is used to determine if the product conforms to design requirements (e.g., drawings, specifications, etc.). A **process audit** is intended to examine a complete process (input, processing, and output) to see if the process is being performed to established process procedures, specifications, and/or standards. At times, process audits are performed to audit process control methods. Process control audits can be used to help determine process control status and process capability.

"Party" Audits

There are first-party, second-party, and third-party audits. **Internal quality audits** are first-party audits that are performed by independent, objective, and qualified personnel within the company. **External quality audits** are second-party audits that are performed by auditors external to the company such as customer representatives (or audits that you perform on your supplier). Third-party audits are conducted by an outside agency.

AUDIT APPROACHES AND PURPOSES

All audits have auditors, an auditee, and a client. The **client** is the organization that requests the audit. The approach to auditing depends considerably on the client's request. Approaches to quality auditing (and their purposes) may vary from company to company because there is a wide selection from which to choose. The following is an overview of different approaches to quality auditing.

Self-audits have been used successfully in preparing for formal audits. In a self-audit, people conduct audits of their own operations for the purpose of making their own internal improvements. Self-audits can be healthy for an organization if they are implemented properly and without imposing blame.

Informal audits are similar to self-audits except that they are conducted by personnel outside the area being audited. Often informal audits are conducted by internal company personnel from another department. They can take an outsider's look at the operation. They can be objective and informal (meaning that findings are not necessarily documented). At times, findings are documented, but only for the auditee's records. The purpose of an informal audit is improvement. It can also be used as a preparatory measure for formal audits.

Formal audits are conducted by internal objective personnel, outside customers, or regulatory agencies. All results are documented and reported to management with formal requests for corrective action.

Third-party assessments are audits performed by outside agencies who do not report to the company being audited or to their customers. Typically, the third-party assessment is conducted by authorities in the field of expertise of the function being audited.

Pre-award surveys are audits of potential suppliers by a customer or an associated regulatory agency prior to awarding a contract. The purpose of a pre-award survey is to assess the supplier's ability to supply the goods and/or services requested by the purchase order.

Follow-up audits are conducted to assess the effectiveness of corrective action that has been implemented as a result of a previous audit.

Surprise audits are usually conducted in areas such as financial management and taxes. There is generally no place for the surprise audit in modern quality auditing approaches. In the old school of auditing, auditors felt that the auditee should not be given time to "cover up" deficient areas. At the same time, some auditees did cover up deficient areas. This was happening because the old school of auditing had the expressed objective of quantity findings regardless of their significance. This approach automatically generated an adversarial relationship between clients, auditors, and auditees, resulting in cover-ups and surprise audits. The new approach to auditing is, and should be, a win-win approach for the expressed purpose of improvement. This objective helps to make the surprise quality audit obsolete.

AUDITORS AND AUDIT TEAMS

An audit team is typically a group of auditors and a lead auditor. The auditors have been trained and are skilled in the auditing process. They understand that the primary purpose of auditing is improvement, and they should have some experience in the specific area they are going to audit. Efforts should be made to select qualified auditors. An objective audit can still be achieved, however, without extensive experience on the auditor's part.

The attitude of an auditor is critical. Auditors should act professionally at all times, keep the primary objective (improvement) in mind, and do their best to help find and correct serious deficiencies when they exist. Auditors should never be biased, subjective, or argumentative. Effective auditors:

1. Have auditing training and experience
2. Have good organization and planning skills
3. Can make decisions
4. Understand human factors and motivation
5. Are unbiased and professional at all times
6. Know the function being audited (technical ability)
7. Have good written and verbal communication skills
8. Are adaptable to changing environments and assignments

The **lead auditor** is usually responsible for the audit team throughout all phases of the audit from audit request to close-out. The following functions are typical responsibilities of the lead auditor:

1. Perform (or participate in) the selection of the auditors.
2. Supervise the audit team and make auditor assignments.
3. Arbitrate between auditors when necessary.
4. Chair the opening meeting and the exit interview.
5. Plan the audit (including preparation of working documents).
6. Prepare the final audit report.
7. Plan the follow-up audit.

The Auditee

The auditee is the organization or function being audited. It shares responsibility for the audit's success and should keep in mind that the primary purpose of the audit is to help improve operations. In general, *before the audit* the auditee will provide

1. Personnel to assist the auditor during the audit
2. Materials or information needed by the audit team, such as manuals, specifications, and procedures
3. An area where auditors can work

During the audit, the auditee will

1. Cooperate with the audit team
2. Review all findings with the auditors to ensure they are correct and complete

After the audit, the auditee will

1. Implement corrective action on all findings
2. Cooperate with the audit team in preparing for the follow-up audit (after corrective action has been implemented)

PLANNING THE AUDIT

Planning for an audit usually begins soon after the client's request for the audit. A wide variety of activities occur during the planning phases. At a minimum, audit planning should be concerned with the following elements:

Objectives and scope need to be defined up front. The objective of the audit is the outcome that the client expects. The scope (extent) of the audit depends heavily on the objective.

Background information is often required to understand the auditee's organization in terms of size, complexity, and so on. This information is required to enhance the understanding of the scope and to plan the length and resources needed for the audit.

Auditor selection has to do with the function being audited, the expertise required, the auditing staff and their qualifications, and other factors. The scope and time frame of the audit helps to identify the number of auditors required. The objectives of the audit help to explain what areas will be audited, the type of audit required, and hence the required expertise of the auditor.

The audit schedule needs to be developed in concert with the client, the auditee, and the lead auditor.

Documents that are required from the auditee should be identified and requested early in the planning phase. These documents could include (depending on the objective and scope of the audit) drawings, specifications, standards, procedures, or other documents pertinent to the function.

"Working" papers used by the auditors (such as audit checklists, assignment sheets, audit report forms, the audit schedule, etc.) need to be assembled or developed during the planning phase. Audit organizations may have prepared checklists and reports, but there are times when special checklists need to be developed.

Preaudit surveys are sometimes necessary to develop a further understanding of the operations and to establish which key personnel will assist in planning the audit. Even if a survey is not required, initial communications still need to occur with the auditee to coordinate the audit.

Approval of the audit plan by the client should be established prior to the start of the audit.

Coordinating the Audit

After approval of the audit plan, initial contact should be made with the auditee to announce the upcoming audit (purpose, authority, and scope), request all applicable documents needed by the audit team, and coordinate the details (such as dates, arrangements, key contact personnel, etc.). The initial contact is usually followed by a formal notification letter announcing the audit (typically 30 days in advance of the audit). The formal notification letter should contain, in general, the information covered by the audit plan. The audit plan may be attached to the letter.

CONDUCTING THE AUDIT

Opening Meeting

The first step in performing the audit is the opening meeting (normally conducted by the lead auditor). The purpose of the meeting is to:

1. Review the purpose of the audit and clarify the audit plan.
2. Establish a working relationship between the auditors and the auditee.
3. Review the ground rules and schedule for the audit.
4. Establish daily debriefings between the audit team and auditee management so that the auditee knows what is going on throughout the audit.

If there are audit findings, someone from the auditee organization should be present when the deficiency is found. The finding should be reviewed by auditors and auditee personnel during the daily debriefing. In any audit, all findings should be known and understood by the auditee.

Performing the Audit

Successful audits are always based on the audit objective and the audit plan. Auditors are seeking facts, not opinions. There are many ways to perform the audit, and different techniques can be used to gather facts. The methods discussed in the following paragraphs can be used by auditors.

Interviewing is one of the methods an auditor can use to obtain baseline information. Interviewing involves some key steps. The most important is to put the person being interviewed at ease. The auditor should explain the purpose of the interview and make sure the interview is nonthreatening.

Interview questions should be formulated so that the answers will give the needed information. At interviews, the auditor analyzes what is happening, makes a tentative conclusion about the next step, and explains the next step to the subject. Interviews do not often result in facts. They should result in tentative conclusions. The auditor must find facts to support those conclusions.

Tracing is one way to audit a process, especially where several process steps are involved. The auditor starts at the beginning of an operation and traces each step to the end (or vice versa). In random tracing, the operations are viewed in random order. During tracing, auditors collect all relevant factors (documents, data, etc.) that support (or refute) the tentative conclusions that have been made.

When gathering facts during an audit, auditors should obtain *objective evidence.* During the audit, auditors are often concerned about how many things, parts, areas, operations, and so forth to sample so that they will have high confidence that the sample represents the population. One of the first rules of auditing is that *sample sizes are very small.* Sometimes sampling needs to be performed with specific confidence intervals. In these cases, refer to statistical textbooks on confidence intervals.

Working papers are documents such as audit checklists, audit reports, structured data collection forms, and so forth. Everything in an audit is documented either by completing working papers or by collecting copies of actual documents. One of the most effective working papers is the audit checklist. Checklists are tools that can help obtain relatively uniform audits among a group of auditors regardless of their experience in auditing or in the area being audited. In many cases, a well-prepared checklist has helped an auditor who knew little or nothing about the area being audited complete a successful audit. Checklists also serve as a detailed part of the audit plan because the checklist leads the auditor to every area involved in the audit objective. Audit reports are typically used by auditors to structure their evidence, record their findings and observations, and remind them of information or evidence that has not yet been obtained.

Daily Debriefings

Daily debriefings should be conducted by the audit team with managers from the function being audited. The purpose of these debriefings is to:

1. Identify significant problems that need immediate correction.
2. Allow auditors to request additional information, if needed.
3. Discuss preliminary findings with the auditee.
4. Allow the auditee to introduce facts about preliminary findings that may support or refute the finding.
5. Review the schedule for the next day of auditing.

The auditor may find insufficient facts or evidence. In these cases, auditees can discuss further facts about the preliminary findings that the auditor may have missed or may not have been exposed to during the audit. Keep in mind that the only thing that counts is the truth supported by evidence. This important part of the audit process helps auditors and auditees to get to the truth. If the truth is negative and a finding results, that should be viewed as an opportunity for improvement. If the truth is positive and no deficiency exists, that should be viewed as a worthwhile exercise.

Exit Interviews

Once the actual audit has been completed, the next step is the formal exit interview. The purpose of the exit interview is to:

1. Present the summary of the results of the audit.
2. Review all findings and observations.
3. Allow the auditee to understand, ask questions, and provide any evidence for or against all findings and observations.
4. Explain the next steps (following the audit) to the auditee.

If the auditee introduces new evidence regarding findings, the audit team should either evaluate the evidence immediately or reaudit the area. If audit findings are refuted, apologies to the auditee are in order and the incorrect finding is removed from the audit report.

Upon completion of the exit meeting, auditee management can begin root cause identification even though the final audit report has not been generated. The reason for this is the fact that the final audit report is guaranteed to precisely reflect the audit findings, observations, and conclusions covered in the exit interview. Since the lead auditor prepares the final report, it is his or her direct responsibility to make sure that the report matches the exit interview results.

AUDIT REPORTING

The final audit report, prepared by the lead auditor, should be nothing more than the formally documented details of the exit interview. The audit report should include a cover letter and the following key topics:

1. Auditee information (e.g., name, address, audit dates, auditors, key auditee personnel contacted, etc.)
2. The objectives and scope of the audit
3. The results of the audit (e.g., rating, acceptability, etc.)
4. A section thanking the auditee for its cooperation
5. The required response date (the date a corrective action response is due from the auditee). Typically, response dates range from 15 to 30 days after the audit. The response is the actual corrective action that will be performed and due dates for when the action will be implemented.
6. The distribution (list of who will receive the report)

Audit reports should be clear, concise, factual, and impersonal. Audit reports should use functions instead of names except where something very positive is being reported and the person is being praised for exemplary work. Audit reports should not report anything that is proprietary in nature unless the distribution lists all have the right to know; otherwise, proprietary items should be reported separately. If legal matters are involved, they should have been reported already to the legal department. Of course, critical or safety-related findings were reported long ago in the daily debriefing and are no surprise now. Any findings that were found and corrected during the audit are also included in the final report.

Audit reports, as mentioned earlier, should cover only the results that were discussed in the exit interview and nothing more. Therefore, it is important to make sure the exit interview is held for as long as necessary to cover every significant finding and observation.

Audit reports should also include a summary of the final results of the audit and the rating achieved (or the acceptability). The summary is one of the most important parts of the report. Once the report is completed, the client should review the report and attach its own cover letter. The final report is then delivered to the auditee.

Acceptability, other than specific customer rankings, is usually stated in one of four different ways:

1. *Unconditional approval* (which is rarely applicable or used) states that the audit was passed with "flying colors" and therefore, the audit is closed.
2. *Conditional approval* indicates that the function audited is acceptable with minor improvements that need to be implemented.
3. *Conditional nonapproval* indicates that the function audited is not acceptable but can be elevated to acceptable if significant and specific improvements are implemented.
4. *Unconditional nonapproval* means that the function or system audited requires major redesigning and further work to be considered for further auditing.

CORRECTIVE ACTION

Corrective action is taken to correct assignable causes that have resulted in deficiencies. Corrective action for audit findings is always the direct responsibility of the auditee (or the function that was audited). Audit reports require a corrective action plan submitted within the time allowed. The action plan, at minimum, should address each finding as follows:

1. State the finding.
2. Identify the root causes.
3. Identify the actions that will be taken to eliminate (and prevent recurrence of) the root causes.
4. Establish target dates when the actions are expected to be implemented.

In most cases, corrective action plans or approaches lack one of the most significant steps in the corrective action process: *follow-up.*

Follow-up is usually the weakest step in correction action. Many believe that simply taking the action is enough. Follow-up ensures that actions taken were effective. In most cases, several probable root causes are determined during the problem-solving process, and then the most probable causes are selected by the problem-solving team. There is no guarantee that the actual root cause of any problem will be properly identified the first time.

In cases where the root cause was not properly identified, the action taken will not correct the problem. When there is more than one root cause, action may be taken on only one of the causes and the problem will be reduced but not solved. Effective follow-up should be performed after every corrective action to make sure that the action taken was effective in eliminating the root causes of the problem. Follow-up can be performed in many ways, including reviewing corrected documents or reauditing a specific area. Not all actions may require a formal follow-up audit.

CLOSING THE AUDIT

The general steps for closing an audit are:

1. Evaluating the corrective action response from the auditee and communicating any problems with the response to the auditee
2. Verifying that the actions in the response have been taken (using appropriate follow-up methods)
3. Closing the audit formally via a letter or memo

Audits should not be formally closed until the auditing agency and the client are satisfied that the audit was effective. A follow-up audit may be in order (which will often result in open findings that have not been fully corrected). At times, auditees are on an audit schedule, in which case the items will be followed up at the next scheduled audit. In other cases, the client may determine that the objective has been met and instruct the lead auditor to close the audit.

REVIEW QUESTIONS

1. The primary purpose of an audit is to
   a. find several findings
   b. achieve improvement
   c. find many observations
   d. convince the auditee to follow procedures
2. Which of the following is considered during the initiation phase of an audit?
   a. gathering facts
   b. needed checklists
   c. auditor selection
   d. purpose and scope
3. A self-audit is
   a. a third-party audit
   b. a first-party audit

  c. a second-party audit
  d. a survey audit
 4. A specific deficiency with regard to a requirement is called a(n) _____ .
 5. An audit of all procedural requirements in a certain area of operations is called a _____ audit.
  a. process
  b. product
  c. survey
  d. system
 6. A customer-focused audit that evaluates a small sample of product for compliance is called a _____ audit.
  a. process
  b. product
  c. survey
  d. system
 7. Which of the following people is responsible for initiating the audit?
  a. lead auditor
  b. chief auditor
  c. client
  d. auditee
 8. Which of the following people are responsible for conducting all audit meetings?
  a. client
  b. chief auditor
  c. auditee
  d. lead auditor
 9. An audit that evaluates all aspects of a particular process is called a _____ audit.
  a. process
  b. product
  c. survey
  d. system
 10. Which of the following is not a good attribute of an auditor?
  a. good communication skills
  b. unbiased and professional
  c. able to find many findings
  d. auditing experience
 11. Who is directly responsible for providing corrective action on all audit findings?
  a. client
  b. lead auditor
  c. auditee
  d. customer
 12. Which of the following is not part of the audit planning phase?
  a. review applicable documents and procedures
  b. prepare needed checklists
  c. develop the audit schedule
  d. conduct the entrance meeting

13. Which of the following is not important when auditors have findings?
    a. they have objective evidence
    b. they provide an audit trail
    c. there are many findings to prove their point
    d. there is a specific requirement supporting their finding
14. For a product audit, sample sizes should be
    a. per Mil-Std-105E
    b. based on the Weibull distribution
    c. very small
    d. none of the above
15. One of the purposes of the daily debriefings is
    a. for auditors to meet
    b. to update the auditee on status and findings
    c. to write the report
    d. to argue specific points or findings
16. Audit reports should be
    a. clear
    b. concise
    c. factual
    d. all of the above
17. Which of the following is not a key element an auditor would expect to see in a corrective action response:
    a. root cause identification
    b. actions to be taken on root cause
    c. who caused the problem
    d. dates for completing the actions
18. Which of the following elements is usually the weakest link in corrective action?
    a. root cause identification
    b. action on root cause
    c. dates for completing the action
    d. follow-up on the action taken
19. At which phase of auditing should the lead auditor fully understand the objective of the audit?
    a. conducting
    b. initiation
    c. planning
    d. closure
20. Who is responsible for selecting qualified auditors?
    a. client
    b. lead auditor
    c. auditee
    d. chief auditee

CHAPTER 5

Inspection Systems and Planning

INSPECTION SYSTEMS

Inspection is an appraisal activity whose primary purpose is to compare the product to applicable specifications and standards and determine whether the product conforms or does not conform to requirements. The inspection system is an integral part of the quality system. Depending on the company and its involvement with the product life cycle, only an inspection system may be required. This is particularly true when the company only makes "products to print" and is not involved with design, testing, or field service.

An inspection system is a combination of:

1. Inspection plans and work instructions
2. Inspection and test equipment
3. Calibration for the inspection and test equipment
4. Inspectors
5. Applicable specifications and acceptance standards
6. Sampling procedures (often included in the planning)
7. Provisions for maintaining records/reports of inspection results

In a typical inspection system all of the above areas have been planned for, implemented, and are functioning properly. There are also commercial, military, and other standards that define inspection system requirements.

Inspection planning, covered later in this chapter, is a critical phase of an inspection system because a good inspection plan provides assurance that inspection activities are performed at the right time, correctly, and consistently. *Inspection and test equipment* are essential because proper inspection requires equipment that is accurate and appropriate for the characteristics being inspected.

Calibration is an important part of inspection because a good calibration system provides assurance that the accuracy of the inspection equipment will be maintained.

Specifications and standards (and control of them) are a vital aspect of assuring that inspectors have an accurate baseline for acceptance. Standards are especially required whenever the characteristics being inspected are cosmetic in nature or may induce errors due to judgment calls by the inspector.

Sampling procedures (often included in the inspection plan) are necessary in order to provide assurance that the correct amount of inspection will be performed and to control **average total inspection (ATI).** Without proper planning and sampling procedures, there is little or no control over the costs of inspection, meeting minimum inspection requirements, reducing risks associated with sampling, and providing assurance that products leaving the area are acceptable.

Inspection records and reports provide a baseline for effective corrective action, monitoring the output of the processes, maintaining historical information for future use, switching sampling inspection levels, and many other important tasks for improving quality. Inspection records should include, at a minimum, the characteristics that were inspected, the quantity inspected, the quantity accepted/rejected, the person who inspected the parts, and information about the part (such as part number, lot/serial number, etc.).

The Changing Role of Inspection

One change to the traditional inspection system (in line with Total Quality Management) is that inspection isn't always performed by inspectors. The most effective inspection is performed at the source where the product is made and as soon as practical after the product has been made. Operators can, and should, perform their own inspection. This is called *in-process inspection* by operators.

The traditional roles of inspectors have changed. In the old days, there may have been roving inspectors who would go from process to process inspecting the work. But it has been found that operators can be the best inspectors when they have had sufficient training and equipment and when they are assigned responsibility for that function. As a result, some inspectors have been switched from their previous function to another inspection function. Some inspectors have become product, process, or system auditors; some have been placed in technical positions for inspection planning purposes; and sometimes the inspection function has been eliminated. Operators make the product and should inspect it so that problems are found and corrected during the process.

How Much Inspection Is Necessary?

The question of how much inspection is best answered in a series of steps. The first step is to classify the characteristics of the product. The typical classifications are critical, major, and minor. *Critical characteristics* are those characteristics that, if defective, could cause bodily harm to the user (or, in military standards, could cause failure of a mission). *Major characteristics* are those that, if defective, could cause loss of function of the product. *Minor characteristics* are those that, if defective, would not affect fit, form, function, safety, or other outcomes classified critical and major. Once critical and major characteristics have been defined, minor characteristics are essentially all other characteristics that remain.

Critical characteristics, in most cases, require a minimum of 100% inspection. Each unit of product is inspected for each critical characteristic. Major and minor characteristics are often subjected to sampling inspection. Refer to Chapter 12, "Lot-by-Lot Acceptance Sampling," for further information about sampling plans and procedures, risks, and the difference in effectiveness between sampling and 100% inspection. In all cases, the inspection plan should state how much to inspect.

INSPECTION POINTS

Inspectors play a key role at specific points in providing the appraisal activity that judges the conformance of products to given requirements. Examples of these points are receiving inspection, source inspection, special inspection operations, and other areas where manufacturing in-process inspection may not be appropriate. Some of these points require professional inspectors (due to the location, equipment, or nature of the job); others are best performed by the operator who runs the process. As mentioned earlier, one of the most important things to keep in mind about inspection is that it should be performed at the source where product characteristics are made.

Identification of these points of inspection is best accomplished using a flowchart of the process of raw materials through finished product.

Source inspection is performed when companies buy products from suppliers and send source inspectors to inspect the product prior to shipment. The role of the source inspector is changing due to various efforts to implement supplier self-inspection (or better yet, control of the supplier's processes). In these cases, the need for source inspection is reduced or eliminated.

Receiving inspection is performed, when necessary, to provide assurance that manufactured parts, purchased parts, and raw materials delivered by suppliers meet purchase order requirements, hence preventing defective products or materials from getting into the assembly or manufacturing process. In receiving inspection, the primary document for requirements is the purchase order. As with source inspection, receiving inspection can be reduced through successful implementation of inspection or process control at the source (the supplier).

First piece inspection is typically performed immediately after the operator has completed, and is satisfied with, the setup of a process. First piece inspection is an important part of the process. The first part is inspected for all characteristics being produced at that process to ensure that the setup is acceptable. Once the first piece has been accepted, the process is allowed to begin production.

In-process inspection, depending on how it is implemented, is very effective because it is performed at the source, the best time to find defects and correct the process. The "old style" in-process inspection was performed either with a roving inspector (from machine to machine) or with batches of products (finished at a specific operation) being sent to the inspection department. The best way to perform in-process inspection is by the operator who runs the process. With proper training, work instructions, measuring equipment, and assignment of responsibility, the operator can be the best inspector.

Final inspection is a process that was originally intended to be the "last chance" to assure product quality prior to shipment (for finished parts or assemblies) or prior to storage (for finished parts to be assembled later). Once again, the "old style" final inspection function included professional inspectors who were responsible for inspecting finished parts and/or assemblies and verifying that all documents were included and in order. With the implementation of in-process inspection by the operator/assembler in manufacturing/assembly processes, final inspection can often be reduced to a verification or product auditing process. Note that final inspection is one of the worst places to find defects in the product because it is so late in the process.

INSPECTION PERSONNEL

The following guidelines apply to professional inspectors or others who perform the inspection function, such as manufacturing operators or assemblers. A successful inspection function depends on the people who perform that function.

1. There are no "gray" areas. Clear up anything subjective with supervisors.
2. Consistently accept good products and reject defective products.
3. Understand all applicable specifications and use only the correct revisions of them.
4. Use calibrated measuring tools and equipment.
5. Keep accurate and complete records/reports.
6. Follow inspection plans and applicable sampling plans.
7. Provide accurate and complete descriptions of defectives on rejection reports.
8. Be open-minded and willing to learn about inspection methods.
9. Assist others, when possible, in taking effective corrective action on defects.

Mechanical inspection functions, regardless of who performs them, require the following knowledge and experience:

1. Knowledge of the role the inspection function plays in product quality systems
2. Knowledge of all applicable standards and specifications and how to interpret them
3. Knowledge of shop-related mathematics
4. Knowledge of acceptance sampling plans
5. Knowledge of applicable dimensional measuring tools, equipment, and methods
6. Knowledge of geometric tolerancing methods (when specifications use this type of tolerancing)
7. Knowledge of applicable records, reports, or charts that are required to report the results of inspection and/or control the process

INSPECTION PLANNING

Inspection planning involves up-front planning for the inspection of the product. Inspection plans can be written by anyone who is capable of preparing them (e.g., understands the quality requirements of the product, the appropriate inspection equipment and methods, sampling plans, etc.). The elements of inspection planning include:

1. *Characteristics to inspect.* The specific characteristics to inspect are determined based on the inspection point (or process where inspection occurs), characteristics finished at that process, characteristics affected by that process, characteristics that cannot be inspected later in the process, and characteristics for which there is "no return" after that process is finished (or subsequent processing may scrap the part).
2. *Quantity of parts to inspect.* The question of how many parts to inspect involves the classification of the characteristic (critical, major, minor), customer's specific requirements, and choices between 100% inspection, applicable sampling plans and acceptable quality

level (AQL) values, or process control requirements. In general, critical characteristics are often required to be 100% inspected (sampling does not apply), major characteristics are often sampled at low AQL values and minor characteristics are often sampled at higher AQL values. Low AQL values cause the sampling plan to be more discriminating than higher AQL values. It is important to note that 100% inspection is not 100% effective (generally because of boredom or fatigue on the part of the inspector). Process control methods generally determine the subgroup sample size for the purpose of control charting, not acceptance sampling. For further information on lot-by-lot acceptance sampling plans, AQLs, and the choices between sampling and 100% inspection, refer to Chapter 12. For information regarding process control samples, refer to Chapter 13.

3. *Inspection equipment or method.* In any department of inspectors (or operators) there is usually wide variation in experience, skill levels, training, and opinions on how characteristics should be inspected. This causes major differences in inspection results, so the inspection plan should identify the equipment and/or method that will be used by all. Planning the specific equipment and/or method provides consistency between inspectors, accuracy and precision of measurements, reduced measurement variation, improved learning curve for entry-level inspectors, and reduced alpha and beta risks from measurement errors. The above is also true in process control planning where measurement error must be avoided so that control charts will show process variation, not measurement variation.

4. *Work area design.* The work area for inspection needs to be designed to make inspection effective. The elements to consider in work area design include appropriate lighting (especially for visual inspection), controlled temperature (for appropriate measurements), workbench layout (for product flow, incoming lots, outgoing lots, and movements of the inspector), readily available inspection equipment, and applicable inspection plans, part specifications, and part standards.

5. *Inspection records.* The purpose of inspection records is to establish historical information for process improvements, product traceability purposes, corrective action, decisions on normal, tightened, or reduced inspection levels, and for many other reasons. Planning for inspection should include the types of records that must be maintained, the format of information on those records, and guidelines for filing records for future reference (near and long term).

6. *Cost of inspection.* While planning for inspection, the planner should consider the cost of inspection, which is often measured as average total inspection (ATI), and the cost of inspection equipment. ATI is the average of the total amount of inspection performed on the product (or in the inspection area). Depending on the way the company decides to measure ATI, the inspection planner should evaluate the plan for its impact on cost. Some of the main elements that have a direct effect on ATI are:
   a. The amount of inspection (100%, large samples, small samples, etc.)
   b. Existing inspection equipment (location, quantity, complexity, etc.)
   c. The cost of new equipment (not already owned by the company)
   d. Existing quality levels (e.g., considerable reinspection required)
   e. The number of inspection points in the process
   f. Duplicate (or redundant) inspections performed

Inspection Plans

Inspection plans should cover, at a minimum, the following topics:

1. What to inspect
2. Which tools and/or methods of inspection to use
3. How many to inspect (sampling, 100%, etc.) and how to select them
4. Which standards and specifications apply
5. What type of inspection records to maintain
6. How to handle and record rejected lots or parts
7. Which workmanship or cosmetic standards apply

Detailed Inspection Instructions

Detailed instructions for the inspector should be prepared in cases where there are complex setups, measuring equipment, methods, or techniques. These instructions may include a customer's special inspection requirements or any special environmental, product, or process concerns. These detailed instructions are shown in the inspection plan so that the inspector is referred directly to them.

Using Flowcharts

When planning for inspection, a useful tool is a flowchart for the product. Flowcharts can help reveal operations or steps where inspection points should be placed, and they can help the planner to avoid redundant inspections. A flowchart reveals the entire picture of the operations that the product undergoes during the process. The flowchart in Figure 5.1 shows four inspection points:

1. Receiving inspection of the shafts and the heads
2. Inspection of the shaft and head assembly after heat treat (which also could have been after the turning process)
3. Inspection of the threads after the deburr process
4. Final assembly inspection

Other inspections, such as in-process inspection by operators at each step, could replace the need for these specific inspection operations. At times, process control requirements require operators to control the process for certain critical or major characteristics.

One can also evaluate points of no return such as the turning operation that must leave stock for the finish grinding operation (or the parts may be scrapped), and the fact that once the parts are heat treated and finish ground, there is no turning back (the parts may also be scrapped if they have no stock for grinding). Note that if the parts were improperly turned on the diameter to be ground or improperly heat treated (too soft, too hard, etc.), finish grinding could cause scrap. For this reason the inspection operation was inserted after the heat treat process. The characteristics inspected at this step would be, at minimum, the hardness of the part and the turned diameter (making sure there is stock for grinding the part later).

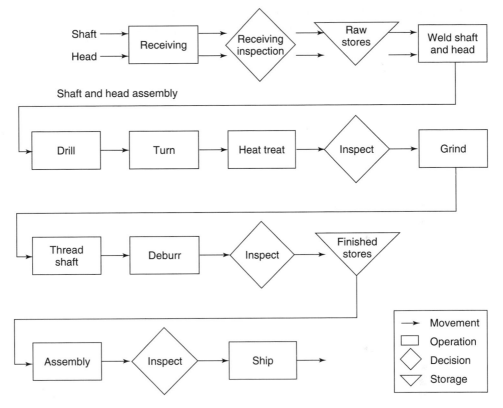

FIGURE 5.1
Flowchart for inspection planning.

Flowcharts should reflect reality by following the part through every stage of the process. If no flowchart exists for the part, one of the easiest ways to prepare one is to physically follow the path of the part and sketch (or annotate) each step of the process. Another method is to review production operation sheets or travelers to see each step of the process.

The Importance of Inspection Planning

The principles of inspection planning apply regardless of the type of product or the quantity to be inspected. They are just as applicable to the truck manufacturer making 30 units per day as they are to a toy manufacturer making hundreds of units per day. Why should inspection be planned? Because it is more economical. The quality professional's job is to keep costs at a minimum. The expense of the inspection planning effort is almost more than offset by decreased failure costs (scrap, rework, etc.) and reduced inspection costs through reduced ATI. ATI is reduced through proper inspection planning because without planning, there is a high probability that each lot and each part within the lot will be overinspected.

Inspection costs can be very high when those involved do not know when, how, what, how much, or where to inspect the product. Inspection planning saves time and therefore saves money. One example of savings is a project producing a complex housing molded by an outside supplier. Before inspection plans were prepared for this part, seven man-hours were spent inspecting the part. As a result of 32 nonrecurring hours spent in planning for inspection for this item, the company was able to reduce inspection time to three hours per part.

Industrial engineers have been planning manufacturing operations since the turn of the century when F. W. Taylor and the Gilbreths (Frank and Dr. Lillian) began developing methods engineering. We in the quality profession have been slow learners. We have been content to let each shop inspector decide what to inspect and how to inspect it. This has caused inconsistency in the characteristics checked and the methods used, not to mention gross inefficiencies and errors.

Some years ago, at one company an inspector worked for several hours to set up a large, complex, machined housing. He then began his inspection while marking on the blueprint those dimensions he had verified. Then he broke down the setup, turned the part over, made his new setup, and began checking dimensions on the alternate side. Then he went over the drawing to determine what dimensions he had missed.

That is when it got interesting! For if he found a characteristic he had forgotten that might require an additional two-hour setup to check, he had to make an important decision. Should he spend the time to do the inspection, or should he assume that the dimension was correct?

The dilemma was eliminated at this company by inspection planning. The planner decided not only what characteristics required inspection but also how to sequence the inspection so that situations like the previous one could be avoided.

Another company did something similar for all in-process inspections and receiving inspection. Taking a page from the industrial engineer, they actually wrote methods sheets for the inspectors. Some of these were extremely detailed. One such set of instructions was a series of 10 photographs showing the method used to inspect assembled components. Under each photo were instructions detailing the characteristics to be inspected. This is an excellent example of documentation accomplished at a minimum of expense.

A side benefit of this type of inspection planning is using it as a training aid for new employees. They can readily see what is required and how to do the job. Not all inspection planning needs to be this detailed, but it must document the proper method of inspecting the product. In every case, these instructions must be available for use by the inspectors.

Another reason why it is important to plan inspection is the need for consistency. Suppose that a complaint has been received from the field about a specific defect that has been shipped. The accepted approach in industry is, of course, to take corrective action to assure that the mistake is not made again. Frequently, a portion of that corrective action is to change the inspection procedure. That information can be communicated in a variety of ways ranging from verbal direction to a written change (written changes are always preferred). At one company, the inspection procedure is revised, and the revised procedure is followed from that point forward. Revised procedures should include the revision letter and date. Older procedures should be removed and replaced.

Intelligent planning of inspection operations eliminates double inspections. If a characteristic is to be checked, it should be done as soon in the manufacturing sequence as possible. Waiting until the product is complete to detect an error that could have been caught earlier and cheaply corrected is an economic disaster. Proper planning allows a selection of the most economical point of inspection.

Establishing Quality Levels

The first step in effective inspection planning is examining how the item is produced and used to establish proper quality levels. A typical method of doing this is to make a simplified block flow diagram of the primary steps in manufacturing the product, as shown in Figure 5.2. The example in the figure was created by a toy manufacturer for a toy car. It is excellent because of its simplicity.

Each block in Figure 5.2 represents a major manufacturing operation. The numbers indicate the acceptable quality level (AQL) of the product (left) for functional characteristics and (right) for cosmetic characteristics. As can be seen, the finished product has a looser AQL than its components because of the influence of the components on the quality of the finished assembly.

In establishing the AQL, start with the finished item and determine from experience and knowledge what AQL is acceptable. Then work backward, keeping in mind what can be economically expected from purchased parts and manufacturing processes. Where conflict exists and it becomes obvious that the process cannot yield the quality required, manufacturing engineering must be advised so that screening or sorting activities can be included in the basic manufacturing process. (Sorting good from bad is a production responsibility, not a quality control responsibility.) The decision that sorting is required must be made early to assure that the expense is calculated in the cost of the product.

It is essential that separate AQLs be established for functional and cosmetic characteristics. Customers normally are less critical of defects of appearance than those relating to function. The extent to which this is true, however, depends upon the type of item being sold. Although both engineering and marketing must be consulted regarding the proper AQL, quality control must make the final decision.

Once AQLs have been established, it is necessary to determine what characteristics require inspection. This is best done by classifying the characteristics as critical, major, or minor, as discussed previously.

Classification of Characteristics

Once characteristics are defined, the easiest method of classifying them is again to start at the end item and work back to the individual parts. As this is done, actual classifications can be marked adjacent to characteristics on the engineering drawings. An obvious but frequently overlooked necessity is that the person doing the classifying understands the application of the product and the definitions of classification of characteristics. A closely toleranced dimension is not automatically a major characteristic. In other words, *tolerances do not equal severity.*

Some companies have the design engineer do the classification. A good quality control engineer can review the designer's decisions to see if they are reasonable. With the classification completed, the inspection plan can be written. Generally, criticals are inspected 100 percent, majors are inspected to a stringent sampling plan (e.g., 1 percent AQL), and minors are inspected using a less stringent plan (e.g., 4 percent AQL).

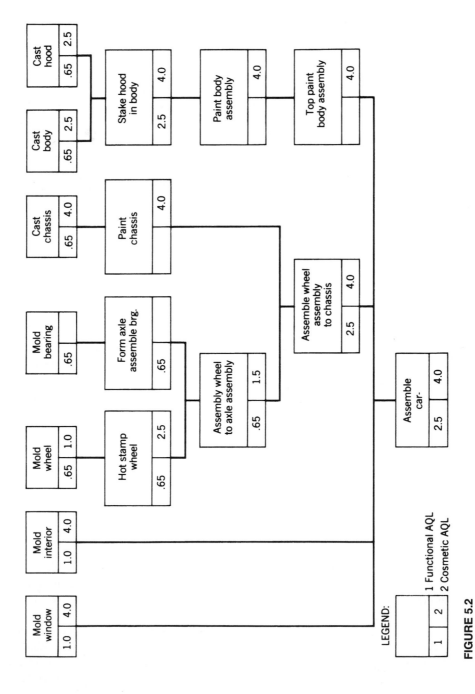

FIGURE 5.2
Block flow diagram.

72

Planning for Purchased Parts

Since buyers have little control over the production of items they purchase, inspection for these parts is designed as an after-the-fact process. An example is a company that uses substantial amounts of molded plastic parts which, for the most part, are made by outside sources using buyer-owned tooling.

In these instances the dimensions to be inspected are selected to show that the tool has been properly operated. Included is at least one major length and width dimension to determine shrinkage after molding. An across-the-parting-line dimension is selected to assure that the tool is properly closed and injection pressure is maintained. Any major characteristic not listed above is also verified. Parts are inspected for flash, short shot, excessive gate, and sink.

Any of these characteristics can change over relatively short periods of time because of the lack of control at the supplier. This concept of inspection is predicated upon inspecting the tooling first (known as a first article inspection). From then on, it is assumed that if the tool is run properly (and this is verified by checking the characteristics previously mentioned) the parts will conform.

Other inspections are also planned into the system. Each month during the time that the tool is operated, parts are subject to more thorough inspection. If the tool is pulled for any reason, an inspection is done not only to verify quality but also to allow corrections to be made while the tool is not needed for production. After tool rework, the tool is inspected for the reworked area. The same philosophy is used in planning parts produced on punch presses and on numerically controlled tools.

In the case of complex, nontooling-produced parts, the quality control engineer is faced with a trade-off between the expense of checking all characteristics and the risk of accepting a nonconforming condition. Other important considerations include how likely it is that the defect will be detected during assembly and, if found, what the rework/repair or replacement cost would be. If the operator can tell quickly during the assembly that the part is defective, there is less of a problem than if the assembly would have to be completed and tested before the defect is found. The quality control engineer is responsible for making that decision.

Inspection plans should be written for all purchased items regardless of complexity. Often, a single plan can be written for a family of products and then modified to fit each product in that family. This type of plan relies heavily on approved samples to display the cosmetic limits of acceptance. These samples often have to be developed as the situations develop. Each time a decision is made regarding the acceptance or rejection of product due to a cosmetic characteristic, a properly identified example of the condition should be added to that part's sample file.

Special sampling plans are frequently a part of the inspection plan for purchased parts. This is essential when a common lot size can be as large as 600,000 parts. Using single sampling, a sample size of 1250 pieces is required. If the part has many dimensions or has a low unit price, obviously some reductions must be made in the sampling.

One company's first choice was to go to double sampling; for this, it has a combined sampling plan that started with single sampling and switched to double sampling when the lot size reached a given quantity (see Figure 5.3). This type of plan technically violated

Lot size	Sample size	Cumulative sample size	0.65 Ac	0.65 Re	1.0 Ac	1.0 Re	1.5 Ac	1.5 Re	2.5 Ac	2.5 Re	4.0 Ac	4.0 Re	6.5 Ac	6.5 Re
			Acceptable quality level—AQL											
2–8	2		↓		↓		↓		↓		↓		0	1
9–15	3										0	1	↑	
16–25	5								0	1	↑		↓	
26–50	8						0	1	↑		↓		1	2
51–90	13				0	1	↑		↓		1	2	2	3
91–150	20		0	1	↑		↓		1	2	2	3	3	4
151–280	20 20	20 40	↑		↓		0 1	2 2	0 3	3 4	1 4	4 5	2 6	5 7
281–500	32 32	32 64	↓		0 1	2 2	0 3	3 4	1 4	4 5	2 6	5 7	3 8	7 9
501–1200	50 50	50 100	0 1	2 2	0 3	3 4	1 4	4 5	2 6	5 7	3 8	7 9	5 12	9 13
1201–3200	80 80	80 160	0 3	3 4	1 4	4 5	2 6	5 7	3 8	7 9	5 12	9 13	7 18	11 19
3201–10000	125 125	125 250	1 4	4 5	2 6	5 7	3 8	7 9	5 12	9 13	7 18	11 19	11 26	16 27
10001–35000	200 200	200 400	2 6	5 7	3 8	7 9	5 12	9 13	7 18	11 19	11 26	16 27		
35001–150000	315 315	315 630	3 8	7 9	5 12	9 13	7 18	11 19	11 26	16 27				
150001–500000	500 500	500 1000	5 12	9 13	7 18	11 19	11 26	16 27						
500001 and over	800 800	800 1600	7 18	11 19	11 26	16 27	↑		↑		↑		↑	

FIGURE 5.3

Combined sampling plan.

some sampling principles, but since Mil-Std-105E provides the operating characteristic curve of single and double sampling, it is possible to determine the risks.

The company's plan saved money and was far superior to allowing the inspector to select his or her own sample size. It was also statistically more valid than using a fixed sample size regardless of lot size. Sampling plans should be detailed in the inspection plan. Quality engineering needs to analyze the economics of the sampling plan (the cost of inspection versus the risk of acceptance of defective product) and specify the best plan to use.

Planning for Manufactured Parts

To plan for manufactured parts, a page can be taken from the industrial engineer's handbook. A flowchart (shown in Figure 5.4) is prepared to show the basic manufacturing steps involved to produce the product. It includes every storage function but ignores all present inspection points.

To start planning the first important rule is to do each inspection as early in the manufacturing as possible. Rule 2 is to do any given inspection only once. Rule 3 is to have the inspection done the same way, regardless of the inspector, by having an inspection plan including careful selection of the measuring instrument. It is far better to take more care doing the first inspection correctly than to have to reinspect that characteristic later in the manufacturing cycle.

Begin at the end! Use the flowchart in Figure 5.4 to identify each characteristic in the finished item that requires inspection. Then decide at what point during manufacturing that characteristic could first be inspected. Mark this information on the flowchart. Continue to do this for every characteristic requiring inspection. Then work backward during the manufacturing cycle doing the same thing at each point along the way. After you are finished, a clear pattern of inspection points will be defined on the flowchart.

Now, starting from the beginning of the manufacturing cycle, detail the inspection points and decide whether it is more economical to move the inspection to a later point in the process. Can more be saved in inspection time than can be risked should the defect be found later in the process? That is the only decision.

After the decision when to inspect is made, the individual inspection plans can be written. At each inspection point, the sampling plan must be defined and the inspection operations sequenced. The sequence must be planned with the same care that the industrial engineer uses in planning manufacturing methods. In fact, advice from the industrial engineering department may be helpful to the inspection planner. Remember, the inspection is probably going to be performed every day, day in and day out, for as long as the product is made. Any savings in labor is one that the company will enjoy from now on.

Inspection Checklists

Checklists offer excellent inspection planning for progressive assembly-line operations (as an example). One company's final assembly checklist system has been instrumental in reducing the average number of defects per assembly by 40 percent while improving the outgoing quality of the product. The assembly line is divided into 11 zones, each with a checklist covering the characteristics requiring inspection (see Figure 5.5).

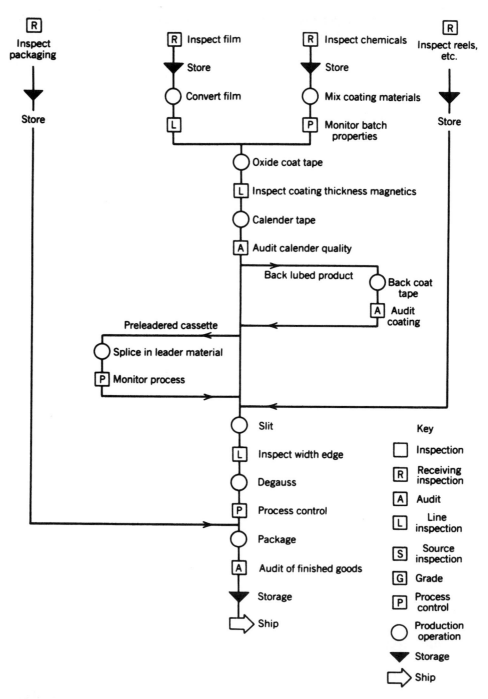

FIGURE 5.4
Process/product flowchart.

Characteristics and Acceptance Criteria	Defect. Class.	Okay	Def.	Comments	Repair	Insp.	Supv'r
1. Radiator: Coolant level and inhibitor added	M						
2. Fuel tank steps: Positioned and secured	M						
3. Sanders: Positioned and secured	m						
4. Lower grille: Positioned and secured	m						
5. Access steps: Positioned and secured	M						
6. Steel hoods: Paint damage, color and cutouts when applicable	M						
7. Fenders: Paint damage, color, positioned and secured	m						
8. Road lamps: Positioned and secured, when required	m						
9. Grille guards: Positioned and secured, when required	m						
10. Exhaust system: Positioned and secured	M						
11. Air cleaner: Positioned and loose assembled	M						
12. Mud flaps: Positioned and secured	m						
13. Swing hoods: Paint damage, color, loose assembled	M						

FIGURE 5.5
Checklist for assembly inspection.

Characteristics on each checklist are listed in the sequence for inspection. A separate set of checklists is completed for each product built. The classification of each characteristic is shown (*M* for major and *m* for minor). If the item is critical it is shown in boldface type.

The inspector is to inspect the characteristics as soon as the production operation is complete. In other words, the inspector works up and down the assembly line within the area, inspecting the assembly as it is built instead of waiting at the end of the line. He or she is encouraged to become a part of the team composed of the assemblers, the repair person, and the inspector. It is best if the inspector can detect a mistake as it occurs and point it out at that time.

Production repair staff are encouraged to assist in the inspection activity (and usually do). It is to their advantage to reduce the amount of mistakes since that directly reduces the amount of repairs to be performed. The most efficient areas are those where the repair people spend the majority of their time training personnel and checking, rather than correcting mistakes.

If an end item is rejected, the defect column on the checklist is stamped "D" and the particular defect is noted in the remarks column. Production makes the repair within the area. Inspection verifies that the defect has been corrected. The inspector records, using tally marks, the number of defects occurring each hour (Figure 5.6).

The information is posted in a particular area of the assembly line. The production supervisor reviews this information and initiates corrective action as required. At final assembly inspection, the inspector reviews all checklists to see that all characteristics have been accepted. The inspector may also perform additional inspections, using a checklist, to assure that other standards are met (an example of other factors could be federal safety standards).

Other Factors in Inspection Planning

Certain human factors should be considered during the inspection planning. For instance, the inspector's workstation must be designed with provisions for holding the product and displaying it during the inspection. Lighting and good air circulation must be provided, as well as a proper place for keeping the inspection equipment, paperwork, inspection tags, and other essentials.

Another consideration is the form used to collect the data generated during the inspection. It is disastrous to inspect without documenting the findings. How else can corrective action be taken? The best form is one that requires minimum time for the inspector to complete, that lays out the data to be readily tabulated and summarized, and that specifies only meaningful data. Whenever possible, expected defect categories should be prerecorded. The inspector can then record the frequency of the defects in each lot inspected. When the columns are totaled, the summary data are ready to publish.

Another consideration during planning is what will be done when defects are found. The major reason for continuous sampling being unsuccessful is that no one plans for rejections. Typically, a continuous sampling plan requires more inspector time when excessive defects are detected than when the line is in control with few defects. The problem is where to get those extra people. Most textbooks say to borrow them from production. However, production has its own problems when defects occur and is least likely to have extra

Characteristics and Acceptance Criteria	8:30	9:30	10:30	11:30	1:00	2:00	3:00	4:00	Total
1. Radiator: Coolant level and inhibitor added									
2. Fuel tank steps: Positioned and secured									
3. Sanders: Positioned and secured									
4. Lower grille: Positioned and secured									
5. Access steps: Positioned and secured									
6. Steel hoods: Paint damage, color and cutouts when applicable									
7. Fenders: Point damage, color, positioned and secured									
8. Road lamps: Positioned and secured, when required									
9. Grille guards: Positioned and secured, when required									
10. Exhaust system: Positioned and secured									
11. Air cleaner: Positioned and loose assembled									
12. Mud flaps: Positioned and secured									
13. Swing hoods: Paint damage, color, loose assembled									
14. Quarter fenders: Paint damage, color, positioned secured									
15. Battery box covers: Positioned and secured									
16. Trailer electrical cable: (In cab) positioned and secured									
17. Seats: Positioned and secured									

FIGURE 5.6
Tally sheet.

people. In short, unless an adequate supply of qualified personnel can be found to handle the line when rejects occur, do not use continuous sampling.

Another consideration is the inspection and handling of rejects on a lot basis. It is easier to handle rejects when using lot inspection, but facilities must be available to hold the rejected lot until production has the time to sort the product. An orderly method must be provided to get this sorted material back to the inspector for verification that the sorting has been done properly.

If the inspection process requires comparison "to approved samples," such samples must be easily accessible to the inspector. If those samples deteriorate easily by handling or exposure, they must be carefully protected. Such is the case with samples showing maximum allowable scratches or with color chips that may fade.

Who Should Plan and When

The person best able to determine the proper method of inspection and the characteristics to check is usually the trained quality planner or engineer. As mentioned earlier, because of the complexity of the product and other factors, the planner must be part industrial engineer to specify methods, part inspector to determine procedure, part design engineer to determine what to inspect, part statistician to determine sampling plans and, most important, part prophet to select the most economical method based upon what will happen during the manufacturing process. A good inspection planner is right most of the time. However, if the inspection planner is right all of the time he or she is probably in the wrong profession. The main idea is that inspection planners should constantly work toward being correct at all times when preparing inspection plans.

In a large firm a separate department may do all the planning. In a small firm, the responsibility may fall to the chief inspector. In every firm where inspection is performed, someone must plan. The greatest economic gains come when planning is done before production begins. It takes the same amount of time then as it does later and will save money in both reduced inspection costs and a better product. If changes are required in the manufacturing process, they can be planned into it rather than having to alter the process to accommodate them later.

REVIEW QUESTIONS

1. One of the basic "tools" used by an inspection planner to plan for inspection is the
   _____.
   a. Pareto chart
   b. caliper
   c. flowchart
   d. AQL
2. Which of the following is not a benefit of inspection planning?
   a. reduced inspection time
   b. consistency among and between inspectors

    c. guaranteed reduction in defects

    d. knowledge of characteristic classification

3. Which of the following is not part of an inspection plan (or inspection instruction)?

    a. the sampling plan

    b. the measurement equipment

    c. what characteristics to inspect

    d. control charts

    e. applicable inspection records

4. Which of the following classifications of characteristics is important with respect to fit, form, and function of the part?

    a. major

    b. minor

    c. critical

    d. incidental

5. (Yes or no) Just because a part dimension has a tight tolerance, does that mean its classification is critical or major?

6. Which of the following classifications has a direct relationship to safety problems that may occur?

    a. major

    b. minor

    c. incidental

    d. critical

7. Inspection is a(n) _____ activity that determines whether a product conforms or does not conform to its requirements.

8. Which of the following elements are part of any inspection system in a company?

    a. inspection plans

    b. sampling plans (or procedures)

    c. calibration of measuring equipment

    d. appropriate inspection equipment

    e. all of the above

9. Which of the following are good reasons to keep inspection records?

    a. for historical process results

    b. for traceability purposes

    c. for decisions on inspection levels

    d. all of the above

10. The most effective inspection that can be performed is in real time and at

    a. every operation

    b. the source

    c. all operations

    d. receipt of product

11. The question of how much inspection has a lot to do with the

    a. time of day

    b. workforce

    c. classification of characteristics

    d. time of the month

12. Source inspection is inspection of a _____ product before shipment.
    a. lot of
    b. supplier's
    c. local
    d. source

13. Which of the following takes precedence with regard to inconsistencies in documents at receiving inspection?
    a. the drawing
    b. the specification
    c. the work instruction
    d. the purchase order

14. Inspectors should be unbiased and accurate. They should consistently _____ good products and _____ bad products.
    a. reject, accept
    b. accept, reject
    c. accept, ship
    d. reject, ship

15. 100% inspection is not _____ effective.
    a. very
    b. at all
    c. 100%
    d. none of the above

Reading Engineering Drawings

Those working in industry must understand engineering drawings. These drawings relate design requirements to operators, machinists, inspectors, assemblers, and many others who use engineering drawings to make, inspect, or assemble products.

Therefore, drawings are the industrial language. This language must be learned. There is a phrase that expresses the importance of reading blueprints:

as designed = as built

It is the intent of an engineering drawing to describe clearly all requirements the product must meet to be a quality product. Many quality problems occur solely because someone did not understand how to read or interpret a drawing. To avoid getting involved in these problems (or causing them), it is necessary to understand how to read and interpret drawings.

Drawings are made up of various "lines" and "views" that are easily learned. Once you learn these, you will be halfway to learning how to read engineering drawings. The other half is understanding the title block, footnotes, change block, tabulations, and in some cases, symbols. The title block gives information about the part including name, part number, engineers, drafter, and tolerances that are applicable. The change block tells the latest revision letter, date, and what was revised. Tabulation is a way of making several similar parts using only one drawing. Footnotes are simply notes that are applicable and necessary for the user to build or inspect the part. Symbols are used on some engineering drawings to save time and space.

LINES ON AN ENGINEERING DRAWING

A basic requirement for reading engineering drawings is understanding the types of lines used and what they are used for. The following shows the type of lines and gives examples of how they are used on engineering drawings.

Object Line (visible line). ━━━━━━━━━ (heavy-solid)   Used to show the outer and inner boundaries of surfaces and features in a particular view.
Example: a washer (Figure 6.1).

FIGURE 6.1
Washer.

Centerline. ———— —— ———— (long and short dashes) Used to show the geometrical center of an object.
Example: a washer (Figure 6.2).

FIGURE 6.2
Washer.

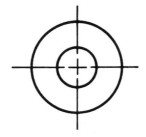

Hidden Line. — — — — — — (short dashes) Used to show surfaces and features that exist, but cannot be seen in that particular view.
Example: a washer (Figure 6.3).

FIGURE 6.3
Washer.

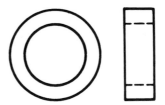

Leader Line. (medium–solid arrow) Used to point (or lead) the reader from a dimension (or note) to the feature or surface where the dimension (or note) applies.
Example: a washer (Figure 6.4).

FIGURE 6.4
Washer.

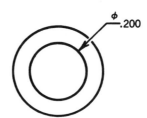

Extension Line. ———————— (light–solid) Used to extend a feature (or surface) from the view so that it can be dimensioned.
Example: a washer (Figure 6.5).

FIGURE 6.5
Washer.

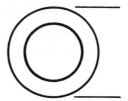

Dimension Line. 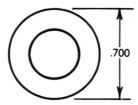 (light–solid arrows)   Used to dimension the part or to show the distance between the extension lines.
Example: a washer (Figure 6.6).

FIGURE 6.6
Washer.

Section Lines. /// (light–solid) 45 degrees. Used to show the surfaces of a part that has been sectioned (or cut).
Example: a washer (Figure 6.7).

FIGURE 6.7
Washer.

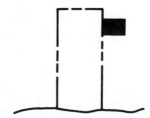

Phantom Lines. ——— — — ——— (one long and two short dashes)   Used in four main ways:

1. To show an existing structure that needs modification.

Example: A box must be put on an existing pole in the ground. The pole is drawn in phantom lines to show that it is already there (Figure 6.8).

FIGURE 6.8
Structure needing modification.

2. To show alternate positions of an object.

Example: A lever that moves on an assembly would be drawn solid in one position and drawn in phantom lines to show its other possible positions (Figure 6.9).

FIGURE 6.9
Alternate positions of an object.

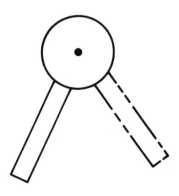

3. To avoid unnecessary detail.

Example: Avoid drawing the coils of a spring when it is not necessary.

4. To show the direction of a unilateral profile tolerance zone in geometric tolerancing.

Cutting Plane (or viewing plane lines).   (medium–broken) Used to show where a part is to be cut, or viewed by the reader.
Example: a washer viewing plane (Figure 6.10).

FIGURE 6.10
Washer.

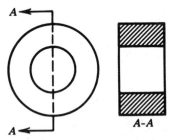

Short Break Line. ‿‿‿‿‿ (medium-jagged)   Used often to break up a particular view and show some of it in section (Figure 6.11).
Example:

FIGURE 6.11
Short break line.

Long Break Line. 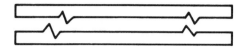 (thin)   Used to break up a long object along its length (Figure 6.12).
Example:

FIGURE 6.12
Long break line.

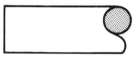

Cylindrical Break Line.   Used to show a break in a cylindrical object (such as a drill rod) (Figure 6.13).
Example:

FIGURE 6.13
Cylindrical break line.

Chain Line. _____ __ _____ (heavy and shaped like a centerline)   See chain line applications on page 150.

ORTHOGRAPHIC PROJECTION VIEWS

On engineering drawings, various views are used to pictorially represent the object. Many different views are used for this, but the main ones are *front, right side, left side, top,* and *bottom.* The most important view of all is the front view. This is the view that is selected to show (in visible lines) the most information about the shape of the part. As you see in Figure 6.14, the front view tells a lot about the actual part configuration because it shows clearly the hole, the slot, and the step.

FIGURE 6.14
(*a*) The part. (*b*) The front view.

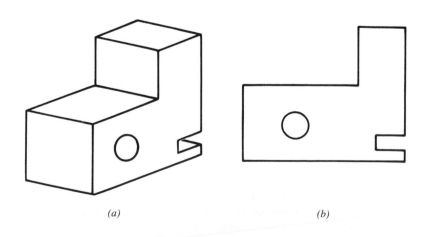

(*a*)   (*b*)

After the all-important front view is chosen, using it as a reference, the right side, left side, top, and bottom views may be used. I say *may be used* because it is important that only the necessary views be used.

You can see in Figure 6.15(*a*) that only the front and right-side views of the washer need to be shown because all other views look exactly like the right-side view.

FIGURE 6.15
Washer drawing with (*a*) correct and
(*b*) incorrect views.

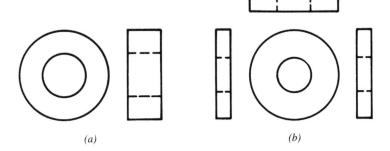

(*a*)   (*b*)

Readers must understand the views on a drawing. They must be able to "fold" them together in their mind and imagine what the real object looks like. For example, Figure 6.16 is a picture of a cardboard box that has been cut. The sides have been folded out in the same way that drawing views are folded from the front view.

FIGURE 6.16
(*a*) A box with sides folded out.
(*b*) The box itself.

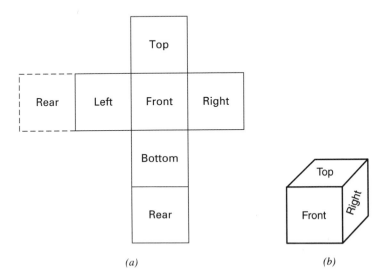

(*a*)   (*b*)

Examine Figure 6.17 (*a*) and try to imagine the solid figure revealed in Figure 6.17 (*b*). It takes practice to learn this; one way to learn is to sketch the views of simple objects.

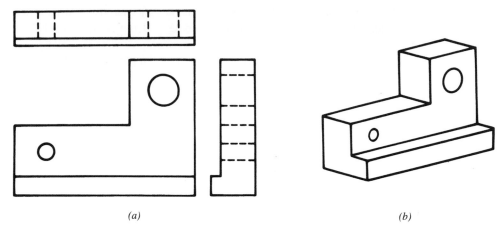

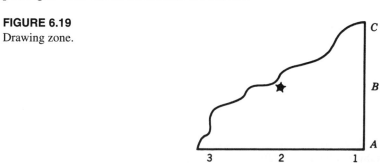

(a)   *(b)*

FIGURE 6.17
(*a*) View of an object. (*b*) The object.

CHANGE BLOCK, NOTES, AND TABULAR DIMENSIONS

Change Block

The change block, usually shown in the upper right-hand corner of a drawing, is where changes in a drawing are recorded. *Revisions* are assigned letters. For example, the first revision (or change) on a drawing would be given the letter *A*, the second *B*, the third *C*, and so on. When the alphabet is used up, changes would start with *AA* and continue. Note: Sometimes the letter *A* is assigned to the initial release of the drawing, so the first revision would be *B*.

Generally, the information given in the change (or revision) block will be as shown in Figure 6.18. The information includes:

FIGURE 6.18
Revision block.

Sym	Zone	Revision	Date	By	Approved

Sym.   This is the revision letter (*A, B, C,* and so on).

Zone.   This is the drawing area in which the change was made (Figure 6.19). For example, Figure 6.19 shows an asterisk in zone *B*2.

FIGURE 6.19
Drawing zone.

Revision.   This will describe the revision that was made: what it was and the specific change that was made. For example, ".510 ± .010 was .510 ± .005."

Date.   This is simply the date (5-4-00, for example) that the revision was made.

By.   The person who made the revision on the drawing.

Approved.   The person who approved the revision.

In the revision block other bits of information may be found, such as the *DCN* (drawing change notice) *number.* This is the document that is written to cause a change in the drawing. Drawings are not just changed. They must be approved by many people in most cases, and drawing change notices are part of the paperwork required to get this approval, after the change is reviewed and found to be acceptable.

Notes

There are two types of notes used on a drawing: *general notes* and *local notes.*

General Notes.   General notes are numbered and are usually located at the bottom of a drawing. They apply to several areas of the drawing and are used to avoid cluttering the drawing with individual notes in all of those areas and to save time.

General Note Example.

1. Finish all over.

This note says that all surfaces are to be finished (or machined). It simplifies the drawing and saves time for the drafter who creates the drawing, workers who read the drawing to find out which parts need to be finished, workers who inspect the finished parts, and others involved in the process.

Local Notes.   Local notes apply to one or more particular characteristics on the drawing. They are usually used with a leader line pointing directly to the characteristic, or they may be flagged, in which case a triangle is used with the note number within.

Local Note Example.

This surface to be medium knurled.
Regardless of the notes used, there is one thing to remember: *Read the notes thoroughly before doing anything.*

Tabular Dimensions

There are times when several similar parts are made with only slight differences. An example of this is when several flat washers are produced, the only difference being the inside diameter.

Part Number	Inside Diameter	Part Number	Inside Diameter
9165-1	.250	9165-6	.500
9165-2	.300	9165-7	.550
9165-3	.340	9165-8	.580
9165-4	.420	9165-9	.630
9165-5	.475	9165-10	.750

FIGURE 6.20
Tabulation block.

These washers usually have the same basic part number but have different dash numbers to identify them. When this is done, tabular dimensioning can be used effectively to save time in preparing a different drawing for each washer (as shown in Figure 6.20).

Now with one drawing and all other dimensions the same, several washers can be made. The only difference between them is the hole size. For example, a 9165-4 washer must have a .420 diameter hole in it.

THE TITLE BLOCK

The title block of a drawing is usually in the lower right-hand corner of the page. It contains information that identifies the part and other particulars about that part. There is no standard arrangement for the information in the title block, but there are standard pieces of information found in it. Some of these are: part name, part number, material to use, next assembly, title block tolerances, draftsman, checker, company name, and scale.

1. *Part name.* The name of the part is always shown in the title block. Sometimes it is shown with a *descriptive modifier,* which is simply a word that follows the name to describe the part further. An example is CAP, HUB (which we commonly call a "hub cap"). The name of the part is CAP and the descriptive modifier is HUB.
2. *Part number.* All parts, assemblies, and subassemblies have a part number that should be distinct from any other in the company.
3. *Material to use.* The title block may also list what material the part is to be made from, such as aluminum, steel, bar stock, and so on.
4. *Next assembly.* Some title blocks may tell you where (or in what assembly part number) the part will be used.
5. *Title block tolerances.* Almost all drawings have title block tolerances. These are usually in the form of a note, as shown in Figure 6.21.

    This means that on all dimensions shown on the face of the drawing that do not have tolerances already specified, use the tolerances stated here. The exception is any BASIC dimension (which is always used with geometric tolerances).

    There are also dimensions without tolerances shown. Three different examples are: .50; .500; and 10°. If you have the .50 dimension, the title block must be used to establish the tolerance. Since the title block says that .XX = ±.03, the .50 is supposed to be .50 ± .03. Therefore, the limits are .53 and .47. The next one is .500 with

FIGURE 6.21
Title block tolerances.

Unless Otherwise Specified
.XXX = ±.010
.XX = ±.03
Angles = ±1 degree

no tolerance shown. Since the title block says that .XXX = ±.010, the .500 dimension is to be .500 ± .010, so the limits would be .510 and .490. The last one is 10°. The title block is used here, and it says that angles are ±1 degree, so that the tolerance band on the 10° angle is 9° to 11°. *Remember:* If dimensions have a tolerance shown, use it. If not, use the title block tolerance.

6. *Drafter.* The title block also shows the name of the drafter who drew the print.

7. *Checker.* The checker checks the drafter's work to make sure there are no errors.

8. *Company name.* The name of the company that makes the part is usually shown in the title block.

9. *Scale.* The title block also shows the scale of the drawing. This is where the actual drawing size and the actual part size are compared to each other. For example:

Full scale (1:1). The drawing is the same size as the actual part.

Half scale (1:2). The drawing is one-half the size of the actual part.

Not to scale. The drawing is not to any particular scale.

Note: One should never scale drawings, but work from the dimensions given.

10. *Size.* Drawing sizes are indicated by letters. The following table indicates letters associated with drawing sizes:

A. $8\frac{1}{2}$in. × 11 in. (or 9 in. × 12 in.)
B.   11 in. × 17 in. (or 12 in. × 18 in.)
C.   17 in. × 22 in. (or 18 in. × 24 in.)
D.   22 in. × 34 in. (or 24 in. × 36 in.)
E.   34 in. × 44 in. (or 36 in. × 48 in.)
J.   36 in. × any length
R.   48 in. × any length

Depending on the company, type of product, and other variables, there could be more information in the title block, but the information above is usually there.

RULES TO FOLLOW WHEN USING ENGINEERING DRAWINGS

Wherever engineering drawings are used, there are some basic rules that should be followed:

1. Never get into the habit of memorizing a drawing regardless of how long you have used it. Always pull the drawing from the file on each job. Engineering drawings do change from time to time, and if you practice memorization, you are likely to make an error due to a change.

2. Unless otherwise specified, always keep only the latest change drawing in the file. When a new drawing is received, tear up the old one or dispose of it in the prescribed manner.
3. To avoid damage to drawings, handle them carefully and fold them properly.
4. Whenever there is a question about anything on the drawing, clear it up with your supervisor. Do not guess at anything.
5. When you suspect something is wrong with a drawing (unusually wide tolerances, unclear statements, missing dimensions, etc.), report it to your supervisor.
6. Always follow drawings to the letter.
7. Never mark on a drawing unless you are authorized to do so. If you are authorized, it is sound practice to sign and date your markings.
8. Make sure you read and understand all notes on a drawing before you begin working.

TOLERANCES

There is variation in every process. No two parts can be made exactly alike. Therefore, all dimensions must have a tolerance. *Tolerancing* simply shows how much variation is allowable in a dimension; it is the area between the maximum and minimum limits. There are various types of tolerances on drawings related to various dimensions, or characteristics. Some of these are discussed here.

Bilateral Tolerance

Variation is permitted in both directions from a specified dimension (see Figure 6.22).

FIGURE 6.22
Bilateral tolerance.

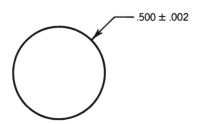

.500 ± .002

Unilateral Tolerance

Variation is permitted in only one direction from a specified dimension (see Figure 6.23).

FIGURE 6.23
Unilateral tolerance.

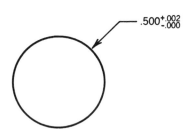

$.500^{+.002}_{-.000}$

Limit Tolerance

Tolerances show the high and low limits of a dimension (see Figure 6.24).

FIGURE 6.24
Limit tolerance.

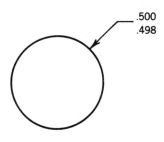

.500
.498

Single Limit Tolerance

A tolerance specifies a maximum or minimum limit only (see Figure 6.25).

FIGURE 6.25
Example of a single limit tolerance.

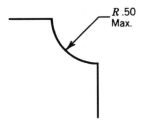

R .50
Max.

Title Block Tolerance

Many dimensions are shown without any tolerance. These are controlled by the title block tolerances at the bottom of the drawing. For example (see Figure 6.26):

$$.xxx = \pm.010 \text{ (a three-place decimal is } \pm.010)$$

$$.xx \ \ = \pm.03 \text{ (a two-place decimal is } \pm.03)$$

FIGURE 6.26
These dimensions are controlled by title block tolerances.

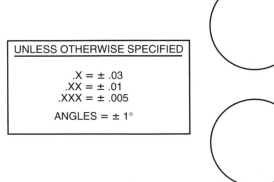

UNLESS OTHERWISE SPECIFIED

.X = ± .03
.XX = ± .01
.XXX = ± .005

ANGLES = ± 1°

.500

.50

ALLOWANCES AND FITS

The following is a small collection of terms and examples that are necessary in the mechanical trades. The list is not all-inclusive but does stress some major points about allowances and fits, and relationships between mating parts.

Clearance Fit (Positive Allowance)

This type of allowance permits two mating parts to be assembled easily (with airspace between them after assembly), as in Figure 6.27.

FIGURE 6.27
Mating parts with clearance fit.

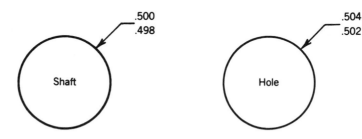

No matter what size the shaft or the hole is, they will easily go together if they are both within their tolerance limits.

Interference Fit (Negative Allowance)

Two mating parts must be forced together (because the male part is larger than the female part), as in Figure 6.28.

FIGURE 6.28
Interference fit of two mating parts.

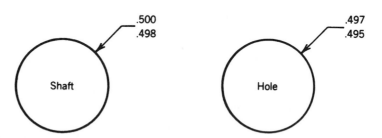

No matter what their sizes (if they are both within their tolerance limits), there will be an interference fit.

Allowance (Positive or Negative)

Allowance is defined as the *intended difference* between the virtual sizes of mating parts (whether it is clearance or interference).

Transition Fit

This is a form of allowance that can be clearance or interference, depending solely on the actual sizes of each part (see Figure 6.29).

FIGURE 6.29
Transition fit of two mating parts.

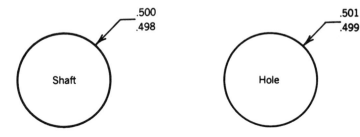

There is interference if the shaft were actually .500 and the hole were actually .499. There is clearance if the shaft were actually .498 and the hole were actually .500.

Line to Line Fit

This is an interference fit where both parts are the same size (see Figure 6.30).

FIGURE 6.30
Line fit where both parts are the same size.

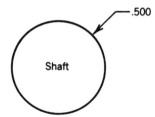
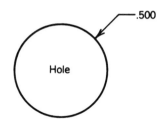

THREADS (NOMENCLATURE)

Screw threads are a very important part of industry. In the old days, threads were not standardized. One manufacturer would make the bolt and another manufacturer would make the nut and they might not fit together. Today, threads are standardized and interchangeable with each other no matter who makes them.

There are names for each geometrical shape and dimension that make up threads. The *unified thread form* is one that is discussed here. There are many other forms of threads such as square, acme, buttress, whitworth, and knuckle, which are not covered in this book.

The drawing in Figure 6.31 shows a sectional view of a thread and names the various parts. The most important parts of the thread that should be known are shown here and are listed as follows.

Pitch.   The distance from a point on one thread to the same point on the next thread.

Crest.   The top of the thread.

FIGURE 6.31
Unified external thread.
(L. S. Starrett Co. Reprinted by
permission.)

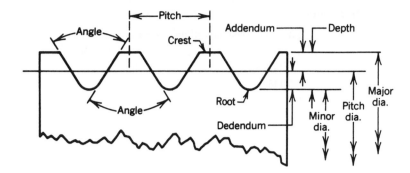

Root. The bottom of the thread.

Angle. The included angle between the two flanks of the thread.

Minor Diameter. The distance from root to root.

Major Diameter. The distance from crest to crest.

Depth. The distance from crest to root.

Addendum. The distance from the pitch diameter to the crest on an external thread.

Dedendum. The distance from the pitch diameter to the root on an external thread.

Pitch Diameter. The imaginary cylinder throughout the thread.

Screw threads are described on drawings as follows: thread nominal size; how many
threads per inch; the type of thread; the class of thread, A external or B internal; and the direction of the thread (left-hand is always stated, and right-hand is understood). Figures 6.32
and 6.33 show examples of thread callouts.

FIGURE 6.32
English thread callouts.

Only if it's a
left-hand thread

$\frac{3}{4}$ – 10 UNC-2A LH

- Class 2 external
- Unified national course
- Threads per inch
- Nominal size

FIGURE 6.33
Metric thread callouts.

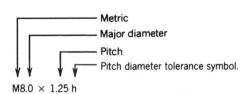

- Metric
- Major diameter
- Pitch
- Pitch diameter tolerance symbol.

M8.0 × 1.25 h

Class Numbers

There are basically class 1, 2, and 3 threads.

- Class 1 is the loosest fit for quick assembly.
- Class 2 is tighter toleranced for torque applications.
- Class 3 is very tightly toleranced for close-fitting requirements.

Pitch (or Lead)

On a single thread, the pitch and lead are the same. The formula for the pitch or lead of a single thread is

$$P = \frac{1}{\text{number of threads per inch}}$$

CHAMFERS

Chamfers are commonly referred to as the "unimportant dimension." However, they can become important for the following reasons: (1) they can break burrs and sharp edges; (2) they aid in easy assembly of component parts; and (3) in some cases, an oversize chamfer, or one with the wrong angle, will cause problems in fit or function of the part. The examples in Figure 6.34 show what kind of callouts may be used for a chamfer.

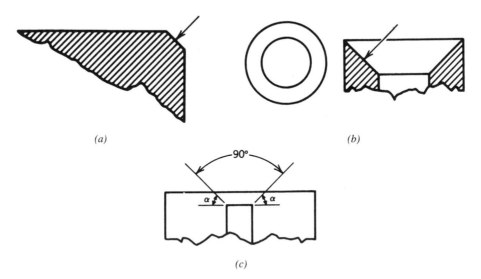

(a)   (b)

(c)

FIGURE 6.34
Various chamfers: (*a*) on an outer edge or outside diameter; (*b*) inside a diameter; (*c*) a countersink.

Figures 6.35 and 6.36 illustrate two different ways to show a chamfer on blueprints and how to measure each. Care should always be taken to ensure that chamfers are *measured as they are shown on the blueprint.*

As Shown on Blueprint	This Means	To Measure It	
(a)	The angle should be 45 degrees and the depth should be 0.050 within certain tolerance limits.	A protractor of some sort. 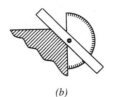 *(b)*	45 degrees can be measured in either direction. 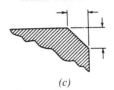 *(c)*

FIGURE 6.35
A bidirectional 45° chamfer.

As Shown on Blueprint	This Means	To Measure It
	The angle of cut should be 45 degrees and the depth must be 0.100 within certain tolerances.	The part is still measured with some type of protractor, but now you must measure it in the specified direction. This is true when it is called out as shown, or with any angle other than 45 degrees. 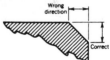

FIGURE 6.36
A unidirectional 45° chamfer.

REVIEW QUESTIONS

1. What is the most important orthographic view of a drawing?
   a. right
   b. left
   c. front
   d. top
2. What type of tolerance is .500−.505?
   a. bilateral
   b. limit
   c. unilateral
   d. single limit

3. Name the fit where two mating parts are easily assembled.
   a. interference
   b. allowance
   c. line
   d. clearance

4. A line that is used to show features or surfaces that exist but are not seen in a particular view is called the _____ line.
   a. border
   b. hidden
   c. center
   d. object

5. Lines that are used to show a surface of a part that has been cut are called _____ lines.
   a. dimension
   b. extension
   c. section
   d. border

6. An alternate position for a switch lever can be shown on an engineering drawing using a(n)_____ line.
   a. border
   b. alternate
   c. phantom
   d. optional

7. A drawing feature that shows only a nominal dimension, without a specific tolerance, would have a tolerance determined by the _____ _____ tolerance.
   a. engineering drawing
   b. title block
   c. change block
   d. optional open

8. A _____ note will typically be located at the bottom of the engineering drawing.
   a. local
   b. foot
   c. special
   d. general

9. One way to use one drawing to establish the configuration and tolerances for many similar part numbers is to use _____ dimensions.
   a. general
   b. different
   c. optional
   d. tabulated

10. The tolerance .750 $\pm^{.000}_{.003}$ is what kind of tolerance?
    a. bilateral
    b. limit
    c. unilateral
    d. single limit

11. The tolerance .500 ± .005 is what kind of tolerance?
    a. bilateral
    b. limit
    c. unilateral
    d. single limit
12. The tolerance .375 Max. is what kind of tolerance?
    a. bilateral
    b. limit
    c. unilateral
    d. single limit
13. The tolerance .675 Min. is what kind of tolerance?
    a. bilateral
    b. limit
    c. unilateral
    d. single limit
14. The tolerance $^{.505}_{.495}$ is what kind of tolerance?
    a. bilateral
    b. limit
    c. unilateral
    d. single limit
15. If a shaft and a hole are exactly the same size, what kind of fit is there between the two mating parts?
    a. clearance
    b. interference
    c. running
    d. line
16. If both mating parts can be such that there will be clearance or interference between them depending on their allowable sizes, what kind of fit is this?
    a. clearance
    b. interference
    c. transition
    d. line
17. If a shaft and a hole were toleranced so that the shaft would always be larger than the hole, what kind of fit is this?
    a. clearance
    b. interference
    c. running
    d. line
18. What is the name of the outside diameter of an external thread?
    a. dedendum
    b. addendum
    c. major
    d. minor

19. What is the name of the distance from the center of one full thread to the center of the next full thread?

   a. crest

   b. pitch

   c. dedendum

   d. root

20. With regard to English threads, the A in the thread classification stands for _____ thread.

   a. internal

   b. single

   c. external

   d. double

7 Geometric Dimensioning and Tolerancing

This Chapter, Chapter 8, and other chapters that use geometric dimensioning and tolerancing symbols and examples have been updated to the latest **ASME Y14.5M–1994** standard. The previous version, **ANSI Y14.5M–1982** is also covered in comparison by a table at the end of the chapter. Although the chapters are not a complete review of the standard, many examples are covered that explain geometric tolerancing symbols, rules, and interpretation.

Geometric dimensioning and tolerancing is used on many engineering drawings. This method of tolerancing uses symbols instead of notes to describe the geometric controls that are necessary for the *fit, form,* and *function* of the part and to control the *relationship* between specific part features. The standard also provides specific rules governing the interpretation and application of these symbols to design drawings. This chapter will review all of the symbols and rules and provide examples of their application. For basic measurement methods and techniques, refer to further coverage of these symbols in Chapters 8, 9, 10, and 11. For detailed coverage of geometric tolerancing methods, applications, and measurement, see reference (1).

The correct application and measurement of geometric tolerances depend on personnel who have been trained to the standard. Geometric tolerancing methods far surpass any other way of describing the function and relationships of a part and providing guarantees that products produced to these requirements will fit and function as they were designed. Geometric tolerancing is, however, a language of symbols and rules that must be learned before they can be applied. For example, size tolerances automatically control the form of features within the stated boundaries of size.

Other rules for tolerance limits apply, such as the rule of **absolute limits.** Absolute limits means that regardless of the decimal places used to specify limits of size, those limits cannot be violated (e.g., a .5, .50, .500, and .5000 maximum dimension can be assumed to have infinite zeros, so any parts exceeding the absolute value are not acceptable). The language of geometric tolerancing, once learned, helps people from all areas of the company (and countries around the world) to communicate and understand design requirements. Figure 7.1 reviews some symbols used in geometric dimensioning and tolerancing. Figure 7.2 shows an example of a feature control frame as it appears on an engineering drawing.

Two other definitions that are important in geometric dimensioning and tolerancing are *feature* and *tolerance zones:*

Feature.   A feature is a physical portion of a part such as a hole, surface, or slot.

Tolerance Zones.   All tolerance zones shown in the feature control frame are total. For example, the position within a .005 cylindrical tolerance zone means that the tolerance zone is a .005 cylinder and the actual centerline of the feature must lie within it. The exact position lies in the center of the .005 zone.

Symbol	Description	Application
A ▲	Datum	Represents physical features (size features or surfaces) that must be used for location in inspection. Any letter (except I, O, or Q) can be used. Use AA, AB, AC, and so on, if the alphabet runs out. Datums must be physical features that can be contacted.
⌀	Diameter	Replaces the word diameter. This symbol precedes all diameters on the drawing and is used in feature control frames when the tolerance zone is a cylinder.
Cl	Datum target (point or line)	A datum target or point on irregular parts such as castings or forgings. Datum targets must be located during machining.
φ6.0 / Cl	Datum target area	A datum target area. This target must be located by a pin locator that is of the specified size and shape shown in the upper portion of the symbol. In this case, a round pin, flat on the end, at 6.0 mm diameter.
.200	Dimension not to scale	Used to identify a dimension that is not to scale on the drawing.
M	MMC modifier	Means that the maximum material condition applies. MMC is the largest external size feature or the smallest internal size feature (within size limits).
L	LMC modifier	Means that the least material condition applies. LMC is the smallest external size feature or the largest internal size feature (within size limits).
P	Projected tolerance zone modifier	Means that the stated tolerance zone is projected above the feature, extending into space. The feature must be measured in comparison to this projected (extended) zone.
1.000	Basic dimension	A perfect dimension with no tolerances. It locates geometric tolerance zones. A basic dimension can also be shown as 1.000 BSC or 1.000 BASIC. Basic dimensions cannot be rejected; they have no tolerance.
(.500)	Reference dimension	For computation purposes only. These dimensions are not to be inspected.

FIGURE 7.1

Symbols used in geometric dimensioning and tolerancing.

FIGURE 7.2
Feature control frame.

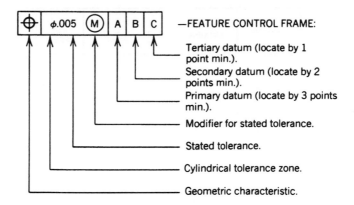

—FEATURE CONTROL FRAME:

Tertiary datum (locate by 1
point min.).

Secondary datum (locate by 2
points min.).

Primary datum (locate by 3 points
min.).

Modifier for stated tolerance.

Stated tolerance.

Cylindrical tolerance zone.

Geometric characteristic.

COMPARISON OF ANSI Y14.5 SYMBOLS

The geometric tolerancing symbols have undergone several changes over the years since their conception. The most recent specifications involved are the 1982 and 1994 revisions. Figures 7.3 to 7.5 show the 1994 revisions in detail. Figure 7.6 is a comparison of the symbols in these two revisions.

OTHER DRAWING SYMBOLS

Some drawing symbols are used to describe certain characteristics that usually require a word to explain. These optional symbols are listed in Figure 7.7.

THE DIMENSION ORIGIN SYMBOL

In certain instances on drawings, there is a need to specify the reference surface of a linear dimension. The dimension origin symbol (Figure 7.8) adequately defines the origin (or reference surface) of a dimension using a circle around its extension line.

The example in Figure 7.9 shows one application for the dimension origin symbol.

As shown in Figures 7.9 and 7.10 the origin of the .500 dimension lies at the shorter surface of the part. Therefore, measurement of the .500 dimension must be made using the shorter surface as a reference and the tolerance of ±.010 surrounds the longer surface of the part. This is true because in the final assembly the shorter surface is the interface (or the surface that makes contact with the mating part).

How to Measure the Part on the Surface Plate

To measure this part you must locate the short surface on the surface plate and indicate it on the long surface as shown in Figure 7.11.

The dimension origin symbol clearly specifies a required "datum" reference for a simple linear dimension without having to use datum symbols or a geometric tolerance.

SYMBOL	AS SHOWN ON DRAWING	TOTAL TOL ZONE	MMC OR RFS?	DATUM USED ?	FUNC GAGE USED?	GOOD PART EXAMPLE	BAD PART EXAMPLE
FLATNESS	⬭ .004	TWO PARALLEL PLANES .004 APART	A SURF. HAS NO SIZE	NO	NO	.004	.004
STRAIGHT-NESS (OF AN AXIS)	φ.500 ± .002 — φ.001	.001 DIA. SAME LENGTH AS SHAFT	Ⓢ UNLESS Ⓜ IS SHOWN	NO	YES IF Ⓜ IS STATED	.001φ	.001φ
STRAIGHT-NESS (SURFACE ELEMENT)	LEADER POINTS TO SURFACE — .003	TWO PARAL-LEL LINES .003 APART	A SURFACE ELEMENT HAS NO SIZE	NO	NO	EACH ELEMENT .003	ANY ELEMENT .003
CIRCULAR-ITY (ROUNDNESS)	SHAFT φ .504 .496 ○ .001	TWO CONCEN-TRIC CIRCLES .001 APART	Ⓢ ALWAYS	NO	NO	.001 PART IS ROUND	.001 PART IS NOT ROUND
CYLINDRIC-ITY	SHAFT φ.500 ± .003 ⌭ .001	TWO CONC CYLINDERS .001 APART	Ⓢ ALWAYS	NO	NO		

FIGURE 7.3
ASME Y14.5 symbols.

SYMBOL	AS SHOWN ON DRAWING	TOTAL TOL ZONE	MMC OR RFS?	DATUM USED?	FUNC GAGE USED?	GOOD PART EXAMPLE	BAD PART EXAMPLE
⊥ PERPEN-DICULAR-ITY	ø.005 ± .003 ⊥ ø.002 C / C	ø.002 / 90°	(S) HERE, (M) MUST BE STATED	YES ALWAYS	YES IF (M) IS STATED		
∠ ANGULARITY	∠ .004 A / 30° / A	TWO PARALLEL PLANES .004 APART / 30°	(M) CAN APPLY TO FEATURE OF SIZE	YES ALWAYS	YES WHEN (M) IS STATED	ENTIRE SURFACE IS INSIDE THE ZONE	ANY PART OF THE SURFACE IS OUT OF THE ZONE
// PARALLELISM	// .003 B / B	.003 WIDE TOL ZONE / B	(M) CAN APPLY TO FEATURE OF SIZE	YES ALWAYS	YES IF (M) USED	THE ENTIRE SURFACE IS IN THE ZONE	ANY PART OF THE SURFACE IS OUT OF THE ZONE
⌒ PROFILE OF A LINE	⌒ .002	.002 WIDE AROUND TRUE PROFILE	A SURFACE HAS NO SIZE	DATUM CAN BE USED OR NOT	NO	.002	.002
⌓ PROFILE OF A SURFACE	BASIC DIMENSIONS ARE REQIURED TO DEFINE THE TRUE PROFILE / ⌓ .001	.001 WIDE ZONE AROUND TRUE PROFILE	A SURFACE HAS NO SIZE	DATUM CAN BE USED OR NOT	NO	ENTIRE SURFACE IS IN	PART OF SURFACE IS OUT

FIGURE 7.4
ASME Y14.5 symbols.

107

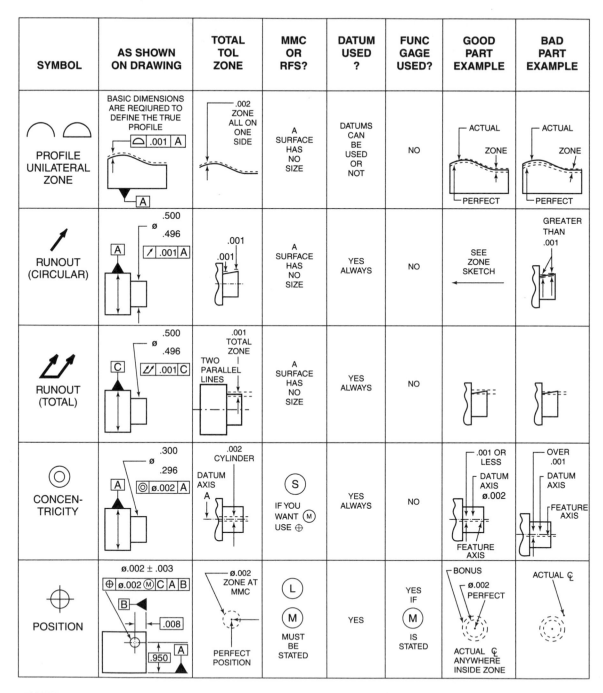

SYMBOL	AS SHOWN ON DRAWING	TOTAL TOL ZONE	MMC OR RFS?	DATUM USED ?	FUNC GAGE USED?	GOOD PART EXAMPLE	BAD PART EXAMPLE
PROFILE UNILATERAL ZONE	BASIC DIMENSIONS ARE REQIURED TO DEFINE THE TRUE PROFILE ⌒ .001 A A	.002 ZONE ALL ON ONE SIDE	A SURFACE HAS NO SIZE	DATUMS CAN BE USED OR NOT	NO	ACTUAL ZONE PERFECT	ACTUAL ZONE PERFECT
RUNOUT (CIRCULAR)	ø .500 .496 A ↗ .001 A	.001 .001	A SURFACE HAS NO SIZE	YES ALWAYS	NO	SEE ZONE SKETCH	GREATER THAN .001
RUNOUT (TOTAL)	ø .500 .496 C ↗↗ .001 C	.001 TOTAL ZONE TWO PARALLEL LINES	A SURFACE HAS NO SIZE	YES ALWAYS	NO		
CONCEN-TRICITY	ø .300 .296 A ◎ ø.002 A	.002 CYLINDER DATUM AXIS A	Ⓢ IF YOU WANT Ⓜ USE ⊕	YES ALWAYS	NO	.001 OR LESS DATUM AXIS ø.002 FEATURE AXIS	OVER .001 DATUM AXIS FEATURE AXIS
POSITION	ø.002 ± .003 ⊕ ø.002 Ⓜ C A B B .008 .950 A	ø.002 ZONE AT MMC PERFECT POSITION	Ⓛ Ⓜ MUST BE STATED	YES	YES IF Ⓜ IS STATED	BONUS ø.002 PERFECT ACTUAL ₵ ANYWHERE INSIDE ZONE	ACTUAL ₵

FIGURE 7.5
ASME Y14.5 symbols.

108

Tolerance type	Name	1982 Standard Symbol	1994 Standard Symbol
Form	Flatness	▱	▱
	Straightness	—	—
	Roundness	◯	◯
	Cylindricity	/◯/	/◯/
Orientation	Parallelism	//	//
	Angularity	∠	∠
	Perpendicularity	⊥	⊥
Runout	Circular runout	↗	↗
	Total runout	↗↗	↗↗
Profile	Profile of a line	⌒	⌒
	Profile of a surface	◠	◠
Location	Position	⊕	⊕
	Concentricity	◎	◎
	Symmetry*	Use ⊕	≡
Other symbols	Datum target	(A1) or (ø.5/A1)	(A1) or (ø.5/A1)
	Dia. symbol	∅	∅
	Datum placement	// ø.002 A	// ø.002 A

*Symmetry symbol replaced by position symbol in the 1982 standard.

FIGURE 7.6
Comparison of symbols, 1982 and 1994 standard.

FIGURE 7.7
Other drawing symbols.

Word	Symbol	Example
Counterbore or spotface	⊔	6 X ⌀.224 ± .005 ⊔ ⌀.375 ± .010
Depth	⊤	⊤ .250 Min.
Quantity	X	6 X ⌀.500 ± .002
Countersink	∨	∨ ⌀.320 ± .010
Square	□	5 X □
All around	⌀↙	⌀↙ ⌒
Conical taper	▷↙	▷↙ .002:1
Flat taper slope	◿↙	◿ .010 ± .003:1
Radius	R	R.250 ± .005
Controlled radius	CR	CR.250 ± .010
Spherical radius	SR	SR.500 ± .005
Spherical diameter	S⌀	S⌀.500 ± .005
Arc length	⌒	⌒ 2.350

MAXIMUM MATERIAL CONDITION—LEAST MATERIAL CONDITION

The maximum material condition (MMC) or least material condition (LMC) of mating parts is very important because these conditions can cause problems between mating parts. The ANSI symbols for them are Ⓜ for MMC and Ⓛ for LMC.

FIGURE 7.8
Dimension origin symbol.

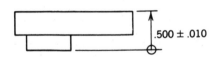

.500 ± .010

FIGURE 7.9
Application for the dimension origin symbol.

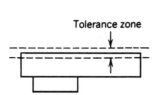

Tolerance zone

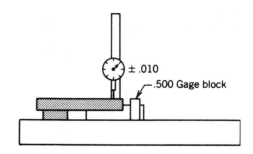

± .010

.500 Gage block

FIGURE 7.10
Tolerance zone.

FIGURE 7.11
Transfer measurement.

MMC. That condition of a dimension where the *most material* allowed (by the tolerance) is still there.

LMC. That condition of a dimension where the most *material allowed to be cut off* (by the tolerance) has been cut off.

For example, we have a shaft .498–.502. Here the MMC is the largest size (.502). The LMC is the smallest size (.498). Now suppose we have a hole .590–.600. Here the MMC is the smallest size (.590). The LMC is the largest size (.600).

A good way to remember this is:

• MMC is the largest shaft and the smallest hole (per the size tolerance).
• LMC is the smallest shaft and the largest hole (per the size tolerance).

■ **Example 7–1**

A shaft is supposed to fit inside a hole with only .001 clearance between them at MMC as shown in Figure 7.12. So

$$
\begin{aligned}
\text{MMC of hole} &= \quad .503 \\
\text{minus MMC of shaft} &= - \underline{.502} \\
\text{clearance} &= \quad .001
\end{aligned}
$$

FIGURE 7.12
Example of fitting a shaft inside a hole.

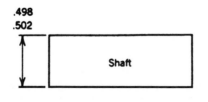

.498
.502

Shaft

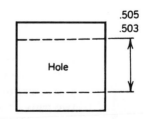

.505
.503

Hole

With only .001 clearance, if the shaft happens to be oversize to .503, it will not fit and will require rework. Or, if the hole happens to be undersize to .502, it will not accept the shaft and will also require rework.

Most products are designed with clearances based on MMC and LMC. As the MMC increases, the clearance decreases. The MMC is very important for assembly because what will not go together cannot be shipped. The LMC is important also because clearances greater than LMC cause parts to be too loose in the assembly.

THE VIRTUAL CONDITION

The virtual condition is a crucial aspect of any mating part size feature. Because the collective effects of size and geometric error contribute to a feature's virtual condition, the virtual condition is the worst case condition of size and geometrical error allowed by the drawing. The virtual size is the collective effects of the actual size and geometric error of the part as it is produced.

■ Example 7–2

A coat hanger wire may have a size tolerance of .050–.060″ diameter. If the wire was made at .060″ diameter and was bent on its axis by .010″, the virtual size of the wire would be .060″ + .010″ = .070″ (as shown in Figure 7.13).

FIGURE 7.13
Example of virtual size.

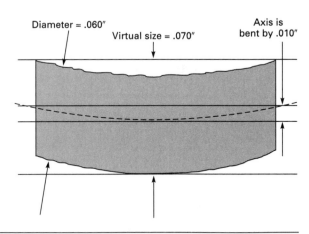

■ Example 7–3

In Figure 7.14, a pin is toleranced at .610–.600 diameter and must pass through a bore that is toleranced .622–.612 diameter. At MMC this is only .002 clearance between the two parts. As shown in Figure 7.15, if the pin was tapered on one end and oversize to .615, the virtual size of the pin becomes .615 also, and there would be problems in assembly of the two parts.

FIGURE 7.14
Pin and bore.

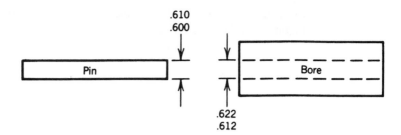

.610
.600

.622
.612

FIGURE 7.15
Effective size due to taper.

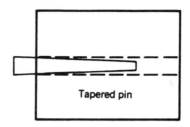

Tapered pin

■ **Example 7–4**

Now let's say that the same pin in Example 7–3 was not tapered, but crooked. The pin measured .600 diameter all along its length. As shown in Figure 7.16, the distance to the highest area of the pin is .623. This pin would never go through any of the mating parts unless they were oversize or a different size, because the virtual size of the pin is now .623 and the largest diameter that the hole can be is .622.

FIGURE 7.16
Virtual size of a crooked pin.

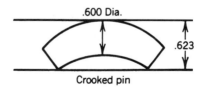

.600 Dia.

.623

Crooked pin

Remember: The virtual size of any dimension is dependent on its size, form, and position (or location).

• Size examples: (1) oversize shaft; (2) undersized hole.
• Form examples: (1) tapered; (2) barrel-shaped; (3) crooked.
• Position example: The feature is out of location (position).

Virtual Size

This is the *effective* size of a feature. For example, if a short shaft is supposed to be .500–.498 diameter, and it is bent by .004, its virtual size is equal to the shaft's MMC (maximum material condition) plus the .004 crookedness. The MMC of the shaft is its largest size (.500), so that the virtual size is

$$.500 + .004 = .504$$

GENERAL RULES FOR ASME Y14.5M–1994 TOLERANCING

American Society for Mechanical Engineers (ASME) specification Y14.5M–1994 covers geometric tolerancing. This tolerancing method is widely used on drawings to maintain control over geometrical problems that are inherent in machined parts. When using the specification, there are some general rules to consider. These rules govern the interpretation of the drawing symbols, datums, and tolerances.

Rule 1. Regarding size tolerances.

Rule 2. Regarding tolerance modifiers.

Pitch diameter rule. Regarding geometric tolerances that are applied to screw threads.

Datum/virtual size rule. Regarding datum features of size that are controlled by geometric tolerances to another datum.

When using geometric tolerancing, these rules must be kept in mind so that the requirements of the drawing are clear. The following is a breakdown of the rules to simplify their meaning.

Rule 1. This rule simply means that unless otherwise specified, the size tolerance of a feature controls form as well as size. No part of a feature shall be undersize or oversize. Size tolerances apply all over the feature. Per this rule, there is a maximum boundary of perfect form at MMC that cannot be violated. This rule controls geometric error within the size boundaries.

Rule 2. Regardless of feature size Ⓢ applies to all symbols unless Ⓜ or Ⓛ have been specified Ⓢ therefore, is understood.

Pitch Diameter Rule. All geometric tolerances applied to screw threads apply directly to the axis of the pitch diameter of the thread unless otherwise specified. Figure 7.17 shows an example of *otherwise specified.*

Note: Geometric tolerances applied to gears and splines must designate the specific feature to which they apply (including the pitch diameter if applicable).

Datum/Virtual Size Rule. A datum feature of size that is being geometrically controlled to another datum has a virtual size. Using this controlled datum as a reference must be at its virtual size (even when MMC is shown next to the datum letter). If the virtual condition is not intended to apply, the designer should apply a zero tolerance at MMC.

FIGURE 7.17
Feature control frame.

INSPECTING SIZE TOLERANCES PER ASME Y14.5 RULE 1

Inspection of size tolerances per ASME Y14.5 specification (Rule 1) is not a simple task. Traditionally, size tolerances have been inspected using various hand tools such as micrometers, vernier calipers, ring gages, and plug gages. The problem is that each of these hand tools only does half the job.

A size tolerance (per Rule 1) invokes two controls on the feature of size: *size* and *form*. Rule 1 states clearly that size features must have perfect form when produced at MMC size.

A shaft produced at MMC (its largest allowable size) must be a perfect cylinder.

A hole produced at MMC (its smallest allowable size) must also be a perfect cylinder.

A thickness dimension between two surfaces produced at MMC size must have sides that are perfectly straight and parallel to each other.

None of these conditions can be verified by using a single hand tool. However, they can be verified using two hand tools.

■ Example 7–5

A plug gage can be used to measure the MMC boundary of perfect form, and a telescoping gage can be used to measure any condition beyond the LMC boundary (Figure 7.18).

FIGURE 7.18
Hole diameter.

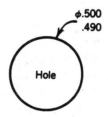

ϕ.500
.490

Hole

■ Example 7–6

A ring gage can be used to measure the MMC boundary of perfect form, and a vernier caliper or micrometer can measure any condition beyond the LMC boundary (Figure 7.19).

FIGURE 7.19
Shaft diameter.

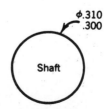

ϕ.310
.300

Shaft

■ Example 7–7

A surface gage and high-limit block stack can set up a boundary of perfect form at MMC where no portion of the surface can exceed. A vernier caliper or micrometer can measure any condition beyond the LMC boundary (Figure 7.20).

FIGURE 7.20
Thickness.

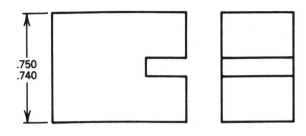

.750
.740

DATUMS

There are certain geometric tolerances that specify a datum surface or feature of size. The feature control frame will contain these datums in the last portion. When three datums are specified, this establishes a *datum reference frame.* The datum reference frame consists of the *primary, secondary,* and *tertiary* datums.

Datum Planes

When datum surfaces are used to establish datum planes, the following rules apply. With respect to functional datum planes:

1. *Primary datum.* This is the supporting datum that must be contacted at the three highest points on the surface. This is usually accomplished by a flat datum locating surface such as surface plates (see Figure 7.21).
2. *Secondary datum.* This is the aligning datum that must be contacted at the two highest points on the surface.
3. *Tertiary datum.* This is the stopping datum that must be contacted at the highest point on the surface.

FIGURE 7.21
Three-plane concept.

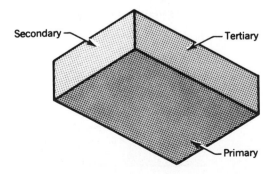

FIGURE 7.22
Three planes on a cylindrical part.

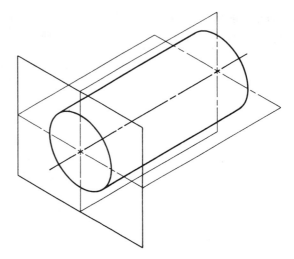

These three planes are mutually perpendicular to each other and establish the datum reference frame.

Note: The same rules for the number of points contacted apply to datum targets except that the points contacted are specified points that are not necessarily the highest ones.

With respect to a cylindrical part the three datum planes are established as follows.

1. *Primary datum (axis).* This is established by two intersecting planes through the parts that are perpendicular to each other (see Figure 7.22).
2. *Primary datum (end surface).* This is established by the three highest points on the end surface.
3. *Secondary datum (axis).* This is established by contacting the extremes of the diameter after the primary (end surface) has been contacted.
4. *Secondary datum (end surface).* This is established by the extremes of the end surface after the primary (axis) has been contacted.
5. *Tertiary datum.* The tertiary datum for a cylindrical part is usually a feature that stops rotation (such as a slot) or one that is used for clocking.

The Selection of Functional Datums

During design of the product, functional datums are selected according to their functional importance, especially when they are interfaces in the next assembly. In some applications, only the primary datum is necessary. How many and what features are designated as datums depend on functional need.

Typically, the primary datum for a hole is that face which is perpendicular to that hole and is the mating surface (or interface). When positioning a hole from the outer edges of a part, the primary, secondary, and tertiary datums are needed. When positioning a pattern of holes to each other, only the primary datum (face) need be specified. This means that the position tolerance applies to the holes from each other and that they must be perpendicular to the datum plane in the same tolerance zone.

Datum axes are typically primary to certain geometric controls such as runout and concentricity. These datum axes are established by *datum features*. They are primary because they establish the functional axis of rotation for the unit. There are also times on a cylindrical part when a functional datum axis may be defined by contacting the diameter (primary) and an end face (secondary).

Who Should Contact Datums?

Inspection.   Functional datums must always be contacted by inspection. A datum plane or axis on a finished part is a functional datum. Nonfunctional datums must also be contacted by inspection. Nonfunctional datums (such as datum targets) are important to the relationships on a part (e.g., casting) for verification of casting relationships but are not functional to the assembly.

Manufacturing.   Functional datums do not have to be contacted in manufacturing (although they are allowed to). Manufacturing can use any means of locating the part that is practical. The important thing for manufacturing to consider is that the end result must be verified with respect to the functional datums. Manufacturing must locate nonfunctional datum targets, however. These datum targets are established for consistency in the manufacturing process. Datum targets are used on parts such as castings, forgings, weldments, and sheet metal parts.

Datums play an important role in manufacturing and design. If they are properly classified (functional/nonfunctional) and located by those concerned, they make measurements on the finished product more functional (therefore, meaningful) and the manufacturing process more consistent part after part.

DATUM TARGETS

Datum targets (symbols shown in Figure 7.23) are used to establish the location of parts that have irregular surfaces to make the initial cut. Castings and forgings are examples of parts where datum targets are applicable. Generally, you locate on datum targets to machine the primary datum plane, then locate on the machined datum plane for subsequent machining.

FIGURE 7.23
Datum target symbols.

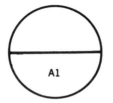 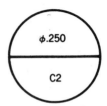

■   Example 7–8

Figure 7.24 shows a casting. As shown, the first machining pass will establish multiple datum A–B. Target symbols applied to the casting drawing for this part may look like the drawing in Figure 7.25. Now, according to the targets, the location of the part on the machine that will make the first cut may resemble the drawing in Figure 7.26.

FIGURE 7.24
Casting.

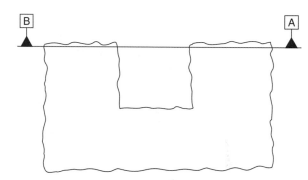

FIGURE 7.25
Targets shown.

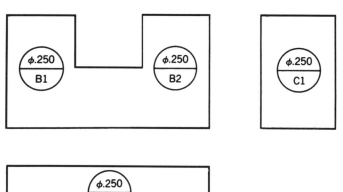

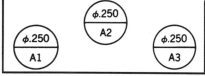

FIGURE 7.26
Path of machining cutter.

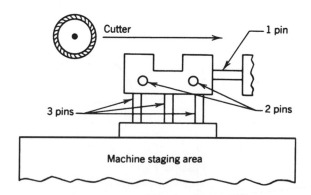

All pins are .250 diameter and flat at the contact area since the target symbol specifies a target area.

Using Datum Targets

When you use datum targets effectively, the initial location of irregular parts is more consistent part after part. This consistency in location of irregular parts for machining can be an asset to producing quality parts in many ways. Some of these ways are listed below.

1. Consistency in location for machining operations. Where to locate on each part is clearly defined.
2. Proper selection of tooling for production.
3. Types of locators to use are defined (flat or rounded).
4. Casting (or forging) dimensional problems (relative to the first cut) are readily discovered.
5. Casting or forging is securely located with respect to subsequent areas that are to be machined.
6. The casting can be inspected in the same way that it will be machined.

Types of Datum Targets

A datum target may take three basic forms: target point, target line, or target area. The symbol describes which one is to be used on the part (see Figure 7.27).

FIGURE 7.27
Target symbols.

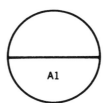
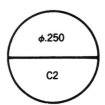

Target Points and Lines.   When a datum target symbol has nothing specified in the upper portion, as shown in Figure 7.28, it means that it is a target point or line. Target points must be located by a spherical (rounded) locator pin such as shown in Figure 7.29.

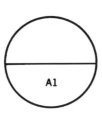

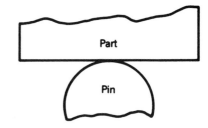

FIGURE 7.28
Target point symbol.

FIGURE 7.29
Target point located.

Target lines are contacted by the outside edge of a round pin, as shown in Figure 7.30.

FIGURE 7.30
Target line located.

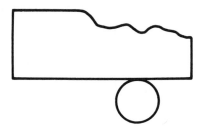

Target Areas. When a target area is specified in the upper portion of a symbol, as shown in Figure 7.31, it means that the specified diameter is the datum target area. Target areas are located by locator pins that are the diameter specified in the symbol and flat on top. An example of this is shown in Figure 7.32.

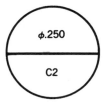

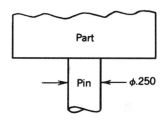

FIGURE 7.31
Target area symbol.

FIGURE 7.32
Target area located.

Where to Locate Datum Targets. Datum targets (points, lines, or areas) are usually shown on a drawing with coordinate dimensions to describe exactly where the targets are located. Figure 7.33 shows an example of how this is done.

FIGURE 7.33
Targets on the drawing.

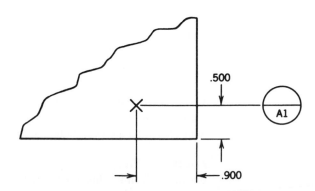

Using datum targets correctly can give you a greater chance of transforming an irregular mass of material into a finished part that is cleaned up on all machined areas.

HOW TO CONTACT FEATURE DATUMS

There are two basic categories of datums: *feature datums* and *datum targets.* Datum targets are covered further in another portion of this book. Feature datums are surfaces, holes, slots, and the like. A feature datum is defined as an actual surface on the part. A size feature datum is defined as an actual size feature (holes, slots, etc.) on the part. A centerline cannot be called a datum.

The ASME standard states that feature datums (surfaces or holes) must be fully contacted prior to the measurement of a geometric tolerance wherever datums apply (the only exception is targets). These feature datums are contacted regardless of their condition (rough, out of round, etc.). This is why a designer must apply a form tolerance on a datum, such as flatness, when it is necessary to make sure that the datum surface is flat enough to reference the measurement.

Simulated Datum Planes and Features

A simulated datum plane is any surface that is accurately flat within gage tolerances. Examples of simulated datum planes are *surface plates, toolmaker's flats,* and *fixtures.* Simulated datum features are any size feature that is to size, round, and/or cylindrical within gage tolerances. Examples of simulated datum features are *gage pins, plug gages,* and *ring gages.* The standard states that datums on the part may be contacted by simulated datum planes to establish the datum plane for the measurement.

Feature Datums (Example: A Surface on the Part)

If a surface or group of surfaces have been identified as a datum on the part they must each be fully contacted against a simulated datum plane prior to any measurement. If the datum is *qualified* by a form tolerance (such as flatness tolerance), the flatness must be measured or verified before the surface is contacted for any other measurement. When the actual part surface is brought in contact with the simulated datum plane, the datum plane has been established and the measurement can begin.

Size Feature Datums (Example: A Hole or Pin)

Size feature datums apply to those measurements that must be related to a size feature on the part. Size feature datums are fully contacted by the most accurate equipment that will do the job.

Datum Is a Hole.   When a hole has been identified as a datum, it must be contacted with the largest true cylinder that will fit the hole (in an RFS situation). This cylinder is often an approved simulated datum feature of size, such as a gage pin or an expanding mandrel.

Datum Is a Pin.   When a pin is a datum it must be contacted with the smallest true cylinder that will fit it. This true cylinder is often a ring gage or a collet.

Inspectors and Datums

It should be clearly understood that whoever does the inspection of the finished product must at all times locate the specified datums. The inspection of the product cannot be from any other feature of size or surface on the part even if it is more convenient. If inspection is performed from datums as shown on the drawing, that inspection will be a correct functional inspection of the part.

Datums That Are Technically Not Acceptable Unless Targeted

Some tools are often used to locate datums that are technically not acceptable unless datum targets are used. The most widely used simulated datum is the V-block. A V-block only makes contact on two "lines" of the surface at one time. This violates the rule of "full contact unless targeted." Another example is the jaws of a rotary table. These jaws only contact at three areas of the datum diameter. The last example (not the least) is the coordinate measuring machine, which only makes contact on points of the datum surface. This can be corrected, however, by bringing the part in contact with a simulated plane, then entering the simulated plane information into the machine.

COORDINATE METHOD OF POSITIONING FEATURES

One of the older methods of defining where a feature must be in position (related to datums, or reference surfaces) is the coordinate method. It is still used in some areas and therefore should be discussed. Coordinate dimensioning usually involves two dimensions locating the position of a feature (in this case, a hole), as shown in Figures 7.34 and 7.35.

Note: This type of tolerancing is not recommended because of its tolerance restrictions. The drawing shown in Figure 7.34 means what is shown in Figure 7.35.

To inspect this part for hole location, you can use the following steps.

Step 1. Measure the size of the hole and record it. For example, we will say that the hole is .402 diameter.

Step 2. Get a pin that will fit the hole with a snug fit and put it through the hole.

FIGURE 7.34
Coordinate ± tolerancing.

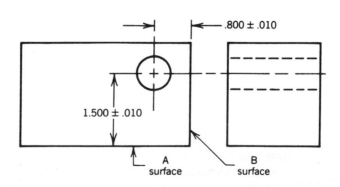

FIGURE 7.35
Square tolerance zone.

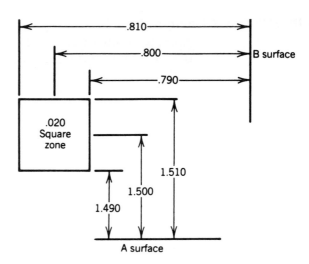

Step 3. Set the part (with surface A down, as shown in Figure 7.36 (*a*) on the surface plate and measure the distance from the plate to the top of the pin. Then subtract one-half of the hole size from that distance. If the answer is between 1.490 and 1.510, the part is acceptable in this direction.

Step 4. Now rest surface B on the surface plate (as shown in Figure 7.36 (*b*) and again, measure the distance from the plate to the top of the pin. Again, subtract one-half of the hole size (which in this case is .201) from that distance. If the answer is between .790 and .810, the part is acceptable.

Remember: The part is good only if *both* of the coordinate dimensions are in tolerance at both ends of the hole.

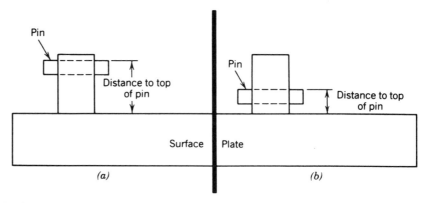

FIGURE 7.36
Measuring hole locations: (*a*) surface A located; (*b*) surface B located.

POSITION TOLERANCING

Positional tolerancing has been widely used since its introduction to manufacturing. It is designed to aid manufacturing in production, which it does in many ways. It allows you a wider tolerance zone to work with than the old coordinate methods. You can have bonus tolerances, depending on feature size and other conditions. Datums are specified for the purpose of location for measurement. There are many more advantages not stated here.

Things to Remember

1. Dimensions used with position will always be BASIC.
2. Position tolerances apply with the feature at MMC, RFS, or LMC—whichever one is stated. RFS modifier is understood.
3. Position means "the theoretical exact location of a feature at its center axis or plane."
4. Position gives you a *circular* tolerance zone instead of a square zone. This circular (diametral) zone gives you 57 percent more tolerance area than the square zone (see Figure 7.37).
5. All the holes together in a bolt circle are called the *pattern*.
6. Some position tolerances apply only to the position of the pattern as a whole, and some apply to the position of the holes to each other.
7. When using position, Ⓜ or Ⓛ must be shown in the frame Ⓢ is understood.

FIGURE 7.37
Circular and square zones. It is best to use all BASIC dimensions and a composite position tolerance.

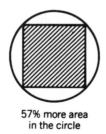

57% more area
in the circle

The example given in Figure 7.38 is a positional tolerance that applies to each hole. There is also a 5.000 and a 4.000 dimension that locates the entire pattern to the edges of the part. This is called *combined position tolerance* (not recommended in the specification).

The drawing in Figure 7.38 means that the centerlines of each of the six holes must be in their proper location within a .010 diameter tolerance zone. Note, however, that the pattern itself must also be within the coordinate tolerances. This applies to the centerlines of the pattern. Before you can inspect this part, you must find a way to convert the hole positions from a bolt circle diameter to linear coordinate dimensions. There are basically two ways to do this. Dimensions and tolerances are shown in Figure 7.38.

One Way to Convert: Right Triangle

In Figure 7.39 you must solve it for both sides, knowing that the angle is 60 degrees because there are six holes divided into 360 degrees and knowing that the hypotenuse of the triangle is one-half of the bolt circle diameter. Then the opposite side length is typical to

FIGURE 7.38
Combined position and coordinate ± tolerances.

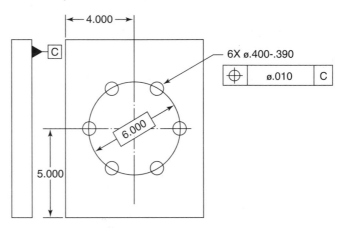

Combined position and ± coordinates

6X ø.400-.390

| ⊕ | ø.010 | C |

hole numbers 1, 6, 4, and 3 vertically, and the adjacent side is typical to numbers 1, 6, 4, and 3 horizontally. The horizontal positions of hole numbers 2 and 5 are exactly one-half of the bolt circle diameter (in this case, 3.000 in.).

An Easier Way: Bolt Circle Chart

Bolt circle charts (Figure 7.40) are available in most manufacturing plants. You simply use the chart by multiplying the actual bolt circle *diameter* times the constant numbers in the table.

Constants for a Six-Hole Circle

A dimensions = .43302 × 6 in.
B dimensions = .250 × 6 in.
C dimensions = .500 × 6 in.

How to Inspect Hole Locations. After you know what the dimensions should be, you simply clamp the part on an angle plate as shown in Figure 7.41 and measure each hole position vertically. Then you rotate the angle plate 90 degrees and measure each hole position that was horizontal. If A and B dimensions all fall within the .010 diameter zone, the part is acceptable.

FIGURE 7.39
Conversion using a right triangle.

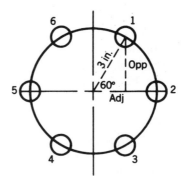

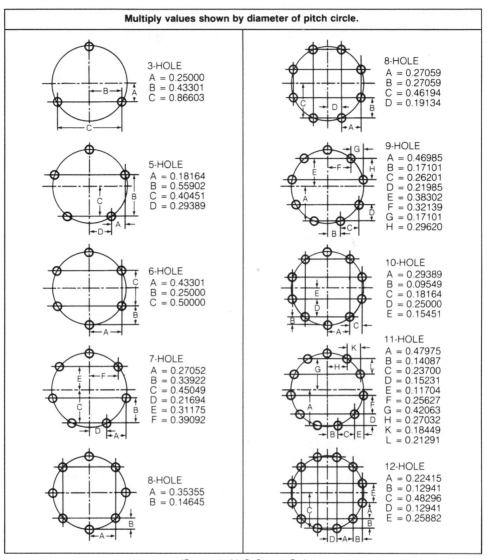

Multiply values shown by diameter of pitch circle.

3-HOLE
A = 0.25000
B = 0.43301
C = 0.86603

5-HOLE
A = 0.18164
B = 0.55902
C = 0.40451
D = 0.29389

6-HOLE
A = 0.43301
B = 0.25000
C = 0.50000

7-HOLE
A = 0.27052
B = 0.33922
C = 0.45049
D = 0.21694
E = 0.31175
F = 0.39092

8-HOLE
A = 0.35355
B = 0.14645

8-HOLE
A = 0.27059
B = 0.27059
C = 0.46194
D = 0.19134

9-HOLE
A = 0.46985
B = 0.17101
C = 0.26201
D = 0.21985
E = 0.38302
F = 0.32139
G = 0.17101
H = 0.29620

10-HOLE
A = 0.29389
B = 0.09549
C = 0.18164
D = 0.25000
E = 0.15451

11-HOLE
A = 0.47975
B = 0.14087
C = 0.23700
D = 0.15231
E = 0.11704
F = 0.25627
G = 0.42063
H = 0.27032
K = 0.18449
L = 0.21291

12-HOLE
A = 0.22415
B = 0.12941
C = 0.48296
D = 0.12941
E = 0.25882

(Courtesy of L.S. Starrett Co.)

NOTES:
1. Pitch circle, hole circle, and bolt circle mean the same thing.
2. This chart is very useful for conveniently converting bolt circle diameters to coordinates for inspection of a bolt circle on the surface plate.
3. Keep this chart handy for use as needed. It is recognized throughout industry as an acceptable tool for use in the shop.

FIGURE 7.40
Bolt circle chart.

127

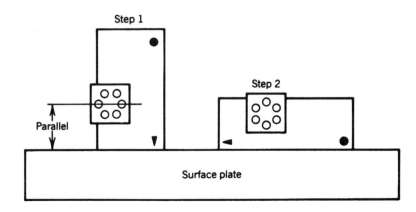

WHEN FEATURES ARE OUT OF POSITION TOLERANCE

Position tolerances ($\oplus$) are used to control the location of features with respect to datum planes or features. One matter of controversy when using position tolerances is how to report those features that are not in the proper position when they are inspected.

The answer to the problem is very simple. All you need to do is report the deficiency according to the drawing requirements. For tolerances of position you have to know two things:

1. The basic coordinate(s) that locate the feature
2. The tolerance zone

■   **Example 7–9**

This example is concerned with a simple part where one hole is drilled that is located by two BASIC coordinate dimensions (see Figure 7.42). The .100–.090 hole must be within a positional tolerance (cylindrical zone) of .005 RFS located by the .500 and 2.000 BASIC dimensions.

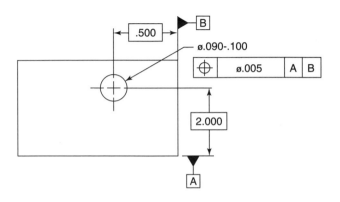

Actual Part Measurements

Let's say that during the inspection, the coordinate dimensions measured as follows.

The .500 BASIC measured .508.

The 2.000 BASIC measured 2.007.

The differences in the actual (measured) coordinates and the BASIC drawing coordinates is 0.008 and 0.007, respectively. Figure 7.43 shows what the true diametral zone of the actual part is.

Reporting the Actual Measurements. The rejection notice should say: "The .100–.090 diameter is out of position tolerance to .021* diameter."

.500 BSC IS 0.508

2.000 BSC IS 2.007

The report above tells everything that is necessary to describe the deficiency (the hole size would have come into play if the position tolerance was at MMC and the part exceeded the bonus tolerance zone).

Hole Patterns on a Bolt Circle

Another example is when there are a multiple of holes located on a bolt circle, as shown in Example 7–10 and Figure 7.44. When holes are located on a bolt circle diameter, each hole has its own specific coordinates. Since the bolt circle diameter is BASIC, so are the coordinate dimensions.

■ Example 7–10

The four holes in Figure 7.44 are to be in position on a 6.000 BASIC bolt circle diameter within a .010 cylindrical tolerance zone (when they are at MMC). As the hole size departs from MMC size (or gets larger), the .010 tolerance zone also gets bigger by the same amount.

The coordinates in Example 7–10 for each hole are 2.121 in. when you calculate them. Since the angle is 45 degrees, both coordinates are 2.121 in. Let's say that we measured all four hole diameters and coordinates and got the following:

Hole 1. Actual size is 0.503
X coordinate is 2.122
Y is 2.120

Hole 2. Actual size is 0.502
X coordinate is 2.120
Y is 2.121

---

*Determined from the table (or formula) in Figure 7.43.

FIGURE 7.43
Actual diametral zone.

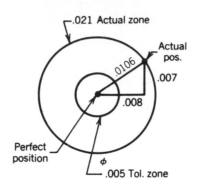

	.001	.002	.003	.004	.005	.006	.007	.008	.009	.010	.011
.020	.0400	.0402	.0404	.0408	.0412	.0418	.0424	.0431	.0439	.0447	.0456
.019	.0380	.0382	.0385	.0388	.0393	.0398	.0405	.0412	.0420	.0429	.0439
.018	.0360	.0362	.0365	.0369	.0374	.0379	.0386	.0394	.0402	.0412	.0422
.017	.0340	.0342	.0345	.0349	.0354	.0360	.0368	.0376	.0385	.0394	.0405
.016	.0321	.0322	.0325	.0330	.0335	.0342	.0349	.0358	.0367	.0377	.0388
.015	.0301	.0303	.0306	.0310	.0316	.0323	.0331	.0340	.0350	.0360	.0372
.014	.0281	.0283	.0286	.0291	.0297	.0305	.0313	.0322	.0333	.0344	.0356
.013	.0261	.0263	.0267	.0272	.0278	.0286	.0295	.0305	.0316	.0328	.0340
.012	.0241	.0243	.0247	.0253	.0260	.0268	.0278	.0288	.0300	.0312	.0325
.011	.0221	.0224	.0228	.0234	.0242	.0250	.0261	.0272	.0284	.0297	.0311
.010	.0201	.0204	.0209	.0215	.0224	.0233	.0244	.0256	.0269	.0283	.0297
.009	.0181	.0184	.0190	.0197	.0206	.0216	.0228	.0241	.0254	.0269	.0284
.008	.0161	.0165	.0171	.0179	.0189	.0200	.0213	.0226	.0241	.0256	.0272
.007	.0141	.0146	.0152	.0161	.0172	.0184	.0198	.0213	.0228	.0244	.0261
.006	.0122	.0126	.0134	.0144	.0156	.0170	.0184	.0200	.0216	.0233	.0250
.005	.0102	.0108	.0117	.0128	.0141	.0156	.0172	.0189	.0206	.0224	.0242
.004	.0082	.0089	.0100	.0113	.0128	.0144	.0161	.0179	.0197	.0215	.0234
.003	.0063	.0072	.0085	.0100	.0117	.0134	.0152	.0171	.0190	.0209	.0228
.002	.0045	.0056	.0072	.0089	.0108	.0126	.0146	.0165	.0184	.0204	.0224
.001	.0028	.0045	.0063	.0082	.0102	.0122	.0141	.0161	.0181	.0201	.0221

ΔY (vertical axis)   ΔX (horizontal axis)

FORMULA:

$$\text{ACTUAL DIA. ZONE} = 2\sqrt{X^2 + Y^2}$$

WHERE X = THE DIFFERENCE BETWEEN THE BASIC AND THE ACTUAL COORDINATE DIMENSION IN "X" DIRECTION
Y = THE DIFFERENCE BETWEEN THE BASIC AND THE ACTUAL DIMENSION IN "Y" DIRECTION

EXAMPLE: The hole is .003 past its basic dimension in "x" direction, and .005 past its basic dimension in "y" direction.
Actual zone = .0117 Dia.

130

FIGURE 7.44
Bolt circle—position tolerance.

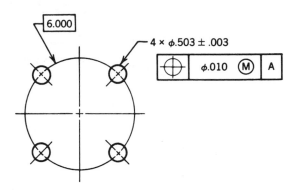

Hole 3. Actual size is .505
       X coordinate is 2.129
       Y is 2.127   — out of position

Hole 4. Actual size is .502
       X coordinate is 2.120
       Y is 2.121

Using your conversion chart, you find that the coordinates for each hole are $\pm$.0035 in each direction. This means that hole 3 is well out of position even when considering the bonus tolerance. Since .500 is MMC size, you can get .005 bonus tolerance to add to the .010 original zone for a total of .015 possible tolerance zone. The actual zone of the hole above is .020, which is much larger than the 0.015 total zone. This part cannot even be reworked because opening the hole size out to LMC size (.506) will only give you a total allowable tolerance of .016. On this part, the rejection notice should read: "The .503 $\pm$ .003 diameter (at hole 3 only) is out of positional tolerance to a .020 diameter (exceeding the total possible tolerance). The 2.121 BASIC coordinates are 2.129 and 2.127."

Now you have reported the problem, and it is understood by those reading the report what is wrong with the four-hole pattern. Again, accurate reporting of parts that are out of position is simply a matter of comparing the drawing requirements with your actual measurements and describing the difference.

CONVERTING POSITION TOLERANCES TO COORDINATE TOLERANCES

On those tolerancing methods that use diametral zones, such as geometric tolerances related to hole patterns, it becomes necessary to convert these diametral zones to coordinate $\pm$ tolerances for measurement. This is especially true when the hole pattern must be measured on a surface plate.

The coordinate tolerances on the hole positions with a diametral zone can be found using the standard chart in Figure 7.45. Reading this chart is easy once you are familiar with it. The chart shows diametral tolerance zones in .001 increments up to .010 diameter. These diameters are shown along the left side of the chart. If your drawing says that the holes are to be within .xxx diameter of position, you first look for this diameter along the left side of the chart.

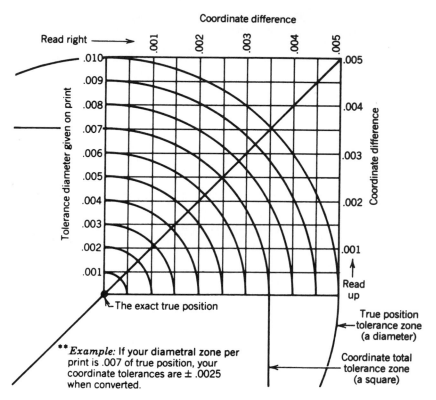

FIGURE 7.45
Sample conversion chart.

Now, note the diagonal line going from corner to corner of the box. This is the point where the coordinate tolerance you need to know meets the diametral tolerance zone you are working with. Once you find the point where the diametral tolerance meets the diagonal line, you simply follow that point to the right and straight up to get the tolerances you are allowed in the vertical and horizontal direction.

Remember: Follow the line from the diametral zone you are dealing with. Stop at the diagonal line. From that point, look right, and straight up for the plus and minus tolerance. For example, a .010 diametral zone has coordinate tolerances of ± .0035. A .044 diametral zone has coordinate tolerances of ± .0155.

You can also use the chart to convert coordinate tolerances to the appropriate diametral zone.

■ **Example 7–11**

The coordinate tolerances are ± .0035 *in both directions.* Find .0035 in the right column and the top row, and follow them to the diagonal line. Then follow the diametral line to find the diametral zone. The answer is the .010 diametral zone.

It is good practice to become well versed in the conversions in either direction using the chart.

Note: This chart shows a maximum capability of only .010 diameter, but you can easily increase its capacity simply by moving the decimal points equally in all of the numbers.

■ **Example 7–12**

Change .010 to .10; then the $\pm$ .0035 becomes $\pm$ .035.

UNDERSTANDING BONUS TOLERANCING

Geometric tolerancing has one major advantage built in—the *bonus tolerance application.* Bonus tolerances can be obtained on geometric controls where the maximum material condition or least material condition is used.

Before going further, a few definitions need to be repeated.

MMC. Simply defined, it is the smallest allowable Ⓜ hole *or* the largest allowable shaft (per the size tolerance).

Original Tolerance. This is the size of the tolerance zone when (and only when) the feature being controlled is at its MMC.

Bonus Tolerance. As the actual feature size departs from MMC, you may add this amount to the original geometric tolerance.

Total Possible Tolerance. This is the sum of the original tolerance (at MMC) *plus* the bonus tolerance (if any).

How the Bonus Tolerancing Concept Works

Figure 7.46 shows two mating parts. Part 1 has a hole in it in a certain position, and part 2 has a pin in it in a certain position. The only important thing in this assembly is that the two mating parts fit together while all four sides are matched evenly.

If the hole position and the pin position are not the same (and they seldom are), a virtual size is created as shown in Figure 7.47. However, if either the hole were larger or the pin were smaller by a significant amount, the parts would still fit together as shown in Figure 7.48. This is how bonus tolerances work. They effectively use all of the possible clearances between mating parts to allow each individual part to be slightly farther away from its location tolerance and still work.

FIGURE 7.46

Two mating parts.

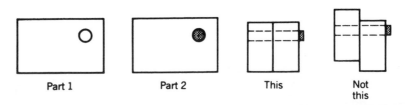

Part 1   Part 2   This   Not
this

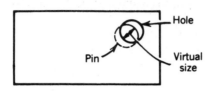

FIGURE 7.47
Virtual size with position tolerance.

FIGURE 7.48
Parts fit together.

How to Calculate Bonus Tolerances. In one statement, you simply measure the feature, find out how far it is from its MMC, and add that much to the stated (original) geometric tolerance. This is your new tolerance (including bonus).

In Figure 7.49, the perfect position would be if the centerline of the hole was exactly .500 from datum B and exactly 1.250 from datum A.

The original tolerance here is .005 MMC if the hole is .100. Let's say that we measured the hole and the actual size was .103. The actual size is .003 larger than its MMC, so you get .003 bonus tolerance on its position.

$$
\begin{array}{r}
.005 \text{ original tolerance} \\
+\ .003 \text{ bonus tolerance} \\
\hline
.008 \text{ total tolerance}
\end{array}
$$

Now you see that due to the maximum material condition concept, you have more tolerance to work with. What started out to be .005 position tolerance became .008 position tolerance simply because of the *actual* clearance between pin and hole.

Restrictions on Bonus Tolerancing

1. You can never go beyond print size dimensions to get more bonus tolerance.
2. Bonus tolerances can only be used in cases where if all the possible bonus tolerance were used, it would not affect the function of the part assembly.
3. Bonus tolerancing applies only on geometric controls where MMC or LMC applies.

FIGURE 7.49
Bonus tolerance.

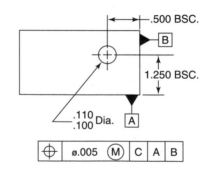

ADDITIONAL TOLERANCE FROM SIZE DATUMS

In applications where a datum feature of size is shown in a feature control frame at MMC, additional tolerance is allowed. This is not the usual type of bonus tolerance, however. In this case a feature or pattern of features is allowed to shift or rotate as the datum feature of size departs from its MMC size. An example of this is shown in Figure 7.50.

In Figure 7.50 datum B is applied at MMC. The .010-in. position tolerance zone on the holes can increase in size as the .100-in. holes depart from their MMC size (.095 in.), and the pattern of holes can shift off centerline or rotate as datum B departs from its MMC size (.902 in.)

Therefore, bonus tolerance from a datum feature of size is not added directly to the position tolerance diameter, but allows the pattern to shift or rotate as the datum feature of size departs from its MMC size. Bonus tolerances do increase the diametral tolerance zone size as the controlled holes depart from their MMC size.

FIGURE 7.50
Bonus and additional tolerance.

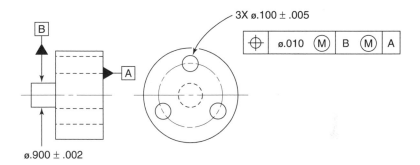

CHAIN, BASELINE, AND DIRECT DIMENSIONING

There are times when a row (or chain) of features must be dimensioned to control their location. An example of this is a set of holes in a straight line as shown in Figure 7.51. Careful selection of the type of dimensioning to use on these holes will assist you in getting the maximum tolerances allowed with a specific buildup of tolerance over the entire chain of holes. Remember that with any type of dimensioning there is no *tolerance buildup* when you use BASIC dimensions. The examples to follow all use linear plus and minus dimensions and have tolerance buildup (or accumulation).

There are three methods in which a "chain" of size features can be dimensioned: *chain dimensioning, baseline dimensioning,* and *direct dimensioning.*

Chain Dimensioning

Figure 7.51 shows an example of chain dimensioning with plus and minus tolerances. When chain dimensions are used, the maximum tolerance accumulation overall is equal to the sum of the intermediate tolerances. Since all three tolerances over distance A are

±0.005, the maximum tolerance for distance A is ±.015 (the sum of the three intermediate tolerances).

FIGURE 7.51
Chain dimensioning.

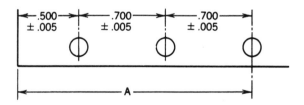

Baseline Dimensioning

Using baseline dimensioning (preferred in most cases), the maximum buildup tolerance for A dimension (Figure 7.52) is equal to the sum of the two dimensions' tolerances that make up A dimension (they are the .500 ± .005 and the 1.300 ± .010, and add up to ±.015). This is the greatest tolerance accumulation.

FIGURE 7.52
Baseline dimensioning.

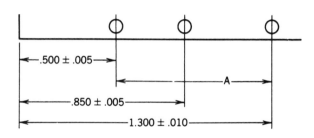

Direct Dimensioning

Direct dimensioning allows you to limit the tolerance buildup in a dimension to the stated tolerance on that dimension. Figure 7.53 shows an example of direct dimensioning where the tolerance on A dimension is as stated.

FIGURE 7.53
Direct dimensioning.

COMPOSITE POSITION TOLERANCES

Composite position tolerances are often used to describe two different controls on a single feature or pattern of features using one geometric symbol for position. Figure 7.54 shows a very basic example of composite position tolerancing. There are three datums (C, D, and E). Datum C is the primary datum for the hole. Specifying datum C normally would control the perpendicularity of the hole where datums D and E control its location.

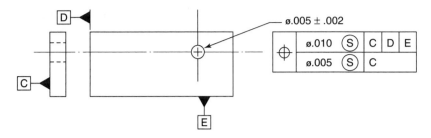

FIGURE 7.54
Composite position tolerance.

In this case the hole can be located within a .010 diametral zone for its location but must be perpendicular to datum C within the .005 diametral zone. To specify this clearly on the drawing, composite position tolerance is used.

There are two different measurements required here. One must first measure the location of the hole regardless of feature size to datums C, D, and E. Then one must measure the perpendicularity of the hole regardless of feature size to datum C alone. If the controls were at MMC, this would call for two different functional gages (or one gage with two different virtual size pins).

Hole Patterns.   When composite position tolerances are used on a pattern of hole, the upper portion of the feature control frame locates the pattern to the outside edges of the part, and the lower portion locates the holes to each other.

Holes in Line.   When composite position tolerances are applied to holes drilled in line, the upper portion of the feature control frame locates each hole to the outside edges of the part, and the lower portion controls the coaxiality of the holes to each other.

POSITION TOLERANCE STATED AT LMC

In some applications position tolerances are stated at LMC. LMC is that condition of a feature of size where all of the material that is allowed to be removed (per the size tolerance) has been removed. The LMC is simply remembered as the largest allowable hole, or the smallest allowable shaft.

The 1982 version of the geometric tolerancing specification allows position tolerances to be stated at LMC condition. Whenever the position tolerance is stated at LMC in the feature control frame, it means that the tolerance shown only applies when the feature is produced at LMC size. As the feature departs from LMC size (toward MMC), the stated tolerance gets larger by the same amount of that departure.

One application for position at LMC could be a boss that takes a drilled hole. If the boss and the hole are both located by the same basic dimensions, and both are in position at LMC, the edge distance (or break out distance) between them can be guaranteed at a certain minimum. This is true because the tolerance applies when the hole is its largest and the boss is its smallest size. Another example application follows.

As shown in Figure 7.55, there is a 1.800 diameter pad on this part that is used as a locator for the part. It must be in position such that it will make contact with a fixed locator on the machine. Position at LMC is applicable here because as the diameter of the pad increases it can afford to be farther out of position, as it will still make contact with the fixed locator on the machine (Figure 7.56).

FIGURE 7.55
Position at LMC.

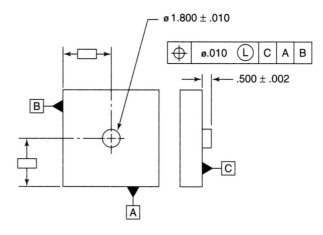

FIGURE 7.56
LMC application.

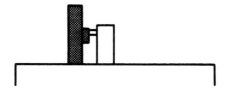

The LMC of the pad is 1.790 at this size and the position tolerance is .010. If the pad is larger than LMC size, you have bonus tolerance equal to that amount. The table in Figure 7.57 shows the position tolerance allowed because of the actual size of a produced pad.

FIGURE 7.57
Table of tolerances.

Actual Feature Size	Position Tolerance Allowed
1.790 (L)	.010
1.791	.011
1.792	.012
1.793	.013
↓	↓
1.801 (M)	.030

ZERO POSITION TOLERANCE AT MMC

Position tolerancing has been around for a long time, but in many cases, the best method of positioning features (zero tolerance at MMC) is not used. One of the main reasons it is not used is that it is misunderstood. The thought of putting zero (or no tolerance) on an engineering drawing strikes fear in the minds of many. Zero tolerance at MMC is not taking tolerance away from the part. In fact, it does not change the tolerance value that is traditionally allowed for a part. The part in Figure 7.58 is an example of zero tolerance at MMC.

In Figure 7.58, the position tolerance allowed for the .300–.295 hole is solely dependent on the actual size of the produced hole. If it is produced at MMC size (.295) it must be in perfect position. That position is described by the basic dimensions (not shown) that locate the hole. The only position tolerance allowed is the amount that is the difference between the actual size and the MMC size. For example, if the hole is produced to .298 diameter, a .003 diametral tolerance zone is allowed for the position of the hole (see Figure 7.59).

FIGURE 7.58
Zero tolerance at MMC.

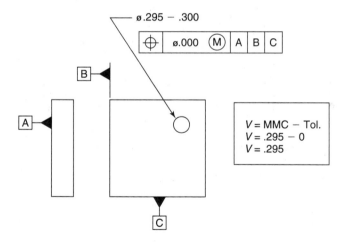

FIGURE 7.59
Table of tolerances.

Actual Hole Size	Position Tolerance Allowed
.295 Ⓜ	0
.296	.001
.297	.002
.298	.003
.299	.004
.300 Ⓛ	.005

The part could be traditionally toleranced as:

Hole size range: .300–.298

Position tolerance = .003 at MMC

Notice that the size range of the hole had to be tightened. The maximum hole size has not changed, and the virtual size of the hole has not changed (it is .295 virtual size either way). Let's also say that a functional gage has been made to check the hole position. The functional gage would have a .295 nominal pin located precisely at the basic dimensions of the hole.

We can also say that on the traditional part some parts were made with undersized holes (.297) but these holes were in perfect position. The functional gage would accept every one of these parts (because they are good parts), but since the hole is undersize, they must be rejected. This is why zero tolerance at MMC is so effective. The functional gage also *doubles as a Go size gage*. No functional parts can be rejected simply because of hole size, as with traditional position tolerancing.

BIDIRECTIONAL POSITION TOLERANCES

In some situations the position of a feature in one direction is more important than its position in another direction. The example in Figure 7.60 shows how this used to be handled by old coordinate (plus and minus) tolerancing methods.

Notice in Figure 7.60 that the hole has a larger location tolerance in Y direction ($\pm$.010) than in X direction ($\pm$.002). This tolerancing method forms a rectangular tolerance zone

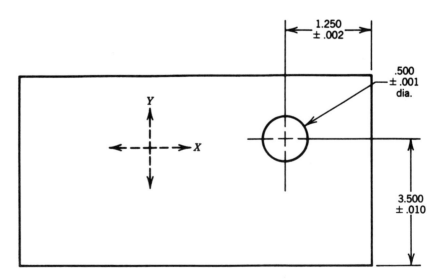

FIGURE 7.60

Bidirectional location tolerance with the old $\pm$ coordinate system.

that is .020 high and .004 wide for the location of the centerline of the hole. Figure 7.61 shows how this rectangular tolerance zone is constructed.

FIGURE 7.61
Rectangular tolerance zone.

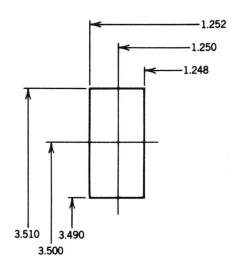

As with all old coordinate methods of tolerancing, this rectangular zone is impractical and restrictive because a round hole is being located with a square-shaped zone. The geometric tolerancing used on this part would show two different feature control frames. The example in Figure 7.62 shows how the part would be dimensioned per geometric tolerancing methods.

FIGURE 7.62
Example of a bidirectional position tolerance.

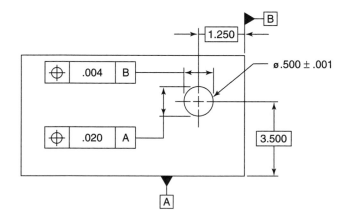

In Figure 7.63 the tolerance zone for each basic dimension is a different width. This forms the same rectangular zone as the old system.

FIGURE 7.63
Tolerance zone.

Now, the feature is controlled in both directions in the same way as the coordinate system. The old system has been converted to geometric tolerancing. The only exception is that we could use MMC now to get bonus tolerance, and the datums are clearly specified (Figure 7.63).

Remember that in the same way the old coordinate system produced a rectangular zone, the new system also created this type of zone. The old system (Figure 7.64) is ±.002, and the new system (Figure 7.65) is a .004 total wide zone.

FIGURE 7.64
Tolerance zone.

FIGURE 7.65
Tolerance zone.

POSITION: FIXED AND FLOATING FASTENER CASES

When using position tolerances on mating parts it is necessary to understand fixed and floating fastener cases.

Fixed Fastener Case. This relationship between mating parts is simply defined by saying that the fastener (or bolt) that connects these parts is restrained in one of them or both. An example is when one of the parts has a tapped hole. This tapped hole would restrict the bolt from moving around. This restriction would mean that the holes must be lined up more accurately to fit.

Floating Fastener Case. This relationship between mating parts is much easier to work with since the fastener (or bolt) is not restricted in either part. Since it is not restricted, the clearance can be put to use in tolerance for the location of each hole position.

Floating Fastener Case

The formula for calculation of the position tolerance using the floating fastener case is

$$T = H - F \quad T = \text{position tolerance diameter}$$
$$H = \text{MMC of the holes}$$
$$F = \text{MMC of the fasteners (bolts)}$$

When you have an established position tolerance and you must calculate the other sizes, the following formulas can be used:

$$H = T + F \quad F = H - T$$

■ **Example 7–13**

$$MMC \text{ (hole)} = .380$$
$$MMC \text{ (fastener)} = .375$$

What should the positional tolerance be for both mating parts?

Solution

$T = H - F$

$\quad = .380 - .375$

$\quad = .005 \; Answer$

■ **Example 7–14: Floating Fastener Case**

$$\text{Position tolerance} = .010$$
$$MMC \text{ (hole)} = .510$$

What is the MMC of the fasteners you should use?

Solution

$F = H - T$

$\quad = .510 - .010$

$\quad = .500 \; Answer$

Fixed Fastener Case

The formula for the fixed fastener case to calculate the position tolerance is

$$T = \frac{H - F}{2} \text{ (the tolerance for both parts)}$$

This formula can be changed to also calculate the hole size or fastener size with an existing position tolerance as follows:

$$H = F + 2T \qquad F = H - 2T$$

■ **Example 7–15**

What is the required position tolerance for each mating part if the hole MMC is .380 and the fastener MMC is .370?

Solution

$$T = \frac{H - F}{2}$$

$$= \frac{.380 - .370}{2}$$

$$= \frac{.010}{2}$$

$$= .005 \text{ position tolerance} \quad Answer$$

Understanding the fixed and floating fastener cases makes position tolerance calculation much easier (using the MMC concept). With the old square tolerance zone for hole positions, this was not possible.

PROJECTED TOLERANCE ZONES

In certain cases a particular geometric tolerance will be shown on a drawing with a ⓟ next to it. This ⓟ is the symbol for *projected tolerance zone*. To introduce projected tolerance zones, we must first understand how tolerance zones work. A tolerance zone for any feature is the same length as that feature. As shown in Figure 7.66, the tolerance zone for this hole is always the same length as the hole itself (unless ⓟ is used).

At times, however, the tolerance zone for a feature must be extended past the feature into space for very good reasons. One of the best reasons is fit. Sometimes it is not enough to control the feature alone but to control the component that goes into it as well.

FIGURE 7.66
Tolerance zone and length of hole are the same. This hole is 1.5 in. long, so the tolerance zone is also 1.5 in. long.

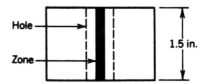

FIGURE 7.67
Part with a .002 cylindrical tolerance zone.

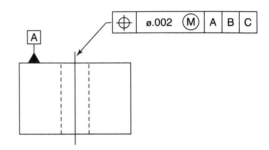

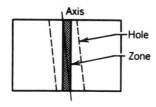

FIGURE 7.68
Tolerance zone of part 1.

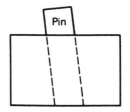

FIGURE 7.69
Part 1 with centerline pin.

For example, part 1 (shown in Figure 7.67) must have a hole put into it that is the correct size and positioned to datum surfaces within a .002 cylindrical tolerance zone. According to this tolerance, you could have a hole that looks like Figure 7.68.

If part 1 must have a press-fit pin installed, the pin will follow the centerline of the hole and look like the drawing in Figure 7.69. The next assembly for part 1 requires that it fit into part 2 and that the mating surfaces close together completely, as shown in Figure 7.70. To ensure that the two parts fit together properly, we must make certain that the press-fit pin is also positioned to its datums. Since the pin extends above datum A by 1.00 in., we will simply project the tolerance zone for the hole above datum A by 1.00 in. This is shown in Figure 7.71.

Now the position tolerance that is applied to the hole is also applied to the pin, and it will easily fit the mating part time after time (as long as both parts are within blueprint tolerance). Projected tolerance zone, $\textcircled{P}$, is a method of controlling a feature in such a way that you also control components that mate with that feature.

FIGURE 7.70
Assembly of parts 1 and 2.

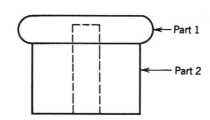

FIGURE 7.71
Projected tolerance zone. The chain line (a thick long dash, short dash) shows the direction of the projected zone. If it were a blind hole, the chain line would not be necessary, and the 1.0 would go next to the circled P.

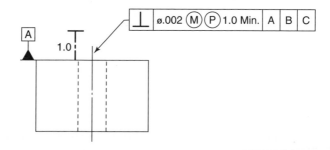

Projected tolerance zones are more restrictive than the original tolerance because the pin sticking up above the surface is being controlled as shown in Figure 7.72. The most important purpose for projected tolerance zones is to make sure that parts will fit together at assembly.

FIGURE 7.72
Press-fit pin also controlled.

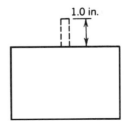

1.0 in.

Another example for projected tolerance zones is to apply them to a part where a bolt must be torqued down with the bolt head meeting the surface flat all around. In this case, if the bolt head does not meet flat with the surface, it could break off when torque is applied. This example is shown in Figure 7.73.

FIGURE 7.73
(*a*) Proper seat for a bolt head.
(*b*) Improper seat for a bolt head.

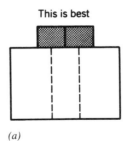
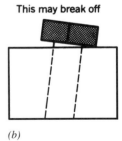

This is best   This may break off

(a)   (b)

Projected tolerance zones should be used only where they are absolutely necessary because they increase the cost of making the part.

Measurement Technique

The measurement technique for tolerances with projected tolerance zones is slightly different in that if you use a pin in the hole, the pin must extend out from the hole in the proper direction and your measurements will be taken on the pin at the distance the projection is indicated. If it is a 1-in. projection, you will measure at the pin and the distance from the end of the hole to 1 in. above it. The measurement takes place only above the hole (in the projected tolerance zone) and not in the hole.

FUNCTIONAL GAGES

Functional gages are designed to inspect the collective effects of the size of a feature and the geometric characteristic controlling that feature. Geometric tolerances can be verified with a functional gage only if they utilize the maximum material condition (MMC) in the feature control symbol on the drawing. These gages are very useful during production and inspection for quick and accurate feature verification; the gaging is also related to fit and function. Note that functional gages only measure the geometric characteristics, not the actual size of the feature being controlled.

Functional gages are attribute gages, meaning that they will only tell you whether or not the part is acceptable. They will not tell you how good or how bad a part is. Generally, functional gages are used for a quick check during production or inspection. When the gage rejects the part, then a "variable" inspection method may be used, if necessary, to find out how bad the part is.

Basics of Functional Gages

Functional gages generally take the shape of the mating part. For example, the feature to be controlled is a shaft (or pin). Here the functional gage would generally take the shape of a ring gage. This is also true for holes. A functional gage to check hole positions would use pins. All functional gages are generally one gage that either goes into the part or does not. They are not two gages (such as Go and NoGo).

Basic Design of Functional Gages

To design functional gages, one must thoroughly understand how to calculate the *virtual size* of any feature (e.g., a hole, slot, shaft, or tab) that is controlled by a geometric tolerance such as angularity, perpendicularity, position, and parallelism.

There are two quick formulas to use in all cases:

1. Virtual size (external feature) = MMC (of the feature) + geometric tolerance
2. Virtual size (internal feature) = MMC (of the feature) − geometric tolerance

■ **Example 7–16: Virtual Size (External Feature)**

A shaft that has a straightness tolerance at MMC of .002 and a size tolerance of .530 ± .003.

$$\text{Virtual size} = \text{MMC} + \text{geometric tolerance}$$
$$= .533 \text{ in.} + .002 \text{ in.}$$
$$= .535 \text{ in.} \quad Answer$$

■ **Example 7–17: Virtual Size (Internal Feature)**

A hole that has a positional tolerance of .005 in. diameter at MMC and a size tolerance of .310 in. ± .003 in.

$$\text{Virtual size} = \text{MMC} - \text{geometric tolerance}$$
$$= .307 \text{ in.} - .005 \text{ in.}$$
$$= .302 \text{ in.} \quad Answer$$

Examples of Functional Gages

Functional Gage to Measure the Straightness of a Pin at MMC. When Figure 7.74 appears as part of a drawing, it means that the axis of the pin must be straight within a diametral tolerance zone of .002 in. diameter only when the pin size is at MMC (which is .502 in.). As the pin gets smaller (until you reach LMC or .498 in.), the straightness tolerance gets larger automatically by the amount that the pin deviates from MMC. The virtual size of the pin is .504 in. (see Figures 7.75 and 7.76).

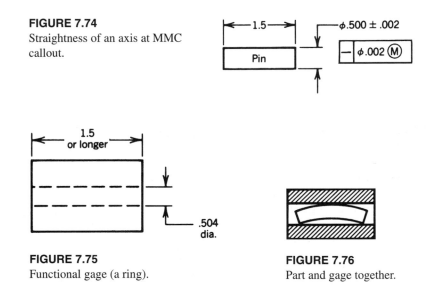

FIGURE 7.74
Straightness of an axis at MMC callout.

FIGURE 7.75
Functional gage (a ring).

FIGURE 7.76
Part and gage together.

Functional Gage to Measure the Position of Three Holes Equally Spaced on a 6-in. Basic Bolt Circle Diameter at MMC. Figure 7.77 shows a part with three holes, which must lie in true position, with each having a diametral tolerance zone of .002 in. (when they are at MMC, which is .398 in.). As the holes get larger, so does the tolerance zone. If

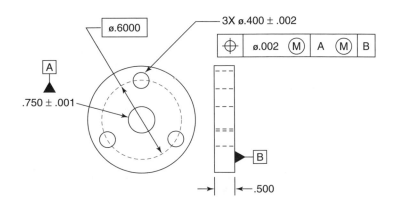

FIGURE 7.77
Bonus tolerance from pattern and datum.

datum A is larger than MMC size, the pattern of holes (Figure 7.78) is allowed to shift or rotate from its center axis by that much. The .002 diameter tolerance zone does not increase in size from the datum bonus tolerance.

FIGURE 7.78
Functional gage for pattern.

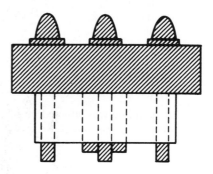

Functional Gage for Perpendicularity of a Hole at MMC.   Figure 7.79 shows that the .490–.500 hole must be perpendicular to datum plane C within a cylindrical tolerance zone of .005 in. diameter when the hole is made at MMC size. The MMC is .490.

FIGURE 7.79
Perpendicularity at MMC.

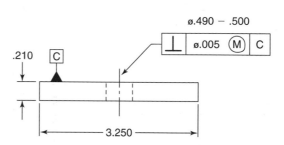

The functional gage for the part must have a virtual size pin that is .485 in. in diameter (which is the MMC of the hole minus the perpendicularity tolerance), and the plate of the gage is large enough to cover datum C completely. The pin must be long enough to go all the way through the hole (see Figure 7.80).

FIGURE 7.80
Functional gage for perpendicularity at MMC.

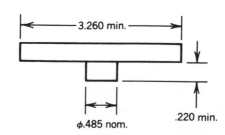

As shown in Figure 7.81, the functional gage surface must mate with datum C (at its three highest points), and the pin must go through the hole at the same time.

FIGURE 7.81
Functional gage and part.

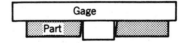

THE CHAIN LINE

The chain line was adopted in the ANSI Y14.5M 1982 revision of the specification. It is similar to a centerline except that it is much thicker (see Figure 7.82).

FIGURE 7.82
Chain line.

Chain Line Applications

Chain lines have the following uses:

The chain line has several uses, most of which are to limit the geometric control of a feature to certain areas (depending on design requirements).

1. *To define a limited length of control* (Figure 7.83). The chain line means that the position tolerance of .004 is limited to control only 1 in. of the length of the controlled diameter. This is true because the rest of the diameter is nonfunctional and need not be controlled.

FIGURE 7.83
Limited length of control.

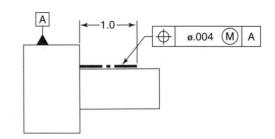

2. *To define a limited area of control* (Figure 7.84). The area shown in Figure 7.84 is the only area on the part that must be flat within .002. The area is bounded by the chain line and filled in with section lines for clarity.

FIGURE 7.84
Limited area of control.

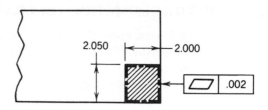

3. *To define partial datums* (Figure 7.85). The drawing in Figure 7.85 means that datum A is not the entire bottom surface. Only 2 in. inward from the end as shown should be contacted to measure the perpendicularity of .003. (This is so because only 2 in. of the bottom surface is contacted by the mating part at assembly.)

FIGURE 7.85
Partial datum.

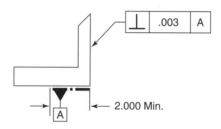

4. *To show the direction of a projected tolerance zone* (Figures 7.86 and 7.87).

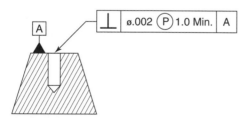

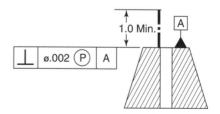

FIGURE 7.86
Projected zone direction—blind hole. There is no need for the chain line; the direction is obvious.

FIGURE 7.87
Projected zone—through hole. Use the chain line to show the direction and height of the projected zone.

Using the chain line a designer can limit the control on a part to exactly where it is needed. As a result, limited control products are more economical to produce without any loss in function. There are many cases in industry where limited control can and should be applied.

CHANGES IN THE GEOMETRIC TOLERANCING STANDARD

There were many changes in the new standard ASME Y14.5M–1994 from the older version (ANSI Y14.5M–1982). This book does not cover all geometric tolerancing symbols and methods, but the contents have been updated to the new standard. The table on p. 153 compares the 1982 and the 1994 standard in terms of some of the changes that were made.

REVIEW QUESTIONS

1. The diameter symbol
   a. precedes all diameters on the drawing
   b. is shown when the tolerance zone is a cylinder
   c. can be used when a datum target area is round
   d. all of the above
2. What is the MMC of the shaft size tolerance .750–.760?
   a. .750
   b. .760
   c. .755
   d. none of the above
3. What is the LMC of the shaft in question 2?
   a. .750
   b. .760
   c. .755
   d. none of the above
4. A _____ dimension has no tolerance. It locates a geometric tolerance zone.
   a. reference
   b. nominal
   c. BASIC
   d. title block
5. If a dimension is underlined, such as .500, what does this mean?
   a. It is an important dimension.
   b. It is a reference dimension.
   c. It is not to be inspected.
   d. It is not to scale.
6. The _____ _____ frame contains the requirements and replaces the old notes for geometric controls.
   a. title block
   b. feature control
   c. change block
   d. none of the above
7. Which of the following is an example of a geometric tolerance zone?
   a. an imaginary cylinder
   b. two parallel planes
   c. two parallel lines
   d. all of the above

TABLE 7–1
A comparison of the 1982 and 1994 standards

GDT SYMBOLS ANSI Y14.5M-1982 VS. ASME Y14.5M-1994

SUBJECT	SYMBOL	ANSI Y14.5-1982	ASME Y14.5-1994
FORM TOLERANCES	FLATNESS	▱	▱
"	STRAIGHTNESS	—	—
"	CIRCULARITY	○	○
"	CYLINDRICITY	⌭	⌭
ORIENTATION TOLERANCES	PARALLELISM	//	//
"	PERPENDICULARITY	⊥	⊥
"	ANGULARITY	∠	∠
RUNOUT TOLERANCES	CIRCULAR RUNOUT	↗	↗
"	TOTAL RUNOUT	↗↗	↗↗
PROFILE TOLERANCES	PROFILE OF A LINE	⌒	⌒
"	PROFILE OF A SURFACE	◠	◠
LOCATION TOLERANCES	CONCENTRICITY	◎	◎
"	POSITION	⊕	⊕
"	SYMMETRY	DELETED Use Position	≡ RENEWED USE
DATUM	DATUM	- A -	Ⓐ ▲
MODIFIERS	MMC	Ⓜ	Ⓜ
"	LMC	Ⓛ	Ⓛ
"	RFS	Ⓢ	No Longer Needed (Understood)
"	PROJECTED ZONE	Ⓟ	Ⓟ
"	TANGENT PLANE	NONE	Ⓣ
"	FREE STATE	NONE	Ⓕ

153

8. Which of the following geometric tolerancing symbols does not use MMC?
   a. parallelism
   b. position
   c. perpendicularity
   d. flatness

9. Which of the following symbols cannot be gaged with a functional gage?
   a. position
   b. parallelism
   c. perpendicularity
   d. flatness

10. Which of the following symbols may (or may not) be applied using one or more datums (in other words, datums are optional)?
   a. position
   b. parallelsim
   c. runout
   d. profile

11. Which of the following is *not* a form tolerance?
   a. flatness
   b. straightness
   c. cylindricity
   d. parallelism

12. Which of the following is not an orientation tolerance?
   a. parallelism
   b. angularity
   c. position
   d. perpendicularity

13. What is the MMC of the hole size .500–.510?
   a. .510
   b. .505
   c. .500
   d. .502

14. What is the LMC of the hole in question 13?
   a. .510
   b. .505
   c. .500
   d. .502

15. What is the virtual condition of a shaft that has a size tolerance of .750–.760 and is allowed to be bent on its axis by .005″?
   a. .745
   b. .750
   c. .760
   d. .765

16. What is the virtual condition of a hole that has a size tolerance of .375 ± .002 and a position tolerance of .005 at MMC?
   a. .373

b. .368

c. .382

d. none of the above

17. Which of the general rules applies when a geometric tolerance is applied to a screw thread?

   a. rule 1

   b. datum/virtual size rule

   c. pitch diameter rule

   d. rule 2

18. Which of the general rules is designed to guarantee the maximum boundary of perfect form of a shaft?

   a. rule 1

   b. datum/virtual size rule

   c. pitch diameter rule

   d. rule 2

19. How many high points, at minimum, are required to contact a secondary functional datum?

   a. 3

   b. 2

   c. 1

   d. 4

20. If the datums are expressed in the feature control frame as follows, what is datum C? Datums: C A B.

   a. secondary datum

   b. tertiary datum

   c. primary datum

   d. clocking datum

21. Which of the following is an example of a datum target?

   a. target point

   b. target line

   c. target area

   d. all of the above

22. Datum targets are mainly used for

   a. function

   b. consistency

   c. convenience

   d. permanent reference

23. Surface plates, gage pins, ring gages, etc., are all examples of what kind of datums?

   a. primary

   b. physical

   c. simulated

   d. targets

24. What is the expected shape of the tolerance zone for locating a hole using the old co-ordinate system?

   a. round

   b. oblong

    c. square

    d. rectangular

25. Which of the following modifiers are applicable using the old coordinate system of locating features?

    a. MMC

    b. RFS

    c. LMC

    d. none apply

26. Convert a cylindrical tolerance zone of .010″ diameter to the even plus or minus coordinates that apply.

    a. ± .010

    b. ± .0035

    c. ± .005

    d. none of the above

27. A hole must be in position within .010″ diameter zone at MMC and the hole size must be .500–.515. What would be the position bonus tolerance allowed if the hole were drilled to an exact size of .502? (Note: Just the amount of bonus tolerance.)

    a. .003

    b. .002

    c. .001

    d. .005

28. In question 27, what would be the total position tolerance allowed for that hole?

    a. .010

    b. .011

    c. .015

    d. .012

29. The following type of dimension has no tolerance, does not allow tolerance buildup, and is used with position and profile tolerances.

    a. perfect dimension

    b. nominal dimension

    c. locating dimension

    d. basic dimension

30. When using composite position tolerances, the tolerance value in the upper portion of the feature control frame controls

    a. hole to hole location

    b. perpendicularity

    c. pattern location

    d. intrinsic location

31. Refer to question 30. What is the tolerance in the lower portion of the composite position control frame used for?

    a. hole to hole location

    b. perpendicularity

    c. pattern location

    d. intrinsic location

32. A hole has a .750–.760 size tolerance and a position tolerance of zero at MMC. If the hole is produced at an actual size of .755, what is the position tolerance allowed?
    a. .010
    b. .005
    c. .012
    d. .011

33. If a designer can allow more tolerance for the location of a hole in one direction than in another direction, the best application of position tolerance would be
    a. composite
    b. feature to feature
    c. pattern
    d. bidirectional

34. What is the fastener case when both mating parts have clearance holes in them?
    a. fixed
    b. tight
    c. floating
    d. bidirectional

35. In the fixed fastener case of mating parts, it is best to use a specific modifier on the position tolerance. That modifier is
    a. MMC
    b. RFS
    c. Ⓟ
    d. LMC

36. A hole must be in position within a .010 diameter cylindrical tolerance zone at MMC, and the hole size range is .375–.385. What would be the nominal size of the pin in a functional gage for this hole?
    a. .385
    b. .375
    c. .365
    d. .380

37. Which of the following special types of drawing lines can be used to show limited control?
    a. section line
    b. border line
    c. limiter line
    d. chain line

38. Whenever a position tolerance for a size feature is shown in the feature control frame without a modifier, this means
    a. bonus tolerance is not allowed
    b. the size feature must be at size
    c. the location tolerance is additive
    d. none of the above

39. Geometric tolerances are designed to help guarantee which of the following?
    a. fit
    b. form

c. function
d. all of the above

40. Which of the general rules applies when a datum size feature is being controlled to another surface and is being used at MMC?
  a. pitch diameter rule
  b. rule 2
  c. rule 3
  d. datum/virtual size rule
  e. rule 1

Graphical Inspection Analysis*

Graphical inspection analysis (GIA), also referred to as *layout gaging,* is an inspection verification technique employed to ensure functional part conformance to engineering drawing requirements without high-cost metal functional gages. With the increased use of dimensioning and tolerancing methods, where we are concerned with concepts such as maximum material condition, cylindrical-shaped tolerance zones, and bonus tolerances, we must use inspection methods that will evaluate parts conformance on a functional basis. Graphical inspection analysis provides the benefit of functional gaging without the expense and time required to design and manufacture a close-toleranced, hardened metal functional gage. Graphical inspection analysis is well suited when small lots of parts are being manufactured; however, the inspection concept applies equally well to large quantity production. Also, inspection can be performed on parts too large and parts too small for conventional, functional receiver gages.

Inspection with functional receiver gages is the most desirable method to verify part conformance for interchangeability. Functional gages are really three-dimensional, worst-case condition, mating parts. If the gage assembles to the part being inspected, we are assured assembly with all conforming mating parts. The drawbacks of functional gages are the expense and time required for design and manufacturing. Graphical inspection analysis is a technique that will provide immediate, accurate inspection results without high cost and time delays.

GEOMETRIC TOLERANCING PRINCIPLES

Graphical inspection analysis is very dependent on the methods and principles used in geometric dimensioning and tolerancing. It is assumed that readers of this chapter are knowledgeable about the concepts and rules of the geometric dimensioning and tolerancing standard as specified in American Society for Mechanical Engineers (ASME) standard Y14.5M–1994. Geometric tolerancing principles will be reviewed to ensure common interpretation for an effective understanding and use of GIA.

Principle 1: Authority Standard

It is imperative that the parts inspector knows which revision of the standard is used as the authority document in the preparation of the drawing. In this chapter we point out various differences between the revisions of the standard (USASI Y14.5–1966, ANSI Y14.5–1973, ANSI Y14.5M–1982, and ASME Y14.5M–1994).

---

* This chapter was written by and figures reprinted by permission of George O. Pruitt, Technical Documentation Consultants. Contents have been edited and updated by the author.

Principle 2: Basic Dimensions

A basic dimension is a theoretical, exact dimension without tolerance. When used in conjunction with position tolerance specification, the basic dimension locates the exact center of the position tolerance zone. The center plane, or axis, of an acceptable feature is allowed to vary within the basic, located, position tolerance zone, as shown in Figure 8.1. Symbolically, the dimensions are enclosed in a box to signify a basic dimension. When features are located by basic chain dimensioning on the drawing, there is no accumulation of tolerance between features (Figure 8.2). Basic dimensions are absolute values and, when added together equal an absolute value. The basic dimensioning locates the position of the tolerance zone, not the manufactured feature locations.

Principle 3: Condition Modifiers

There are three material condition modifiers used with position tolerancing: maximum material condition (MMC), regardless of feature size (RFS), and least material condition (LMC). The symbol for maximum material condition is Ⓜ . The 1982 standard required modifiers Ⓜ, Ⓢ, or Ⓛto be specified for position tolerances. When the MMC symbol is associated with the tolerance or a datum reference letter in the feature control frame, the specified tolerance applies to the feature *only* if the feature is manufactured at its maximum material condition size.

Let's review the maximum material condition as it applies to position tolerance. The maximum material condition is the condition of a feature when it contains the maximum amount of material or weight within its allowable size limits. For example, a .500 ± .010 diameter pin will be at maximum material condition only if the machinist manufactures the pin at .510 diameter—a .509 diameter pin would not be at maximum material condition. Another example is: A .525 ± .010 diameter hole in a part would be at maximum material condition, or its greater weight, if drilled at .515 diameter. In other words, maximum material condition is maximum allowable shaft size or minimum allowable hole size.

As shown in the feature control frame (Figure 8–3), the .028 tolerance value applies *only* if the feature with which the frame is associated is produced at MMC. What happens

FIGURE 8.1
Positional tolerance zone.

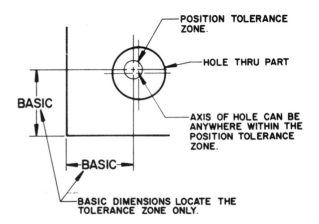

FIGURE 8.2
Chain dimensioning versus baseline
dimensioning.

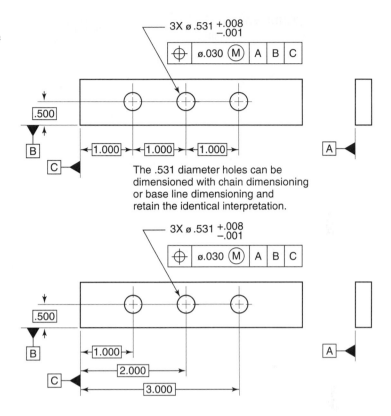

The .531 diameter holes can be
dimensioned with chain dimensioning
or base line dimensioning and
retain the identical interpretation.

if the associated feature or datum departs from MMC? Principle 5 will discuss additional
tolerance allowances.

If no modifier is specified, (S) is understood to apply. When the RFS applies to the tolerance or datum reference, the specified tolerance applies to the location of the feature regardless of the feature's size. In Figure 8.4, the .028 tolerance applies if the hole is manufactured at .515, .535, or any other value in between. The .028 is the total location tolerance
for the feature regardless of its size.

FIGURE 8.3
MMC feature control frame.

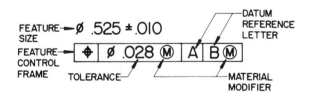

FIGURE 8.4
RFS feature control frame.

.525±.010 DIA HOLE

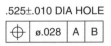

FIGURE 8.5
Feature control frame.

The symbol for least material condition is L. The least material condition concept was not introduced until the approval of ANSI Y14.5–1982. Least material condition is the condition of a feature when it contains the least amount of material or weight; for example, smallest shaft size and largest hole size—just opposite of the MMC concept. When a feature control frame references LMC to the tolerance or datum reference letter, the specified tolerance *only applies* when the feature is produced at the LMC size. Additional discussion of what happens when the feature or datum size departs from LMC will be discussed in Principle 5.

Principle 4: Feature Control Frame

The *feature control frame,* formally called *feature control symbol,* tells the drawing user many things. The shape and size of the tolerance zone is specified, the surfaces and order of precedence for setup of the part is dictated, and the material condition modifiers are assigned to the tolerance and datum reference letters.

As shown in Figure 8.5, the user is aware that the .014 tolerance value applies only when the feature being verified for locations is produced at its maximum material condition size. Also, in Figure 8.5, the diameter symbol (ø) preceding the tolerance indicates to the user that the .014 tolerance zone will have a cylindrical shape. Absence of the diameter symbol would indicate that the tolerance zone shape would be the area between two parallel planes or two parallel lines .014 apart. No symbol is specified to indicate the latter tolerance zone shape. The 1966 standard required the abbreviation DIA for diameter tolerance zone and R for radius tolerance zone to be included in the feature control frame. The 1966 standard also required the specifications of TOTAL for total wide tolerance zone and R for one-half of total wide tolerance zone.

The order of specified datum references in the feature control frame is very important. The primary datum is shown at the left; the least important datum is shown at the right. This datum precedence allows manufacturing and inspection to determine part orientation for their respective functions. The part is fixtured to allow a minimum of three points of contact on the primary datum surface, minimum two points of contact for the secondary datum feature, and minimum one point of contact for the tertiary datum feature.

The 1966 revision of the ANSI Y14.5 standard required the datum reference letters to precede the tolerance compartment of the feature control frame, while the 1982 and 1994 revisions require the datum reference letters to follow the tolerance compartment. Either method was accepted in the 1973 revision of the standard. The left-to-right datum procedure concept is identical in both methods (Figure 8.6).

Principle 5: Bonus Tolerance and Additional Tolerance

As discussed in Principle 3, when a tolerance or datum reference letter is modified with the MMC symbol, the specified tolerance in the feature control frame *only applies* to the feature location when the feature is manufactured at its MMC size. As the feature departs from

FIGURE 8.6
Datum precedence.

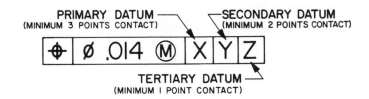

PRIMARY DATUM
(MINIMUM 3 POINTS CONTACT)

SECONDARY DATUM
(MINIMUM 2 POINTS CONTACT)

⌖ | ∅ .014 Ⓜ | X | Y | Z

TERTIARY DATUM
(MINIMUM 1 POINT CONTACT)

MMC size, the position tolerance is increased. The amount that the feature deviates from MMC size is added to the position tolerance specified in the feature control frame. This extra tolerance is called a *bonus* tolerance. As shown in Figure 8.7, the .014 diameter position tolerance applies when the hole is drilled at .515 diameter, the MMC size. If hole 2 were drilled at .525, a departure of .010 from the MMC size of the hole, we would gain a bonus tolerance of .01. The .010 bonus tolerance is added directly to the original .014 position tolerance to give an allowable positional tolerance of .024 (see the table in Figure 8.7). The axis of hole 2 must be within the .024 diameter tolerance. The allowable positional tolerance zone size for each hole must be determined in conjunction with the actual manufactured hole size.

We can also gain extra locational tolerance as a datum feature of size, referenced in the feature control frame as MMC, departs from MMC size. This added tolerance is not called a bonus tolerance and is treated somewhat differently than a bonus tolerance. In Figure 8.8, we could gain up to .005 bonus tolerance as the .260 diameter hole departs from MMC. Additional tolerance could also be gained as the datum B hole departs from MMC. The added tolerance as datum B departs from MMC does not add directly to the original position tolerance as a bonus tolerance would. The datum additional tolerance must be applied to the hole pattern as a *group*. For example, the added tolerance could allow the four-hole pattern to shift to the right as a group. No additional hole-to-hole tolerance gain is realized within the four-hole pattern.

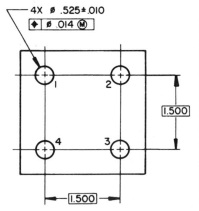

FEATURE SIZE	BONUS TOLERANCE	TOTAL POSITIONAL TOLERANCE
.515 (MMC)	.000	.014
.516	.001	.015
.517	.002	.016
.525	.010	.024
.533	.018	.032
.534	.019	.033
.535 (LMC)	.020	.034

FIGURE 8.7
Bonus tolerances.

FIGURE 8.8
Additional tolerance from a size datum.

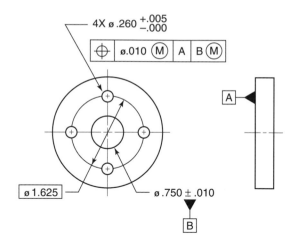

The rules and principles of MMC bonus and additional tolerance also apply to tolerances and datum references that are modified at LMC. The bonus and additional tolerances are determined from the feature's departure from LMC size.

The 1966 and 1973 revisions of ANSI Y14.5 allowed the MMC symbol to be implied by the drawing user when the material condition modifier had been omitted from the position feature control frame. It should be noted that this rule is contrary to international practice, which implies RFS in the same situations. The 1982 revision of ANSI Y14.5 requires the feature control symbol to be completed with the appropriate modifier (no implied modifier). The 1994 standard, then, adopted international practice, which implies RFS when no modifier is specified.

Principle 6: Datums

Datum specification in the feature control frame is of great importance to the drawing user. Proper functional datum selection allows the part to be fixtured for manufacturing and inspection, as it will be assembled as a finished product. Consistency with quality control will be assured when manufacturing uses the specified datum features in the proper order of precedence for machining operations.

The 1966 and 1973 revisions of the ANSI Y14.5 standard allow the use of implied datums by not specifying datum reference letters in the feature control frame. This required the print users to make certain assumptions in the course of manufacturing and inspection, such as which features to fixture upon and in what order of precedence. The use of implied datums can lead to problems if everyone concerned does not make the same assumptions. Designers and drafters should be encouraged to specify all necessary datums to provide uniform drawing interpretation. The 1982 revision of the standard has discontinued the use of implied datums.

The three-datum plane reference frame is ideal for ensurance of common drawing interpretation. For noncylindrical parts the manufacturing/inspection fixture shown in Figure 8.9 can be constructed to ensure uniformity during manufacturing and inspection operations.

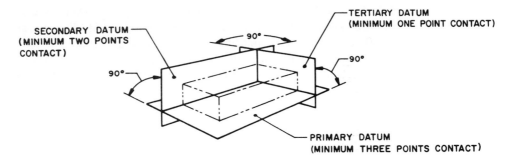

FIGURE 8.9
Three-plane reference frame for noncylindrical features.

All related measurements of the part originate from the fixture datum planes. For cylindrical parts the three-plane reference frame is more difficult to visualize. The primary datum is often described as a flat surface perpendicular to the axis of the cylindrical datum feature, as shown in Figure 8.10. This axis can be defined as the intersection of two planes 90 degrees to each other, at the midpoint of the cylindrical feature. The tertiary datum is used if rotational orientation of the cylindrical feature is required because of the interrelationship of radially located features. The tertiary datum is often a locating hole, slot, or pin. If no angular located features are involved in the hardware requirements, the tertiary datum is omitted.

There are two types of datum features: datums of size and nonsize datums. Nonsize datums are established from surfaces. A datum surface has no size tolerance because it is a

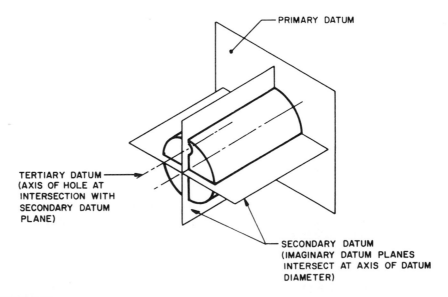

FIGURE 8.10
Three-plane reference frame for cylindrical features.

---

166 Chapter 8

plane from which dimensions or relationships originate. Datums of size are established from features that have size tolerance, such as holes, outside diameter, and slot widths. The centerplane or axis of datum feature of size is the actual datum. For example, a .250 ± .005 hole specified as a datum would be a datum feature of size (±.005 tolerance on feature size). The centerline or axis of the manufactured hole is the actual datum.

Principle 7: Position Tolerance

A position tolerance is the total permissible variation in the location of a feature about its exact position. For cylindrical features, such as holes and outside diameters, the position tolerance is the diameter of the tolerance zone within which the axis of the feature must lie. The center of the tolerance zone is located at the exact position. For other than round features, such as slots and tabs, the position tolerance is the total width of the tolerance zone within which the center plane of the feature must lie. The center of the tolerance zone is located at the exact location. Position tolerance zones are three-dimensional and apply to the thickness of the part. As shown in Figure 8.11, a diameter position tolerance zone is really a cylindrical tolerance zone through the part within which the feature axis must lie. This cylindrical tolerance zone is 90 degrees basic from the specified datum reference surface. Note that the tolerance zone will also control perpendicularity of the feature within the position tolerance requirement.

FIGURE 8.11
Position: cylindrical tolerance zone.

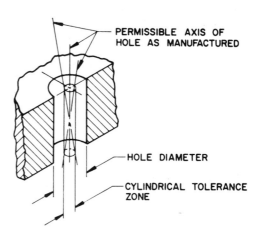

PERMISSIBLE AXIS OF HOLE AS MANUFACTURED

HOLE DIAMETER

CYLINDRICAL TOLERANCE ZONE

Composite position tolerancing provides a method of allowing a more liberal locational tolerance for the pattern of feature as a group, and then controlling the feature-to-feature interrelationship to a finer requirement. Each horizontal entry of the feature control frame shown in Figure 8.12 constitutes a separate inspection operation. The upper entry specifies the positional requirements for the pattern as a group, and the lower entry specifies the position tolerance within the pattern.

FIGURE 8.12
Composite feature control frame.

⊕	⌀ .060 Ⓜ	A	B	C
	⌀ .028 Ⓜ	A		

Principle 8: Screw Thread Specification

Where geometric tolerancing is expressed for the control of a screw thread or where a screw thread is specified as a datum reference, the application shall be applied to the pitch diameter. If design requirements necessitate an exception to this rule, the notation MINOR DIA or MAJOR DIA shall be shown beneath the feature control frame or datum reference as applicable.

Principle 9: Virtual Condition and Size

Virtual condition is the worst possible assembly condition of mating parts resulting from the collective effects of size and the geometric tolerancing specified to control the feature (see Figure 8.13). Virtual condition is primarily a tool used by product and tool/gage designers to calculate basic gage element size or to perform tolerance analysis to ensure assembly of mating parts. The following formulas can be used to determine virtual conditions:

External features = MMC size + tolerance of form, attitude, or location
Internal features = MMC size − tolerance of form, attitude, or location

The virtual condition is calculated from information specified on the engineering drawing (such as size and geometric tolerancing), but virtual size is determined by the parts inspector from the actual measured size and location, or attitude configuration of each feature. The virtual size of features must be considered for accurate GIA verification.

FIGURE 8.13
Virtual condition of mating parts.

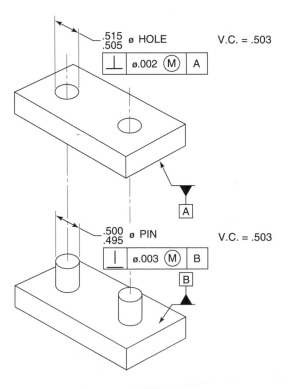

GRAPHICAL INSPECTION ANALYSIS PROCEDURE

■ **Example 8–1***a*

The performance of graphical inspection analysis (GIA) is a relatively simple, four-step procedure. The first step of the process requires plotting the basic location of the feature to be inspected in relationship to the datum features as described by the engineering drawing. This plot is called the *data graph* and is drawn on 10 × 10 graph paper at any desirable scale; 8 1/2 × 11 paper can be used for simple parts, whereas larger sheets will be required for complex parts. We will use the part shown in Figure 8.14 as an introductory explanation of a GIA procedure.

Figure 8.15 shows the data graph for Example 8–1*a*. The basic feature locations are plotted on the graph in relationship to the datum features at any convenient scale. We shall refer to this scale as the *configuration scale.*

Step 2 of the analysis requires plotting the locations of the actual inspected features on the data graph in relationship to the basic plot. These data are established by the parts inspector through surface plate inspection techniques as shown in Figure 8.16, or with the coordinate measurement machine as shown in Figure 8.17. The data are recorded on the appropriate inspection report form (Table 8–1) to become part of the inspection report.

To establish accurate data, the part shown in Figure 8.14 must be fixtured for surface plate inspection as specified in the feature control frame. When set up for inspection, the part must contact the angle plate at a minimum of three points for the primary datum (datum A) and a minimum of two points of contact on the surface plate for the secondary datum (datum B), as shown in Figure 8.18.

The tertiary datum surface (datum C) is established by placing the second angle plate against the original plate as shown in Figure 8.19. Since the part being inspected has a symmetrical configuration, the machinist must mark the surfaces used to establish datums A, B, and C for the placement of the holes. This will ensure consistency between manufacturing and quality control. Notice that the orientation of the two angle plates and the surface plate

FIGURE 8.14

Example 8–1*a* part.

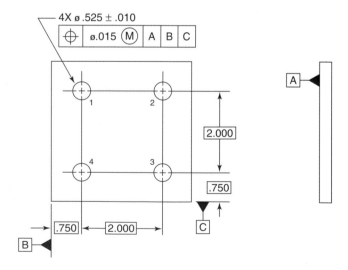

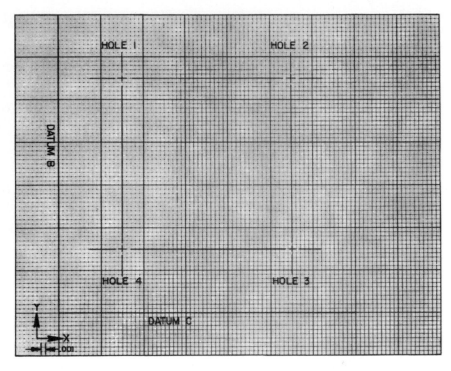

FIGURE 8.15
Data graph for Example 8–1a.

FIGURE 8.16
Surface plate inspection setup.

FIGURE 8.17
Coordinate measuring machine
inspection setup.

TABLE 8–1
Inspection data for Example 8–1a.

FEATURE NUMBER	FEATURE LOCATION						FEATURE SIZE			POSITION TOLERANCE			
	X AXIS			Y AXIS						MATERIAL CONDITION	SPECIFIED	BONUS	TOTAL
	SPECIFIED	ACTUAL	DEVIATION	SPECIFIED	ACTUAL	DEVIATION	SPECIFIED	ACTUAL	DEVIATION				
1	.750	.754	.004	2.750	2.751	.001	.515	.531	.016	MMC	.015	.016	.031
2	2.750	2.760	.010	2.750	2.750	.000	.515	.533	.018	MMC	.015	.018	.033
3	2.750	2.762	.012	.750	.748	.002	.515	.535	.020	MMC	.015	.020	.035
4	.750	.753	.003	.750	.749	.001	.515	.530	.015	MMC	.015	.015	.030

FIGURE 8.18
Surface plate setup (primary, secondary datums).

FIGURE 8.19
Surface plate inspection setup.

has created the three-plane reference frame shown in Figure 8.19. The part will be clamped to the angle plate to restrict its movement when performing inspection measurements. Snug-fitting gage pins are placed in each hole, and a measurement is established from the surface plate to the top of the pin as shown in Figure 8.20. One-half of the gage pin diameter is subtracted from the measured value to establish the dimension from datum surface

FIGURE 8.20
Surface plate inspection from datum B.

B (surface plate) to the center of the hole. Measurements are recorded for each hole in this setup. The angle plate is rotated 90 degrees (without loosening clamps) to establish similar measurements from datum C as shown in Figure 8.16.

A three-plane reference frame must also be employed when using the coordinate measurement machine to verify hole locations of the part shown in Figure 8.14. The tabletop establishes the primary datum (Figure 8.21a) and the two parallels provide contact surfaces for the secondary (Figure 8.21b) and tertiary datums (Figure 8.21c).

The appropriate measurement probe is placed in the coordinate measurement machine and calibrated to a zero reading at the intersection of the parallels that establish datums B and C. The probe is placed in each hole of the part to verify the actual measurements from datums B and C as shown in Figure 8.17. These data are recorded on the inspection report form.

Step 2 of the GIA process for Example 8–1a can be completed with the data tabulated in Table 8–1. The feature location deviations from the basic specified location will be in the order of magnitude of thousandths of an inch. We will let each grid square of the graph equal 0.001 in. to allow sufficient accuracy for the gaging operations. We will refer to this scale as the *deviation scale*. The location deviations are plotted for each feature as shown in Figure 8.22.

In step 3, a transparent *tolerance zone overlay gage* will be generated (see Figure 8.23) by placing a transparent material over the data graph. With a compass, the allowable position tolerance zone will be drawn at each of the basic locations for the specified tolerance zone plus any bonus tolerance. The bonus tolerance data have been recorded in the feature size deviation block of the inspection report shown in Table 8–1. The tolerance zones will be generated at the same scale (deviation scale) used to plot hole locational deviations in step 2.

In step 4, the transparent tolerance zone overlay gage is superimposed over the data graph and rotated and/or translated to provide alignment with the datum feature. We have an acceptable part if the actual feature axis of the four holes falls within the allowable tolerance zones at proper setting of the overlay (shown in Figure 8.24).

The part drawing for Example 8.1a is well done, in that datum surfaces have been specified, and the feature control frame has indicated a datum setup precedence. Both manufacturing and inspection will locate the part with minimum three-point contact of surface A, minimum two-point contact of surface B, and minimum one-point contact of surface C. The manufacturer marked the surface used for datums A, B, and C in drilling operations to ensure consistency with inspection. The MMC modifier associated with the .015 position tolerance zone in the feature control frame has alerted manufacturing and inspection to the bonus tolerance possibilities as the hole size departs from .515 diameter. In this example the manufacturer used the largest standard drill available to take advantage of this additional production tolerance.

Inspection data for a number of parts can be plotted on the same data graph, as shown in Figure 8.25. Notice that the tolerance zone overlay gage includes numerous tolerance zones to allow evaluation of the features as they depart from MMC. Each feature would be checked for conformance at its allowable position tolerance.

Another technique of graphical inspection which employs common axis data plotting and a tolerance zone overlay of prepared concentric circles can be used for feature location verification of simple parts as shown in Example 8.1a. However, this technique does not

FIGURE 8.21
Setting up the three-plane reference
frame on a coordinate measuring
machine.

FIGURE 8.22
Data graph for Example 8–1a.
Locations are plotted.

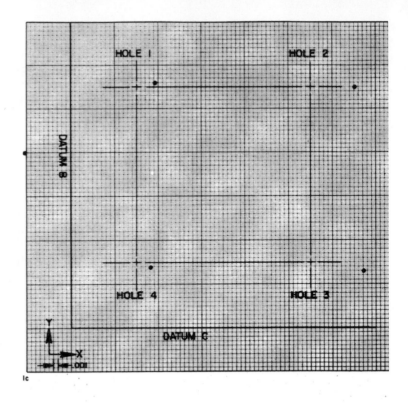

FIGURE 8.23
Example 8–1a tolerance zone
overlay gage.

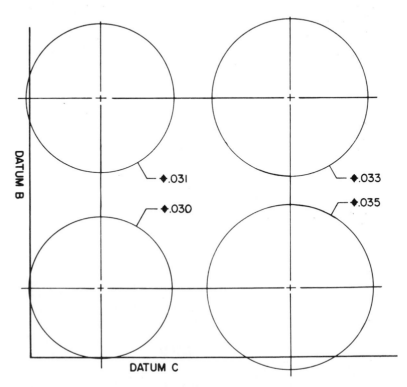

174

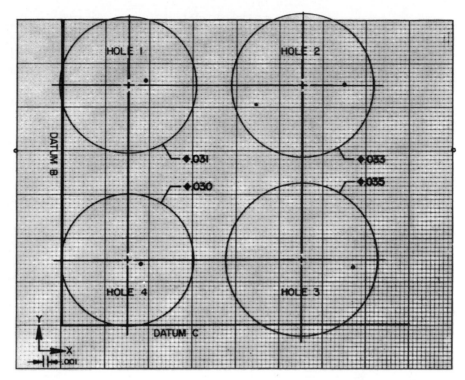

FIGURE 8.24
GIA of Example 8–1*a*.

lend itself to the evaluation of parts with secondary and tertiary datums of size or where angular orientation is concerned. Therefore, we will only demonstrate the graphical inspection technique where data plotting resembles the geometry of the part.

■ **Example 8–1*b***

In Example 8–1*b*, Figure 8.26 is the composite position tolerancing method of dimensioning. Example 8–1*b* does not combine the coordinate dimensioning system with the position tolerancing system. The .750 dimensions are specified basic.

Each entry of the composite feature control frame must be verified separately. This is an expensive gaging operation when using functional receiver gages, since separate gages must be designed and manufactured for each entry. The upper entry has specified selected datum features and order of precedence for common interpretation. The lower entry is independent of any datum location requirement. The four-step GIA verification process can be performed with the inspection data shown in Table 8–2.

Steps 1 and 2 require plotting the actual locations of the feature in relationship to their basic locations (Figure 8.27). In step 3, tolerance zone overlay gages are prepared for both the upper and lower entry of the composite position feature control frame.

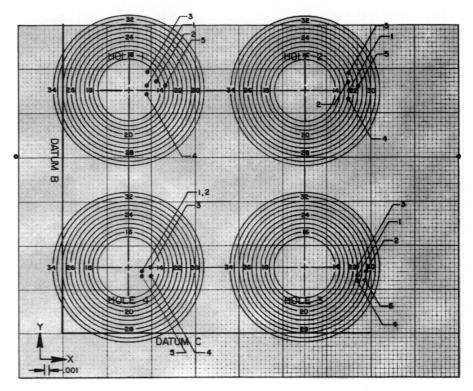

FIGURE 8.25
Multiple-part verification.

FIGURE 8.26
Example 8–1*b* print.

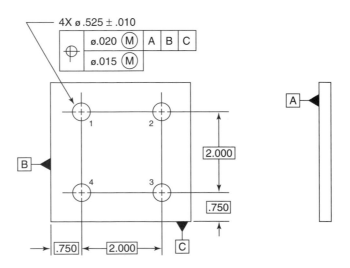

TABLE 8–2
Inspection data for Example 8–1*b*.

INSPECTION DATA

FEATURE NUMBER	FEATURE LOCATION						FEATURE SIZE			POSITION TOLERANCE			
	X AXIS			Y AXIS									
	SPECIFIED	ACTUAL	DEVIATION	SPECIFIED	ACTUAL	DEVIATION	SPECIFIED	ACTUAL	DEVIATION	MATERIAL CONDITION	SPECIFIED	BONUS	TOTAL
1	.750	.754	.004	2.750	2.757	.007	.515	.515	.000	MMC	.020	.000	.020
1										MMC	.015	.000	.015
2	2.750	2.755	.005	2.750	2.743	.007	.515	.516	.001	MMC	.020	.001	.021
2										MMC	.015	.001	.016
3	2.750	2.745	.005	.750	.743	.007	.515	.516	.001	MMC	.020	.001	.021
3										MMC	.015	.001	.016
4	.750	.746	.004	.750	.757	.007	.515	.519	.004	MMC	.020	.004	.024
4										MMC	.015	004	.019

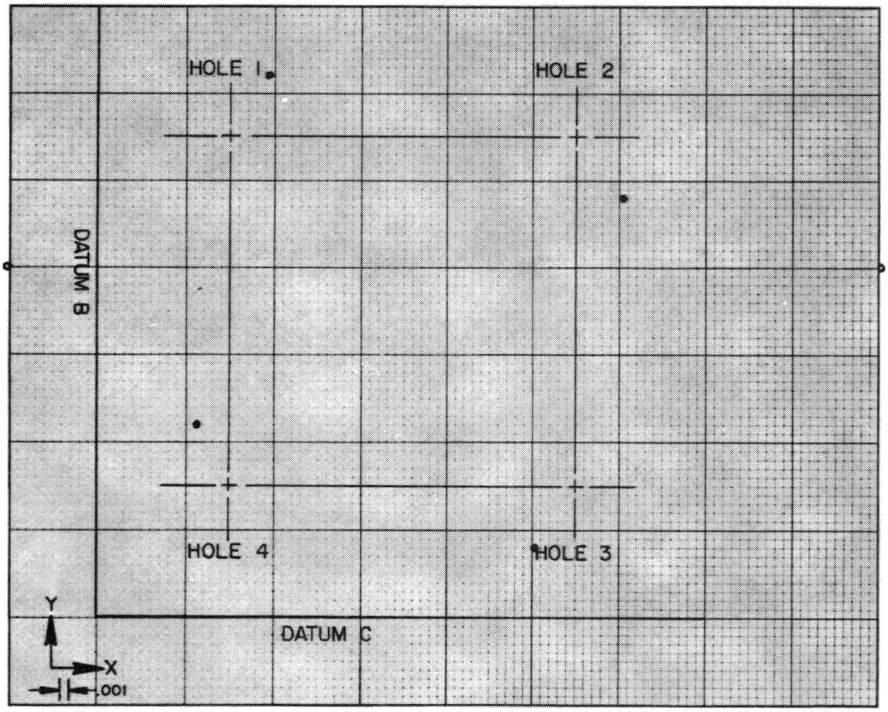

FIGURE 8.27
Data graph for Example 8–1*b*.

Notice the presence of datum feature lines in the upper tolerance zone overlay gage (Figure 8.28), which allow overlay alignment with the specified datum surfaces on the data graph as specified by the feature control frame. The lower tolerance zone overlay gage (Figure 8.29) is not datum related and is therefore allowed to float at will to establish hole-to-hole locational requirements. Step 4, Figure 8.30, requires the plotted hole located to be

FIGURE 8.28
Upper tolerance zone overlay gage.

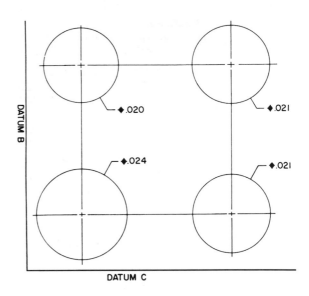

FIGURE 8.29
Lower tolerance zone overlay gage.

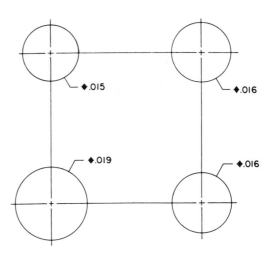

within the tolerance zones of both the upper and lower tolerance zone overlay gages to meet drawing requirements. This part is acceptable.

WORK PROBLEMS

Example 8–2 is provided so that you can gain experience in the concepts of graphical inspection analysis. Refer to the part drawing, inspection data, and pictures of the inspection process for the information required to generate data graphs and tolerance zone overlay gages (a completed example is provided).

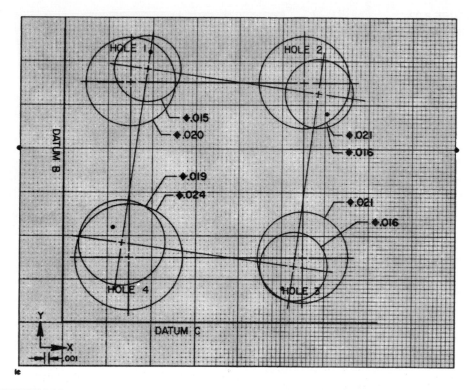

FIGURE 8.30
Data graph for Example 8–1,*b*/tolerance zone overlay gage.

■ **Example 8–2**

Example 8–2 provides a part with radially located holes related to a datum of size as shown in Figure 8.31. The inspection procedure shown in Figure 8.32 was used to generate the inspection data shown in Table 8–3.

FIGURE 8.31
Example 8–2 print.

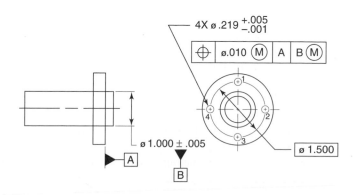

FIGURE 8.32
Inspection procedure for
Example 8–2.

TABLE 8–3
Example 2 Inspection Data

											INSPECTION DATA		
FEATURE NUMBER	FEATURE LOCATION						FEATURE SIZE			POSITION TOLERANCE			
	X AXIS			Y AXIS			SPECIFIED	ACTUAL	DEVIATION	MATERIAL CONDITION	SPECIFIED	BONUS	TOTAL
	SPECIFIED	ACTUAL	DEVIATION	SPECIFIED	ACTUAL	DEVIATION							
1	.000	.000	.000	.750	.754	.004	.218	.219	.001	MMC	.010	.001	.011
2	.750	.743	.007	.000	+.007	.007	.218	.220	.002	MMC	.010	.002	.012
3	.000	-.006	.006	.750	.745	.005	.218	.219	.001	MMC	.010	.001	.011
4	.750	.757	.007	.000	+.007	.007	.218	.219	.001	MMC	.010	.001	.011
DATUM B	—	—	—	—	—	—	1.005	.995	.010	MMC	—	—	—

Steps 1 and 2 of the GIA procedure require the generation of the data graph and the incorporation of the inspection data (see Figure 8.33). Notice from the inspection data that hole 1 has been used by the inspector to establish a starting point. Step 3 requires generation of the tolerance zone overlay gage for the holes (Figure 8.34) and a tolerance zone overlay gage (not shown) for any tolerance gained as the datum feature of size departs from MMC size.

Step 4 shows the hole tolerance zone overlay gage rotated and translated within the added tolerance, resulting from datum B departure from MMC in an attempt to accept the hole locations (Figure 8.35). This part would not be accepted if verified by a technique that was incapable of evaluating the additional tolerance allowed as datum B departs from MMC.

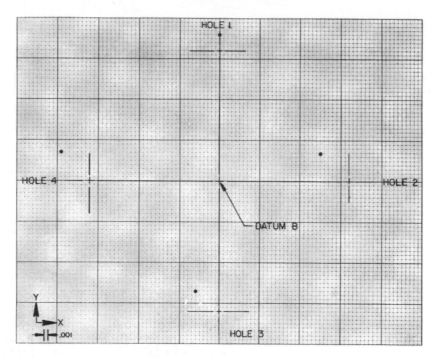

FIGURE 8.33
Data graph for Example 8–2

FIGURE 8.34
Example 8–2 tolerance zone overlay
gage.

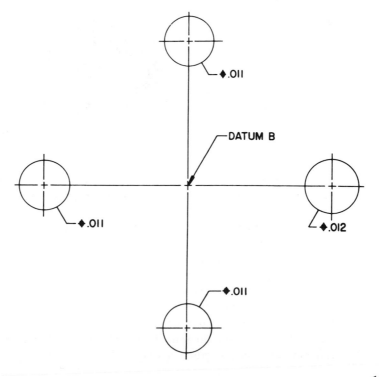

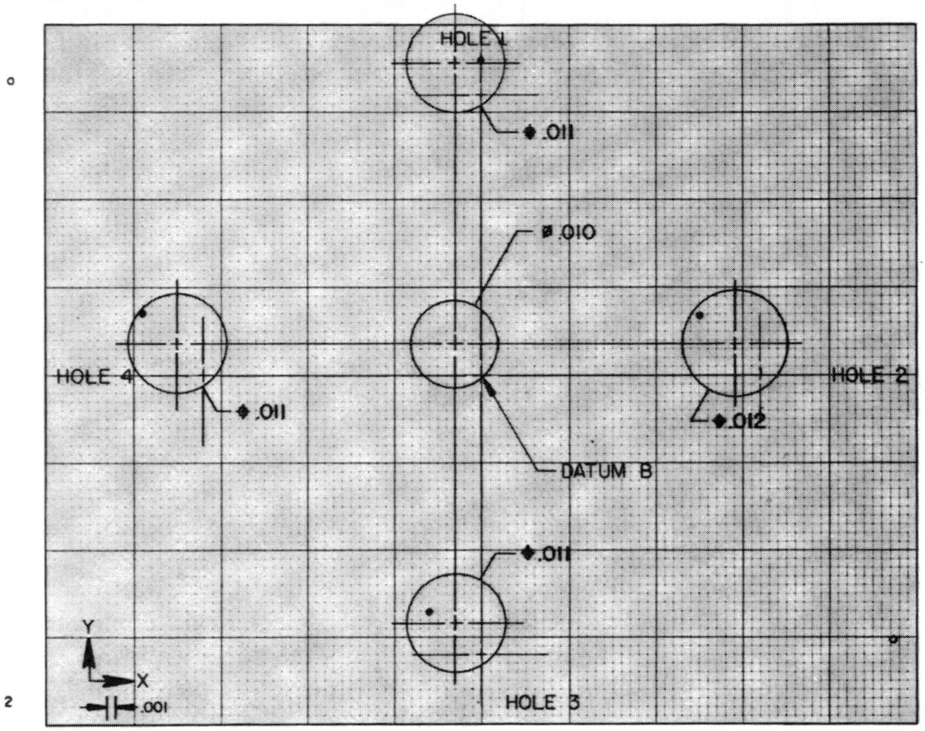

FIGURE 8.35
Example 8–2 data graph/tolerance zone overlay gage.

GRAPHICAL INSPECTION ANALYSIS ADVANTAGES

There are many advantages to the graphical inspection analysis concept for hardware verification. A partial list includes the following.

1. *Functional acceptance.* Most hardware is designed to provide interchangeability of parts. As machined features depart from their MMC size, location tolerance of the features can be increased while maintaining functional interchangeability. Graphical inspection analysis techniques evaluate these added functional tolerances in the acceptance process.
2. *Reduced cost and time.* The high cost and lead time required for the design and manufacture of a functional gage can be eliminated with the GIA technique. An immediate, inexpensive functional inspection can be conducted by the parts inspector at his or her workstation.
3. *No gage tolerance or wear allowance.* Functional gage design allows 10 percent of the part tolerance to be used for gage manufacturing tolerance. Often, an additional wear allowance (of up to 5 percent) will be designed into the functional gage. This could allow

up to 15 percent of the parts tolerance to be assigned to the functional gage. GIA techniques do not require any portion of the product tolerance to be assigned to the verification process. The GIA concept does not require a wear allowance since there is no wear.

4. *Functional verification of MMC, RFS, and LMC.* Functional gages are designed primarily to verify hardware designed with the MMC concept applied. In most instances, it is not practical to design functional gages to verify hardware designed with RFS or LMC tolerance requirements. With GIA techniques all three material modifiers can be verified with equal ease.

5. *Verification of any shape tolerance zone.* Virtually any shape tolerance zone (round, square, rectangular, etc.) can be easily constructed with graphical verification methods. On the other hand, hardened steel functional gaging elements of nonconventional configurations are difficult and expensive to produce.

6. *Visual record of tooling capabilities and problems.* Capabilities of machines can be evaluated through the GIA inspection data. For example, the data graph for the five hole patterns shown in Figure 8.25 would indicate a machine problem in the *X* axis. There appears to be a .004-per-inch offset in the one axis. Tooling wear and misalignment can also be detected during the production operation with periodic GIA verification of parts.

7. *Visual record for material review board.* Material review board (MRB) meetings are postmortems that examine rejected parts. Decisions on the disposition of nonconforming hardware usually are influenced by engineering rank and perseverance rather than engineering information and evaluation techniques. GIA can provide a visual record of verification methods that can be evaluated with tolerance zone overlay to ensure part function.

8. *Minimum storage required.* The inventory and storage of functional gages can be a problem. Functional gages can also rust and corrode if not properly stored. The GIA graphs and overlays, similar to drawings, can be stored in drawers.

Accuracy

The overall accuracy of graphical inspection analysis is affected by such factors as the accuracy of the graphs and layouts, the accuracy of the inspection data, the completeness of the inspection process, and the ability of the drawing to provide common drawing interpretations.

An error equal to the difference in the coefficient of expansion of the materials used to generate the data graph and tolerance zone overlay gages may be encountered if the same materials are not used throughout. Papers will also expand with an increase in humidity and should be avoided. Mylar is a relatively stable material and, when used for both the data graph and tolerance zone overlay gage, the expansion/contraction error will be nullified.

The GIA procedure outlined in this chapter requires the generation of the tolerance zone overlay gage using the grid of the data graph to determine the proper zone size. Error in the grid printing accuracy will be canceled with this procedure.

Layout of the data graph and tolerance zone overlay gages will allow approximately a .010 error in the positioning of lines. This error is minimized by the scaling factor selected for the data graph. If a 10×10 to the inch grid is used, with each grid representing 0.001 in., a scale factor of 100-to-1 will be provided. With the following formula we can calculate the actual error from line positioning:

$$\frac{\text{line position error}}{\text{scale factor}} = \text{actual error}$$

The .010 assumed line position error will equal an actual error of .001 in. This error can be minimized by consistently working to one side of the line or by increasing the scale factor.

A certain amount of error is inherent in all inspection measurements. The error factor is dependent on the quality of the inspection equipment, facility, and inspection personnel. Most inspection operations will produce an error of less than 5 percent.

Accuracy of graphical inspection analysis operations is dependent on the quality of the inspection report. Inspection reports must contain adequate information to ensure common understanding and provide complete variables data. Drawings that specify properly selected datums and provide datum precedence in the feature control frame will allow common drawing interpretations. When the inspector must select datum features and datum precedence because of the incompleteness of the drawing, the datums and setup procedure must be included in the inspection report.

Complete inspection data must also be provided to ensure GIA accuracy. For example, the inspection data included in this chapter have disregarded the perpendicularity condition of the inspected features. As discussed in Principle 7, position tolerance controls the location of feature through the part's thickness. Figure 8.36 shows the inspector inspecting both sides of the feature axis. This information would allow the establishment of both ends of the axis of the data graph for each hole axis and joined by a line to indicate that they are the axis for a single hole (Figure 8.37). This procedure creates the effect of a three-dimensional gage at virtual condition (see Principle 9).

Figure 8.37 shows a data graph, for the part shown in Figure 8.14, where the location of the features are plotted at their virtual size (see the inspection data in Table 8–4). The complete axis of each hole must be within the tolerance zone overlay gage to be an acceptable part, as shown in Figure 8.38. This method of verification is imperative if accurate, meaningful results are to be achieved with graphical inspection analysis.

A datum feature of size that is interrelated by a separate tolerance of attitude or form and is referenced within the same feature control frame must also be evaluated at its virtual size (often referred to as General Rule 5). The examples shown previously for parts that referenced datums of size did not depict complete part definition, since the attitude requirements of the datum feature of size have been omitted. When we added this additional control required for drawing completeness, we must consider the virtual size of the datum feature when evaluating the allowable additional tolerance that can be gained as the datum feature departs from MMC. The parts inspector will determine the datum feature virtual size (by measurement) and subtract this value from the datum feature virtual condition (calculated from the drawing specifications) to determine the additional tolerance allowance.

FIGURE 8.36
Inspecting perpendicularity.

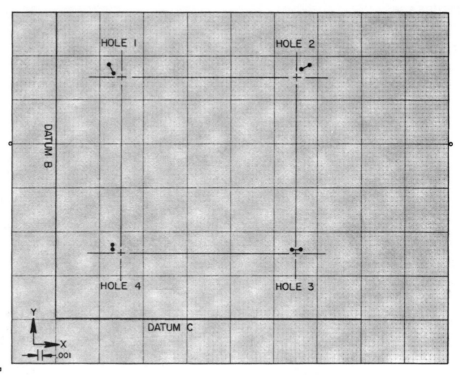

FIGURE 8.37
Three-dimensional hole verification (two coordinates and perpendicularity).

TABLE 8–4
Virtual size data for Example 8–1a.

							Inspection Data						
Feature Number	Feature Location						Feature Size			Position Tolerance			
	X Axis			Y Axis						Material Condition	Specified	Bores	Total
	Specified	Actual	Deviation	Specified	Actual	Deviation	Specified	Actual	Deviation				
1	.750	.748	.002	2.750	2.751	.001	.515	.516	.001	MMC	.015	.001	.016
		.747	.003		2.753	.003							
2	2.750	2.751	.001	2.750	2.752	.002	.515	.516	.001	MMC	.015	.001	.016
		2.753	.003		2.753	.003							
3	2.750	2.749	.001	.750	.751	.001	.515	.517	.002	MMC	.015	.002	.017
		2.751	.001		.751	.001							
4	.750	.748	.002	.750	.751	.001	.515	.516	.001	MMC	.015	.001	.016
		.748	.002		.752	.002							

COMPUTER INSPECTION ANALYSIS

Computer programs can be written to process the X and Y coordinate measurements and tolerance from inspection data and provide the required rotation/translation of the data to determine functional acceptance of parts. The computer can print out data indicating accept-

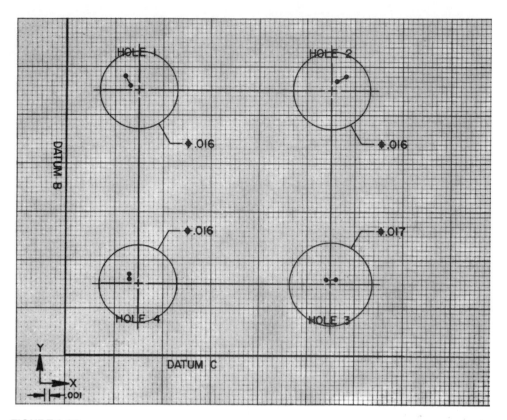

FIGURE 8.38

Example 8–1a virtual size data graph/tolerance zone overlay gage.

ance or rejection, or produce the data in graphical form, resembling manually prepared graphical inspection analysis. Recommended procedures can be provided for rejected parts to allow rework and ultimate part acceptance.

REVIEW QUESTIONS

1. A basic dimension is a theoretically exact dimension with no _____.
2. A tolerance stated at MMC applies only when the feature controlled is at _____ size.
3. (True or false) Bonus tolerance comes from controlled features, and additional tolerance comes from size datums.
4. The plot of basic locations of features related to the datum features in GIA is called a _____ graph.
5. The scale used for the GIA technique is referred to as the _____ scale.
6. (True or false) The determination of whether features are out of position tolerance is based on the amount the location departs from basic dimensions.
7. The GIA technique superimposes a tolerance zone _____ gage for measurement.

8. Name two advantages of GIA over hard functional gages.
9. The recommended material to use for the overlay gage is _____.
10. (True or false) An inspection report (variables) is needed in order to perform the GIA technique.
11. Another name for GIA is _____ gaging.
12. The authority standard for position tolerancing is ANSI or ASME _____.

Common Measuring Tools and Measurements

DIMENSIONAL MEASUREMENT

Dimensional measurement is a comparison to some specific standard. Gages and measuring equipment that are used to gage (or measure) a part feature are comparing that feature to the standard that was used to calibrate the gage.

MEASURING TOOLS

Some very basic measuring tools are covered in this chapter. Although it does not discuss all of the basic measuring tools, it does lead you through the ones that are commonly used. Helpful hints are given on the many causes of errors in measurement and certain ways to avoid such errors. It is worthwhile to study the possible errors related to each tool and keep them in mind when using the tools.

References and Measured Surfaces

One of the first things to consider with these measuring tools is which surface of the tool is the reference surface, and which surface is the measured surface. The reference surface is fixed (not movable), and the measured surface is movable. Proper measurement with these tools requires that you fix the reference surface of the tool on the part and take your measurement at the movable (measured) surface.

Reading the Tools

It is also necessary to become proficient at reading the various tools, for example, *micrometers, indicators,* and *verniers.*

There are different micrometer models (such as those that discriminate to .001 in. and those that discriminate to .0001 in.). There are also different *vernier scales* (such as each line division being spaced at .025 in. or .050 in. apart).

Selecting the Proper Tool for the Measurement

The rule of thumb (known as the 10% or 10 to 1 rule) is to select a measuring tool that is ten times more accurate than the total tolerance to be measured, or the tool can discriminate to one-tenth of the total part tolerance. For example, if you are measuring a part with a .010 in. total tolerance, use a tool with .001 in. discrimination. If the part has .001 in. total tolerance, use a tool with .0001 in. discrimination capability.

Unnecessary Accuracy

It is important to select the proper accuracy of the tool for the measurement. An example of unnecessary accuracy is measuring a casting with a tolerance of ±0.030 in. with a dial snap gage graduated in .0001 in. Unnecessary accuracy can take considerable time in inspection of parts that is just not necessary or economical. A more appropriate accuracy of the tool would be graduated in .001 in.

Transfer Tools

It is also necessary to understand *transfer tools*. These are tools that cannot be read directly. They make contact with the part (on the dimension being measured), are then locked in the measured position, and are measured with another tool (such as a micrometer). Therefore, the dimension has been "transferred" from the part, to the transfer tool, to the scale on the tool used to make the measurement.

Attribute Gages

Attribute gages are simply gages that are designed to either *Go* or *Not Go* in the dimension of a part. Attribute gages tell you if a part is "good" or "bad" but do not provide a reading to tell you how good or how bad. There are many attribute gages. Two of the most widely used are *plug gages* and *ring gages*. The plug gage is designed with two plugs (one at either end). One is the Go member and the other is the NoGo member. If the Go member goes into the hole, and the NoGo member does not, then the hole is within its limits, and acceptable. The ring gage does about the same thing but gages an outside diameter on a Go–NoGo basis. Attribute gages are used for quick verification of component parts, but limit you severely in your capabilities of measurement. They will not tell you how good or how bad a part is. They also cannot be adjusted in any way. They can only be reworked into another size gage.

Not being able to tell how good or how bad the part is can become troublesome during production or inspection. The operators cannot tell when their machines are going out of tolerance; it is not possible to maintain control charts for averages and range. Inspection must use another method of measurement when rejecting a part to get its actual size (actual size is the measured size with a variable tool).

Variable Gages

These gages provide actual size information (in the form of discrete values). This can be very helpful in spotting the need for adjustments. Variable gages allow inspectors to measure and record actual sizes of parts. Variable gages, for the most part, use indicators and travel mechanisms to show actual movement so a dimension can be compared to the master from which the gage was set.

Accuracy and Precision

Accuracy is the difference between the average of several measurements made on a part and the true value of that part. *Precision* means getting consistent results repeatedly. Accuracy and precision are very important in all measurements. Both accuracy and precision can be adversely affected by measurement errors. Certain types of measurement errors only affect accuracy; other types of errors only affect precision. Some measurement errors affect both accuracy and precision. The difference between accuracy and precision can be explained by the following example.

A dimension is measured by comparing (or transferring) the height of a gage block stack to the part dimension. If the observer uses the wrong gage blocks in the stack, the measurements will be inaccurate, although they will be repeatable. Therefore, the measurements made using this gage block stack will be precise, but not accurate. On the other hand, if the gage block stack were stacked properly, the measurements will be accurate and precise.

Measuring Pressure (Feel)

Measuring pressure is a large factor when considering accuracy. If the pressure is too heavy or too light, the measurement will not be accurate. A key to measuring pressure is to attempt to use a *constant* pressure, which is preferably the same pressure that was used to calibrate the tool. This is where ratchet stops and friction thimbles come in handy. Strive to develop the capability to achieve constant and proper pressure in your measurements.

Some measuring tools require the person using them to obtain a feel for the proper measuring pressure. You should not use a micrometer as a C-clamp, nor should you have it too loose on the part. Measuring pressure should be light enough to avoid squeezing the part or damaging the tool, yet heavy enough to make good contact with the part surfaces. At no time should the pressure be heavy enough to secure parts in the tool. Most often, the feel for hand tools improves with experience. For good practice with feel, use a pressure micrometer. This micrometer has an indicator for proper measuring pressure and it can be used to increase your consistency with feel.

Care of Tools

Hand tools are all susceptible to damage, rust, wear, miscalibration, and other problems that can cause them to fail in their function and reduce the accuracy of the measurement. Precision tools are generally sturdy but they do require careful handling and proper care.

A light film of oil will protect tools that are susceptible to rusting. Avoid rough handling and dropping tools. If a tool is dropped, check it for damage and calibration. Examine tools frequently for wear on the measuring surfaces and try to keep them calibrated. Keep tools protected in a toolbox and avoid storing them on top of each other. Know the particular precautions necessary to keep your tools in good condition. Well-cared-for tools and effective inspectors are an unbeatable combination.

MEASUREMENTS AND ACCURACY

When making measurements, keep one goal in mind—to be as accurate as possible. Due to many variables, no measurement in the shop is ever perfectly accurate; we can only come very close. The accuracy of any measurement made in the shop is largely dependent on a few select variables. These variables are *calibration, manipulation, reading, workpiece geometry, measuring pressure,* and *dirt and burrs.*

Calibration

If a measuring tool is not properly calibrated, the accuracy of the measurement will be poor. Many conditions can cause a tool to be out of calibration, including wear, dropping the tool, improper standard used, and improper pressure (too much or not enough). Measuring tools should be checked periodically for calibration to make sure that this variable is not a problem. Also, a gage that is found "out of tolerance" should be corrected and a review should be performed to assess the effect on products already inspected by that gage.

Manipulation

The persons using the tool must know how to manipulate (use) it to perform the measurement. They must be aware of those measurements that require centering or avoid cocking (or tilting) the tool the wrong way. An example of manipulation error is when a micrometer is used to measure a diameter. The micrometer faces (anvil and spindle) must be 180 degrees apart so that the true diameter is measured. Manipulation errors are easily made unless the inspector understands that it is essential to locate the tool properly on the part.

Reading

Regardless of the other variables, if a person is not well versed in how to read the scale of the tool, accuracy cannot be achieved. Reading errors that are made (on any scale) are of two basic types: *lack of knowledge* and *parallax*. A person must know how to read the scale of the tool, and even then, must avoid parallax error. Parallax error occurs when a person takes a scale reading from an angle instead of looking directly at the scale. Looking at any scale from an angle will cause the person to make an error because the person cannot see the reading exactly. An example of parallax error is a person on the passenger side of an automobile who tries to read the speedometer.

Workpiece Geometry

Many measurements are not accurate because the workpiece has excessive geometry errors. The piece may be tapered or out of round, may have surface waviness (irregularities), or be barrel-shaped, hourglass-shaped, and so forth. If the person making the measurement does not understand these conditions and does not search for them, accuracy cannot be achieved.

Dirt and Burrs

Dirty measuring tools or parts, or parts with machining burrs are always a problem. Tools should always be kept clean, and parts being measured should be cleaned before any attempt is made to measure them. Dirt and burrs will cause errors in measurements as follows. On an external feature (such as a shaft), dirt and burrs will cause your measurement to be larger than the actual size of the shaft. On an internal feature (such as a hole), dirt and burrs will cause your measurement to be smaller than the actual size of the hole.

Many other variables can cause inaccuracy. One example is heat, which will cause a part or a tool to expand in size. Understanding these possible errors and how to avoid them will give you a much better chance of obtaining an accurate measurement. Remember, accuracy means getting an unbiased true value. This definition boils down to "Accuracy means the ability to measure the true size of an object as closely as you possibly can."

TYPES OF MEASUREMENT ERRORS

There are several different types (or categories) of measurement errors. The following are some types and examples:

- **Observational**
     Misreading the gage
     Parallax error
- **Manipulative**
     Holding the gage incorrectly
     Not locating datums properly
     Mounting the part on the wrong datums
     Not aligning the gage properly
- **Bias**
     Rounding off (on purpose)
     Gage inaccuracy (often called bias)
- **Gage errors**
     Precision loss (e.g., a sluggish or sticking indicator)
     Accuracy loss (e.g., incorrect gage block stack, bent micrometer frame)
     Out-of-calibration error
- **Part error**
     Within-the-piece variation (e.g., taper, roundness)
     Dirty parts
     Poor surface finish
     Flaws

THE STEEL RULE

A common measuring tool that is used in the shop for a variety of measurements is the steel rule. The standard rule is graduated on both sides in inches and fractional parts of inches. These graduations are shown in Figure 9.1.

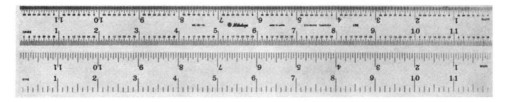

FIGURE 9.1
Steel rules. (M.T.I. Corp. Reprinted by permission.)

The finest divisions that are on a steel rule (sixty-fourths on the one at the top of Figure 9.1) establish its *discrimination*. This is the greatest accuracy that can be achieved on this particular steel rule. Some steel rules discriminate to $\frac{1}{10}$ in. and $\frac{1}{100}$ in.

Use of the Steel Rule

To avoid errors, the steel rule must be used properly. The first thing to consider is where a steel rule can be used and where it cannot. Obviously, it cannot be used on any dimension that is tighter in tolerance than its discrimination. For example, you would not use the steel rule to measure a dimension that has a tolerance of .010 total, because the best discrimination of this tool is $\frac{1}{64}$ (or almost .016 in.). However, if you are measuring a casting or forging with a tolerance of, for example, .060 total, the steel rule will do very well.

Proper Reference Is Important.   Steel rules should be referenced in one of two ways: either the edge of the rule should be firmly fixed, or your measurement should begin with the 1-in. line. These two conditions are shown in Figure 9.2.

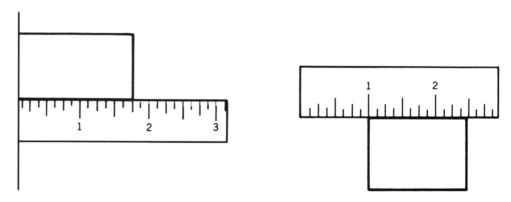

FIGURE 9.2
Referencing the steel rule.

Hook Rules.   Some steel rules have a hook built in the reference end that is designed to provide an accurate reference on the scale when applied to a part (Figure 9.3). The hook must be watched for looseness or wear, to avoid errors. The hook is designed to line up with the end of the steel rule (or 0 in.). Applications for the hook rule are shown in Figures 9.4 and 9.5.

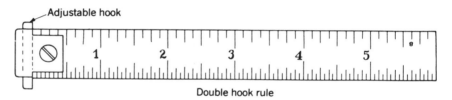

FIGURE 9.3
Hook rule. (L. S. Starrett Co. Reprinted by permission.)

FIGURE 9.4
Hook rule referenced. (L. S. Starrett
Co. Reprinted by permission.)

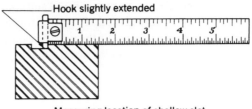

Measuring location of shallow slot

FIGURE 9.5
Measuring over a round corner.
(L. S. Starrett Co. Reprinted by
permission.)

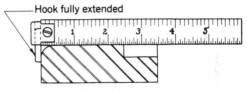

Measuring over round corner

Steel rules are a necessary measuring tool but must be used as they were intended—
for measurements within the limits of their discrimination.

SPRING CALIPERS

Spring calipers are adjustable calipers that are used to transfer a dimension to a line-
graduated instrument (such as the steel rule). There are inside and outside spring calipers
as shown in Figure 9.6.

FIGURE 9.6
Spring calipers. (L. S. Starrett Co.
Reprinted by permission.)

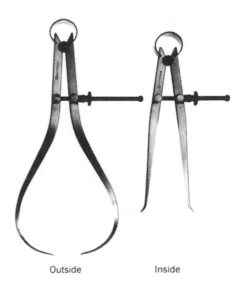

Outside   Inside

Inside calipers are used for measuring inside dimensions and outside calipers are
used on outside dimensions. Spring calipers are a transfer tool. First, you locate them on

the feature to be measured and adjust them by feel as the contacts rub against the surfaces of the dimension. Then you transfer the measurement to the line graduated tool to find the actual size of the part. Spring calipers are generally used on wide tolerances, such as castings or forgings, and generally involve the use of a steel rule for the measurement (see Figure 9.7).

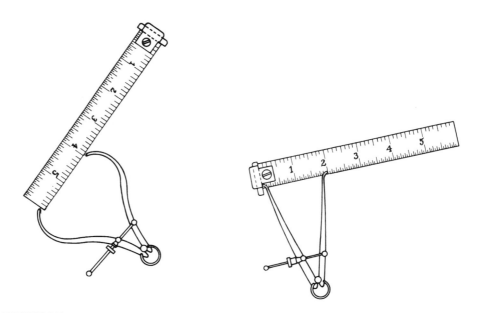

FIGURE 9.7
(*a*) Setting an outside caliper. (*b*) Setting an inside caliper. (L. S. Starrett Co. Reprinted by permission.)

GAGE BLOCKS

Gage blocks are used in almost every shop that manufactures a product that requires mechanical inspection. They have many purposes. Some of these are: to set up a length dimension for a transfer measurement; to set (or calibrate) fixed gages, such as snap gages; and to be used with special attachments for various applications. Gage blocks come in three standard shapes: *round, square,* and *rectangular* (Figure 9.8).

Gage blocks are usually made of hardened tool steel or Tungsten carbide and their surfaces are square, parallel, and flat to very accurate tolerances. Some gage blocks are made of ceramic materials.

There are many methods to clean gage blocks before use; one is with kerosene and a chamois or a gage cleaning solution provided by gage block suppliers. It is best to filter the kerosene through the chamois.

Gage blocks are used by "wringing" them together to obtain the length you need for the measurement. Wringing or rubbing them together squeezes the air out from between

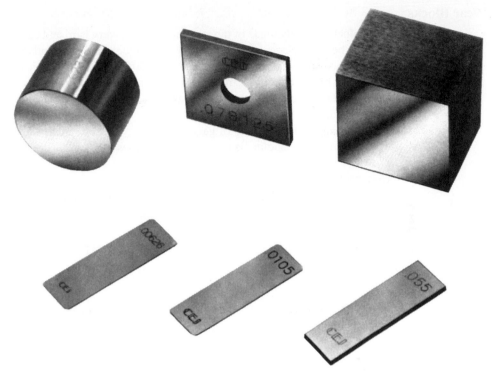

FIGURE 9.8
Gage blocks. (Federal Products Corp. Reprinted by permission.)

them, and since their surfaces are accurately flat, a "cementing" action occurs. Therefore, the blocks will stick together. The correct way to wring blocks together is shown in Figure 9.9. Position the blocks as shown in (a), and with a clockwise, twisting action, slide them together. The blocks, once wrung together, are called a stack.

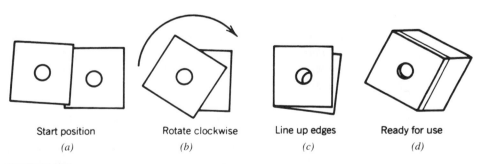

Start position	Rotate clockwise	Line up edges	Ready for use
(a)	(b)	(c)	(d)

FIGURE 9.9
Wringing gage blocks.

Wear Blocks

Wear blocks (usually made of tungsten carbide) are very hard and can withstand considerable use before replacement is necessary. Wear blocks are frequently used on both ends of a gage block stack. They come in two sizes. .050 in. and .100 in. thick. Their purpose is to protect the rest of the blocks from wear. It is important that wear blocks are placed on the ends of a stack, and always the same side out. Remember to include their length in the stack height needed.

To maintain the accuracy of gage blocks, they must be treated as carefully as a gage; therefore, delicate handling is important at all times.

General Guidelines

1. Avoid using the gage blocks if they are not required for the accuracy of the measurement.
2. If there are no wear blocks in the set then pick two small blocks and use them as wear blocks (or order wear blocks).
3. Always use the fewest number of blocks in a stack as possible.

Gage Block Sets

A typical gage block set (Figure 9.10) contains various increments of block sizes. Listed below is an 81-piece set.

Ten-Thousandths Blocks	Fifty-Thousandths Blocks	
.1001	.050	.500
.1002	.100	.550
.1003	.150	.600
.1004	.200	.650
.1005	.250	.700
.1006	.300	.750
.1007	.350	.800
.1008	.400	.850
.1009	.450	.900
		.950

One-Inch Blocks	One-Thousandths Blocks				
1.000	.101	.111	.121	.131	.141
2.000	.102	.112	.122	.132	.142
3.000	.103	.113	.123	.133	.143
4.000	.104	.114	.124	.134	.144
	.105	.115	.125	.135	.145
	.106	.116	.126	.136	.146
	.107	.117	.127	.137	.147
	.108	.118	.128	.138	.148
	.109	.119	.129	.139	.149
	.110	.120	.130	.140	

FIGURE 9.10
Square gage block set. (M.T.I. Corp.
Reprinted by permission.)

Note: Usually included in the set are two wear blocks. These blocks are either .050 in.
or .100 in.

For practice in the selection of blocks, try this example. Notice that the example selects
blocks by eliminating values from right to left.

■ **Example 9–1**

Stack up 3.6824 in.

Solution   3.6824
          − .1004   (use this block)
            3.582
          − .132   (use this block)
            3.450   (use .450 and 3.000 blocks)

MECHANICAL INDICATORS

Mechanical indicators are contact instruments that are a widely used *variable* tool in manufacturing and inspection areas. They are called a variable tool because of their ability to
detect the actual variation (or difference) between a dimension and a reference standard. Indicators are used in many applications such as making surface plate measurements and
checking parts on the machine during production. Indicators can be used to make direct
measurements or comparison measurements. Figure 9.11 shows an example of a mechanical indicator.

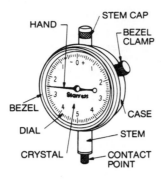

FIGURE 9.11
Parts of a dial indicator. (L. S. Starrett Co.
Reprinted by permission.)

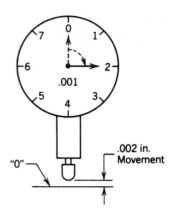

FIGURE 9.12
Dial indicators amplify distance.

Amplification

An indicator is a tool that amplifies the actual movement of the tip to the dial face so that this movement can be measured (see Figure 9.12). The discrimination of the common indicators used today is .00005 in., .0001 in., .0005 in., and .001 in. You can determine an indicator's discrimination by looking at the dial face. For example, if the indicator has .001 in. discrimination, the dial face would show (have lines at) .001 in. On this indicator, the distance between each line on the dial face represents .001 in. of movement (or travel) at the tip. Because of their internal mechanisms, an advantage of using dial indicators is *constant measuring pressure.* This adds to the accuracy of any measurement. Note also that an indicator with a yellow face is a metric indicator.

Types of Dials

Two main types of dials are used on indicators: *balanced* and *continuous* dials (see Figure 9.13).

FIGURE 9.13
Types of dial faces: (*a*) balanced;
(*b*) continuous.

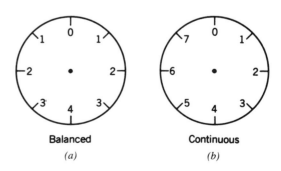

A balanced dial simply means that the indicator can be read directly in either direction from 0. Balanced dials are used mostly to measure dimensions with bilateral tolerances. The continuous dial starts at 0 and continues in number sequence all the way around until you reach zero again. This type of dial is used mostly to measure dimensions with unilateral tolerances.

Revolution Counters

Some indicators are equipped with another small built-in indicator called a *revolution counter* (see Figure 9.14). It tells how many full revolutions the large indicator needle has made. In some applications, use of the revolution counter is very helpful in keeping track of the total travel during a measurement. The total revolutions of an indicator, of course, depend on the measuring range of the indicator.

Choosing Which Indicator to Use

There are two important items to consider when choosing what indicator to use for a given measurement: *range* and *discrimination*. The range of an indicator is simply the total travel (or linear movement of the tip or plunger) of which the indicator is capable. An example is

FIGURE 9.14
Dial indicator with a revolution counter. (M.T.I. Corp. Reprinted by permission.)

Revolution counter

a .0001-in. indicator with a .004-in. range. In this example, .0001 in. is the discrimination and .004 in. is the range. This indicator could be used on a part tolerance of ±.001 or so, but you could not use it on a tolerance of ±.005 (it is outside the .004-in. range). Be sure you choose an indicator that has the range that will envelop the entire tolerance of the dimension you are measuring, and that will be accurate enough (have adequate discrimination) to do the job.

The Rule of Discrimination

It is bad practice to use an indicator on a tolerance where you would have to read between the lines on the dial face. An example might be where the tolerance on the dimension is ±.0002 in. and an indicator with only .001-in. discrimination is erroneously chosen. Since this indicator can only measure to .001 accuracy, you cannot possibly measure a tolerance of .0002 in. without trying to estimate by reading between the lines. Do not estimate—use an indicator that will do the job. Refer to the (10% to 1) rule covered earlier in this chapter.

Indicator Travel

When using any indicator, always make sure that you have enough plus and minus travel to measure the entire tolerance you are trying to measure (Figure 9.15). Many errors occur because the indicator was set up to travel in only one direction (not both) from the nominal size of the part (0 on the indicator dial).

FIGURE 9.15
Travel on an indicator.

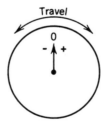

Contact Tips

Dial indicators can be used with a variety of contact tips for many different types of measurement (see Figure 9.16). You should use the proper tip for the feature you are measuring. Shown in Figure 9.17 are examples of correct and incorrect choices of contact tips.

Checking for Repeatability and Accuracy

Indicators should be checked frequently for repeatability and accuracy. Repeatability can be checked by measuring the same gage block over and over. Accuracy within the indicator's range can be checked by using a series of gage blocks within the range of the indicator; the discrimination (tolerance) of the blocks should match the discrimination of the indicator. You should also be very careful not to use an indicator that is erratic (jerks or is sluggish) in its movement.

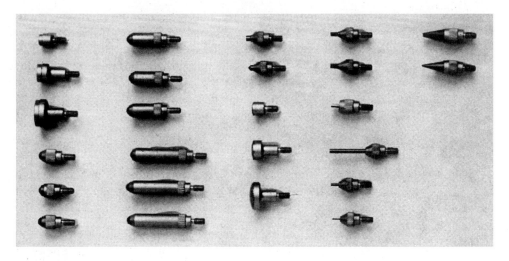

FIGURE 9.16
Various dial indicator tips. (M.T.I. Corp. Reprinted by permission.)

FIGURE 9.17
Proper selection of indicator tips.

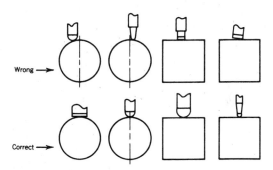

Care of Indicators

Indicators should be treated like delicate precision instruments.

1. Never bump or bang them around.
2. Avoid dropping them.
3. Never blow them with air pressure.
4. Store them in a safe, dry place.

Errors in Using Indicators

Below are listed just a few of the many possible errors in using mechanical indicators.

1. *Cosine error* (discussed later). This error is caused when an indicator tip is at an angle to the surface being measured.
2. *Mistaken travel.* A person using an indicator that makes more than one revolution must be aware of how many revolutions the pointer has made when making the measurement.

3. *Parallax.* This is caused by not looking straight (90 degrees) at the face of the indicator when reading it. The indicator pointer can look as if it is right on a line from the side view when it really is not.

4. *Travel.* Indicator pointer should be moved at least one-fourth revolution from dead position when used. Dead position is the position of the pointer when the indicator is not being used.

COSINE ERROR IN INDICATORS

When using indicators in making linear measurements, it is necessary to understand and avoid cosine error. Cosine error is simply defined as that linear measurement error that occurs when an indicator tip is at an angle to the surface of the workpiece being measured. Note, at this point, that the position of the body of the indicator makes no difference. It is the indicator tip itself, when at an angle, that causes cosine error. As shown in Figure 9.18, indicator tips should be parallel to the workpiece, because the proper movement of the tip is when it moves directly up and down (or in a plane perpendicular to its axis). *Movement in any other direction has the same effect as using a tip that is too short.*

Example of Cosine Error

In Figure 9.19 you can see that the indicator tip is at an angle to the surface being measured and therefore causes cosine error. Figure 9.20 shows the correct methods (remember, only the tip must be parallel).

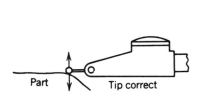

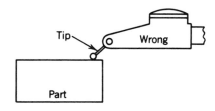

FIGURE 9.18
Indicator tip movement.

FIGURE 9.19
Angle between the indicator tip and the workpiece—cosine error.

FIGURE 9.20
No cosine errors here.

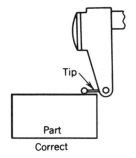

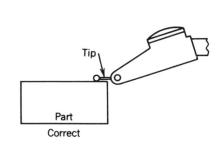

Correcting for Cosine Error

There are times when, because of the shape of a part, you must use the indicator tip at an angle to perform the measurement. At these times, there is a formula for correcting for the cosine error if it is necessary to do so.

Cosine error can be calculated. The answer is expressed as a percentage. This is a percentage of the reading seen when making the measurement. With cosine error, the reading will be too high, and you compensate by decreasing the reading by whatever percentage you get by using the formula.

First, know the meaning of the following symbols:

- ψ means exact angle of the tip.
- Delta (Δ) means the change (in percent) that the reading must be decreased.

The formula is

$$\Delta(\%) = \left(\frac{1}{\text{cosine } \psi} - 1 \right) \times 100$$

This reads: *Delta is equal to 1 divided by the cosine of the angle, minus 1.* That answer is multiplied by 100.

■ Example 9–2

Your measurement is made with the indicator tip at an angle of 15°, and the reading (variation) on the indicator shows .010. The error is

$$\Delta(\%) = \left(\frac{1}{\text{cosine } 15°} - 1 \right) \times 100$$

$$= \left(\frac{1}{.9659} - 1 \right) \times 100$$

$$= (1.035 - 1) \times 100$$

$$= .035 \times 100$$

$$= 3.5\% \text{ error}$$

Now you must multiply the reading (variation .010) times .035 (or 3.5 percent written as a decimal) to get the actual error. Then subtract the actual error from the original .010. So

$$.010 \times .035 = .0035 \qquad \text{(the cosine error)}$$

and

$$.010 - .00035 = .0097 \qquad \text{(the true reading or variation)}$$

This tells you that instead of .010 variation as shown on the indicator dial, there was only .0097 variation. The .00035 error occurred because of the indicator tip being at an angle. The cosine error here is therefore .00035.

Note: In most measurements cosine error is relatively negligible, but it is still a measurement problem of which we should be aware.

Remember: *The larger the tip angle, the larger the error becomes.*

TIR, FIR, AND FIM

Many dimensions are shown on drawings as TIR, FIR, or FIM. These abbreviations stand for

TIR   total indicator reading
FIR   full indicator reading
FIM   full indicator movement

Although according to the American National Standards Institute (ANSI) Y14.5, the preferred term is FIM, TIR and FIR are still used in industry and will probably continue to be used for years to come. The important thing to know is that all three of these mean the same thing. They mean the total distance that the indicator point travels from the start of the measurement to the finish.

■ **Example 9–3**

As shown on the print callout (Figure 9.21), *B* surface is to be parallel to *A* surface within .010 TIR (Figure 9.22).

FIGURE 9.21
Feature control frame.

FIGURE 9.22
TIR: Parallelism.

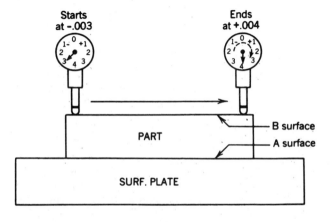

Note: In this case, the part must still be within its thickness tolerance and parallel within .010 TIR at the same time.

Full indicator movement (FIM) is −.003 in. to +.004 in., a total of .007 in.

One thing to remember when making a measurement calling for TIR, FIR, or FIM is to watch the extreme minus and the extreme plus amount on the indicator during the measurement.

■ Example 9–4

Indicator reading begins at −.002 and ends at +.002 (TIR is .004).

Indicator reading begins at 0 and ends at +.002 (TIR is .002).

TIR is not a plus or minus value. It is simply a total value.

An easier way is to set your indicator purposely at 0 in the beginning so you can observe the movement in one direction more conveniently.

TIR (FIR, or FIM) measurements apply to many different characteristics. Most of them are geometric tolerances (see Chapter 7) such as

- Runout (circular and total)
- Parallelism
- Flatness
- Straightness
- Angularity
- Perpendicularity

DIAL INDICATING GAGES

Dial indicating gages are often referred to as *variable gages*. This is so because they are gages that use a dial indicator to gage a part on a variable basis (variable meaning that the gage will detect the actual size of the part, not just whether or not it is in or out of tolerance).

Dial indicating gages work on a *comparison basis*. The gage is usually preset to indicate zero at the nominal size of the part with a set master. Then it is applied to the part, and the difference between the master size and the actual part size is shown on the indicator. For example, you have a dial indicating snap gage that is preset on zero with a .5000-in. cylindrical master. You use a cylindrical master because you are going to measure a shaft diameter with the gage. The dial snap gage is then applied to the shaft you are measuring, and the indicator reads a plus .003 in. Knowing that zero set on the gage is the same as .5000 in., the actual size of the part is .003 in. larger, or .503 in. If the dial read a minus .003 in. (for example), the actual size of the part would be .5000 in. − .003 in. = .497 in.

As you can see, dial gages are simply hand-held mechanical comparators. Unlike the micrometer (where you can read the actual size directly), the dial gage only shows the difference between the master size and the actual size. You must use simple mathematics to add or subtract the reading from the set master size.

Cautions When Using Dial Gages

1. *Allow for indicator travel.* There must always be enough plus and minus travel from zero set to allow for the tolerance limits you are measuring. It would be poor practice to measure a dimension with a tolerance of .005 in. total with a gage that has a total travel of .002 in. There must be enough travel in the gage to indicate the actual size of the part, no matter what that size is within the tolerance band.

2. *Use the proper discrimination.* The dial gage used should have an indicator that has the necessary discrimination to measure the part. For example, a shaft with a tolerance of .5000 in. ± .0005 in. *should not* be measured with a gage that discriminates to .001. It should be measured with a gage that discriminates within .0001 in. or less. Therefore, the gage must have a ten-thousandths indicator on it and at least enough travel to measure the part within the tolerance band.

3. *Know the set master size.* Most set masters have the size marked on them and, ideally, they are the nominal size of the dimension being checked. For example, if the nominal size of the dimension is .500 in., the set master should be .5000 in. This avoids confusion when using the gage. Always be sure of the set master size before setting the gage.

4. *Watch indicator travel when the gage is applied to the part.* Sometimes, when a part is out of specification, the indicator can make a complete revolution quicker than the eye can see. In this situation, the part can seem to be within tolerance when, in fact, it is considerably undersize or oversize. Take care to ensure that the same amount of pointer travel from null position, when the gage is set, is duplicated when the gage is applied to the part.

5. *Reference the gage.* Whatever the design of the gage, there is always a reference surface. This reference surface should be kept clean and free of damage. When the gage is being used, take care to make sure the reference surface is securely located and in the proper place. The reference surface of any gage is one that is not movable, whether it be a bar of some kind, an anvil, or pins. The only movable part of a dial indicating gage is the indicator rod itself.

6. *Beware of indicator problems.* When using a dial indicating gage, regardless of how often it is calibrated, always be on the lookout for indicator problems. Indicator pointers that move in a sluggish manner, jerk from one point to another, or stick are a sign of indicator malfunction, and the gage should be checked out prior to use. Always make certain that the indicator will repeat.

7. *Avoid parallax error.* When using a dial indicating gage or any other instrument that has a moving pointer and line divisions, make sure you are looking directly (90°) at the dial during the measurement. Avoid looking at the dial from an angle at all times. Look at the dial directly, so that you can accurately line up the pointer to the scale line. If necessary, use a mirror to see the dial directly.

8. *Use revolution counters on long-range dial gages.* If you use a dial gage that has the ability (within the indicator) to measure long ranges (such as 2 in., etc.), it is better to use an indicator with a revolution counter. The revolution counter involves a small pointer on the indicator face that tells you how many revolutions the large pointer has made during

a measurement. Since the large pointer moves very quickly during the measurement, carefully count the revolutions to avoid making a serious error in the measurement.

9. *Always be aware of what is minus and what is plus.* All indicating gages have a *null position.* This is the "free" position of the pointer when the indicator is not being used. Generally, this position is at nine o'clock on the dial face. Depending on the type of gage and the dimension being measured, a plus or minus value can be on either side of the zero setting. While setting the gage with the set master, you must compare the measuring jaws of the gage to the direction of the indicator movement. An example is a dial snap gage. As the jaws open (or the gap gets larger), the indicator moves clockwise from null position. In this case, a reading clockwise from zero is a plus reading and must be added to the set master size to obtain the actual size of the part. In some dial indicating gages, it is exactly the opposite. Just remember to compare the motion of the measuring jaws to the motion of the indicator pointer, and this error can be avoided.

10. *Make sure that all clamps are tight.* All indicating gages have some sort of clamping device to hold the indicator to the gage. Make sure that this clamp is not loose, or your measurement could be in error. Hand-tight will usually suffice.

Examples of Dial Indicating Gages

There are a wide variety of dial indicating gages on the market today and this book does not intend to cover them all. There are, however, a few types that are seen more often on the shop floor. Some of these are as follows.

Dial Indicating Micrometers.   The dial indicating micrometer has a constant pressure anvil that transmits movement to the dial face, which is built into the frame (refer to Figure 9.23). The dial indicating micrometer is the same as any other micrometer with the exception

FIGURE 9.23
Dial indicating micrometer. (Federal Products Corp. Reprinted by permission.)

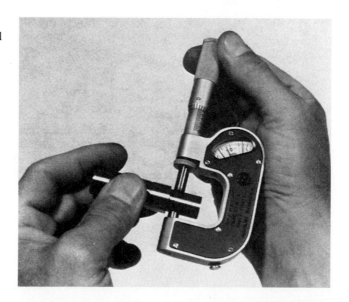

of this feature. When using the dial indicating micrometer, you read the micrometer on the thimble and sleeve for a .001-in. increments, then read the .0001 in. directly from the dial face. This micrometer is very useful in making accurate measurements because of:

1. Constant measuring pressure (in the anvil)
2. Direct reading in the ten-thousandths range

The dial indicating micrometer can also double as a snap gage for outside measurements. You preset the desired size, lock it in, and use the dial for comparisons.

Dial Calipers.   The dial caliper is very useful for quick and accurate measurements (within the capability of the tool). The dial is there to represent the vernier scale, so that the user does not have to find the coincidence line (see the discussion of vernier equipment on p. 230). The dial caliper is a direct reading tool. You simply read the main scale and add what you see on the dial indicator.

Take care to use the dial caliper slowly so that the teeth on the track are not damaged or jumped. Figure 9.24 shows a typical dial caliper. Figures 9.25 and 9.26 are further examples of dial calipers.

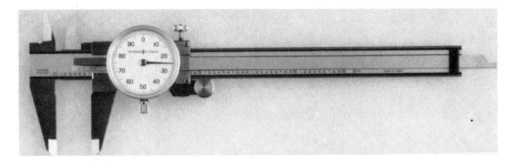

FIGURE 9.24
Dial caliper. (Scherr Tumico. Reprinted by permission.)

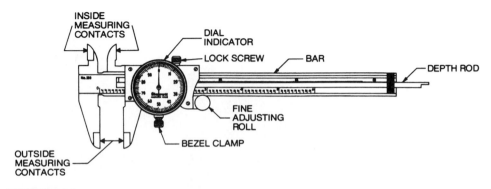

FIGURE 9.25
Parts of a dial caliper. (L. S. Starrett Co. Reprinted by permission.)

FIGURE 9.26
Dial caliper in use. (L. S. Starrett
Co. Reprinted by permission.)

OUTSIDE
MEASUREMENT

INSIDE
MEASUREMENT

DEPTH
MEASUREMENT

Dial Depth Gage.   There are two basic types of dial depth gages. The first type is shown in Figure 9.27. This dial depth gage employs a base and a drop indicator with a revolution counter. This gage does not use a set master but still requires calibration. This is a *direct-reading dial depth gage.* It can be calibrated with various stacks of gage blocks throughout its measuring range. Once calibrated, you can apply this gage to the part and use the indicator itself to establish the depth of the part.

The second type of dial depth gage is one that compares the depth of a part to the depth of a set master. The gage and master are shown in Figure 9.28. This gage generally employs

FIGURE 9.27
Dial depth gage. (Federal Products
Corp. Reprinted by permission.)

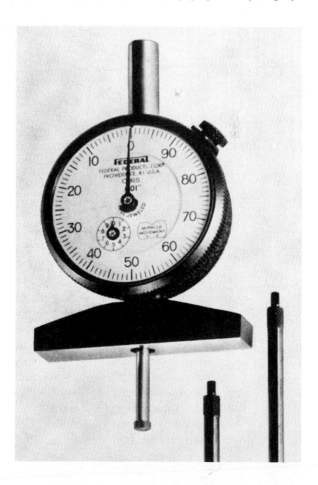

an indicator with a *balances dial* face (one that reads directly in the plus or minus direction), and the user must compare the plus or minus reading to the set master depth.

Dial Bore Gages.   All dial bore gages employ some kind of mechanism that translates movement from the contact points to the indicator. A variety of different contact points are used in dial bore gages. These contact points may expand because of a tapered shaft inside, or may utilize an internal level mechanism. The mechanisms that translate movement in a dial bore gage are not covered in this book.

There are two basic methods in which dial bore gages make contact: (1) two-point contact and (2) three-point contact. Whenever possible (because of the size of the bore) three-point contact is preferred. This is so because three-point contact centralizes the gage in the bore automatically.

FIGURE 9.29
Positioning a dial bore gage. (M.T.I.
Corp. Reprinted by permission.)

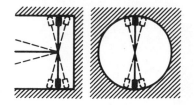 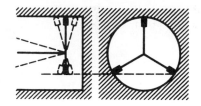

Centralizing the gage avoids *chordal* error. This error is often found when using two points of contact to measure any inside diameter. Two-and three-point contact are shown in Figure 9.29.

When using a dial bore gage, take care to insert the gage into the bore so that the bore is not damaged. Rock the gage to find the true diameter. Some dial bore gages (such as comptor gages) have a built-in *trigger* to retract the contacts for easy entry into the bore. Once inside the bore and rocked into position, the difference between bore size and the size of the set master can be seen on the dial face. The set master for a dial bore gage is usually a ring gage. Care must also be taken to centralize the gage in the set master. Dial bore gages can be very accurate, depending on the indicator used, the set master, and the manipulation of the gage inside the bore (see Figure 9.30).

Dial Snap Gages.   Dial snap gages come in a variety of different shapes and sizes, but all share the same purpose—to measure an external feature on a variable basis. Most dial snap gages are used to measure the following:

1. Outside diameters
2. External pitch diameters of threads
3. External groove diameters
4. Other external features

Some examples of dial snap gages are shown in Figure 9.31.

Dial snap gages employ a reference anvil and a measuring face. The reference anvil is fixed in the frame and should always be fixed (not moved) during the measurement. The measuring face is equipped with spring-loaded tension and transmits the movement to the dial indicator. There is one more feature of a dial snap gage that should be discussed: the centralizing anvil. The centralizing anvil is usually movable to allow you to adjust it to any size part that you are checking with the snap gage. Its purpose is to be a positive stop for the part when it is placed in the gage. This anvil should be set so that the part being meas-ured rests on it in such a way that the actual diameter of the part is centralized between the measuring faces of the gage.

Using the dial snap gage is comparing the size of the set master to the actual part and reading the difference on the dial face (plus or minus).

In-Process Dial Gages

Dynamic measurement is measurement of a moving feature. There are various kinds of dial gages that are used directly on the machine during production (see Figure 9.32). These gages

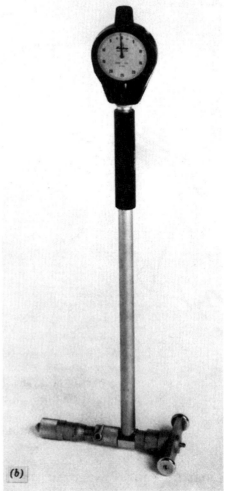

FIGURE 9.30
Two models of dial bore gage. (M.T.I. Corp. Reprinted by permission.)

are similar to other dial gages with the exception of mounting brackets and special indicators. The special indicators have *dampers* built into them so that they can indicate a piece that is being machined and will not have erratic pointer movements. During machining there is considerable vibration, and the damper on the indicator is necessary to enable the operator to read the indicator. The mounting brackets are there so that the gage can be mounted to the machine in such a way that its measuring faces can rest on the part during machining.

In-process gages are very useful because they tell the machine operator when the part is at its desired size. Some in-process gages are equipped with electrical connections that will shut the machine off automatically when the part is to size. In-process gages should be checked periodically for calibration and reset, if necessary, to avoid running defective parts.

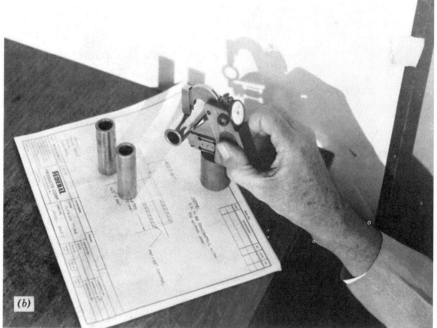

FIGURE 9.31a and b
Dial snap gage in use. (Federal Products Corp. Reprinted by permission.)

FIGURE 9.31c
Positioning the part in the dial snap
gage. (Federal Products Corp.
Reprinted by permission.)

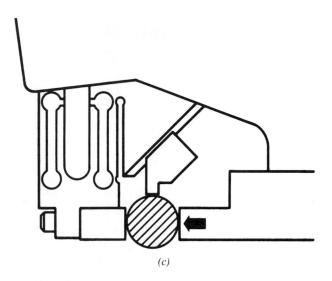

(c)

FIGURE 9.31d
Dial snap gage. (Federal Products
Corp. Reprinted by permission.)

(d)

FIGURE 9.32
In-process gage in use. (Federal Products Corp. Reprinted by permission.)

Comparing Fixed Gages to Dial Indicating Gages

Almost all fixed and dial indicating gages are used during production to determine whether parts are to the size specified (within tolerances). There are some major drawbacks and benefits to using either type of gage. Some of these are listed in Table 9–1. Remember that fixed gages are only used for *one* dimension and tolerance.

The drawbacks and benefits noted in Table 9–1 should be weighed against your needs in gaging. In my opinion, the indicating gage is the best gage to use (even though it is more

TABLE 9–1
Drawbacks and Benefits of Fixed and Indicating Gages

Characteristics	Fixed Gage	Indicating Gage
Cost	Less expensive	More expensive
Ease of use	Easier to use	More difficult to use
Maintenance costs	Less cost	More cost
Calibration time	Less time	More time
Information per piece inspected	Less information	More information

costly). There are several reasons why an indicating gage is suggested for use in most applications of gaging in the shop.

1. The indicating gage gives you more information about the part being inspected. It not only tells you whether the part is good or bad, but it tells you *how good* or *how bad* the part is.
2. Indicating gages allow you to inspect a variety of different dimensions and tolerances. To plot an average and range chart, variable gages must be used (see Chapter 13 on statistical process control).
3. The information given by an indicating gage can help the operator tell when the machine is nearing the high or low limit of the tolerance.
4. The indicating gage enables the operator to make more precise setups than when using the attribute (or fixed) gage. When the setup is precise, there is greater chance of producing good-quality parts in a long production run.
5. The indicating gage is useful when parts must be rejected. Most material review boards require a rejection that clearly states how bad the part is, so a decision can be made on what to do with the part.

Fixed gages do find considerable use in the shop and at the assembly line, but remember that they only tell you that the part is in tolerance or out, and nothing more. If your only need is to tell whether or not the part is in tolerance, fixed gages should be used because they are less expensive and easier to use and maintain.

Variable gages (such as dial indicating gages) are recommended for use because of the information they can supply about the part. Even though variable gages are a little more expensive to purchase and maintain, they can often save you considerable costs in preventing defective products.

MICROMETERS

One of the most widely used basic tools in the shop is the micrometer (or "mike" as it is sometimes called). There are many kinds of micrometers, but one thing that is common to all of them is the way they are read. Before going further, study Figures 9.33 and 9.34.

FIGURE 9.33
Micrometer. (Scherr Tumico.
Reprinted by permission.)

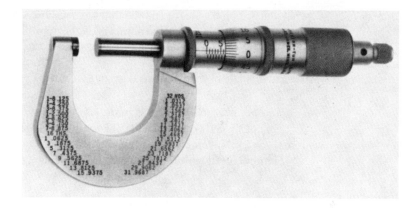

FIGURE 9.34
Micrometer reading.

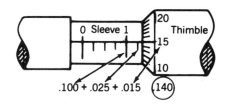

Reading the Micrometer

The scale of a micrometer ranges from 0 to 1 in. and is usually graduated in "thousandths" of an inch (think of 1 in. as $\frac{1000}{1000}$). The sleeve of the mike shows the numbers 0 1 2 3 4 5 6 7 8 9 0. Each of these numbers represents 0.100 (or 100 thousandths) in travel. In between each number, there are four spaces each representing .025 (or 25 thousandths) of an inch. So, you can see that the micrometer *sleeve* divides 1 in. into .025 increments.

Now look at the thimble. Each line on the thimble represents .001 and there are numbers every five lines (or .005). One complete turn (revolution) of the thimble is equal to .025. So each revolution of the thimble (.025) is equal to one division on the sleeve (.025). Following the same logic, one revolution of the thimble moves the thimble one line along the sleeve. When the edge of the thimble lines up with one of the divisions on this sleeve and the thimble shows 0 in line with the horizontal line on the sleeve, you simply add what you see on the sleeve.

■ Example 9–5

The edge of the thimble is past the .100 mark and in line with the first line after the .100 mark. This reading would be .100 + .025 = .125.

The example in Figure 9.34 shows what happens when the 0 on the thimble does not line up with a division on the sleeve. In this example, you can see that the edge of the thimble is past one of the .025 marks, but not quite to the next .025 mark. Also, the horizontal line on the sleeve is lined up with the 15 mark on the thimble. To read this, you simply add what you see on the sleeve and the thimble, as shown. Keep in mind that every line on the thimble is equal to .001 and the numbers on the thimble are in increments of .005. The reading, as shown in Figure 9.34, would be

$$.100 + .025 + .015 = .140 \text{ in.}$$

Another way to look at it is in dollars. If 1 in. = $1000, each number on the sleeve (1, 2, 3, etc.) is $100, each small division on the sleeve is $25, and each division on the thimble is $1. Now the previous reading is $140.

Figure 9.35 shows the parts of the micrometer.

FIGURE 9.35
Parts of a micrometer. (L. S. Starrett Co. Reprinted by permission.)

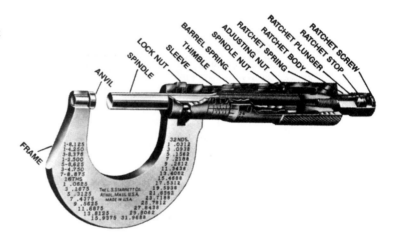

Types of Micrometers

There are many types and sizes of micrometers, but all of them have one thing in common. They all have a 0 to 1 in. barrel (or sleeve). The following are a few examples (there are also digital micrometers; see Chapter 11):

Depth micrometer.   The micrometer shown in Figure 9.36 reads the same way as the one in Figure 9.38 but caution must be used because the thimble hides the scale. You must know the scale well to read the "mike" in Figure 9.37.

Hub micrometer.   This micrometer is very useful in tight places where the conventional micrometer cannot reach because of its frame (see Figures 9.38 and 9.39).

V-anvil micrometer.   This "mike" is very good for measuring diameters or detecting out-of-roundness in a part (Figure 9.40).

Inside micrometer set.   Sets of inside micrometers come in handy for measuring a variety of inside diameters (Figure 9.41). They can be used to measure ranges from about 2 in. to 12 in. by changing the rods (Figure 9.42).

FIGURE 9.36
Depth micrometer in use. (M.T.I.
Corp. Reprinted by permission.)

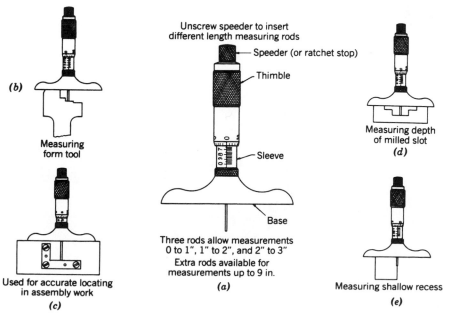

(b)

Measuring
form tool

Used for accurate locating
in assembly work
(c)

Unscrew speeder to insert
different length measuring rods

— Speeder (or ratchet stop)

— Thimble

— Sleeve

— Base

Three rods allow measurements
0 to 1″, 1″ to 2″, and 2″ to 3″
Extra rods available for
measurements up to 9 in.
(a)

Measuring depth
of milled slot
(d)

Measuring shallow recess
(e)

FIGURE 9.37
Various uses for a depth micrometer. (L. S. Starrett Co. Reprinted by permission.)

FIGURE 9.38
Hub micrometer. (M.T.I. Corp. Reprinted by permission.)

FIGURE 9.39
A hub micrometer measures a
dimension not easily accessible.

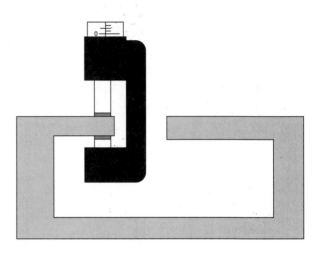

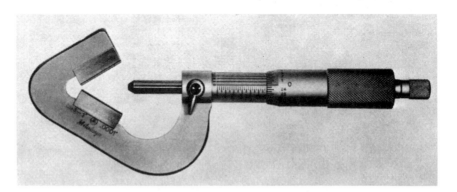

FIGURE 9.40
V-anvil micrometer. (M.T.I. Corp. Reprinted by permission.)

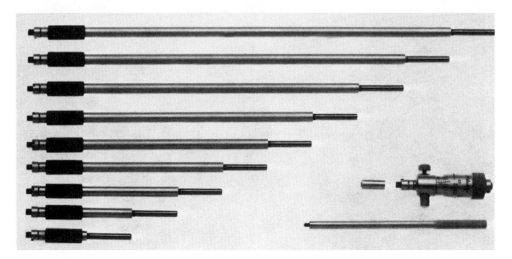

FIGURE 9.41
Inside micrometer set. (M.T.I. Corp. Reprinted by permission.)

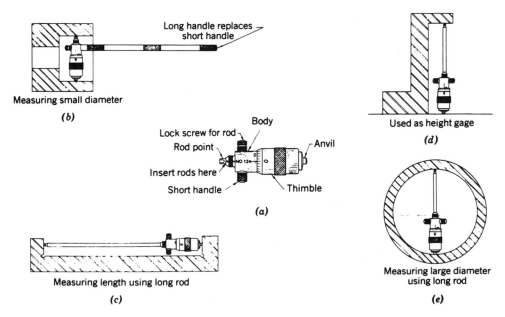

FIGURE 9–42
Inside micrometer measurements. (L. S. Starrett Co. Reprinted by permission.)

FIGURE 9.43
Multianvil micrometer. (Brown &
Sharp. Reprinted by permission.)

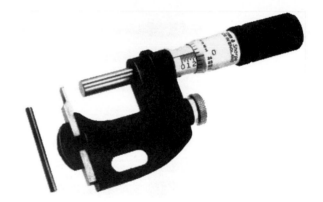

Multianvil micrometer.   This tool has flat and round changeable anvils. You can remove the anvil to measure the height of a step (Figure 9.43).

There are many more specialized micrometers that range from 0 to $\frac{1}{2}$ in., 0 to 1 in., up to very large sizes.

The Proper Way to Care for a Micrometer

The correct way to care for a micrometer is shown in Figures 9.44 to 9.46. Keep it clean, oiled, and safe. Treat it like an expensive tool, because it is.

FIGURE 9.44
Proper way to hold a micrometer.
(L. S. Starrett Co. Reprinted by
permission.)

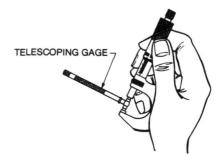

TELESCOPING GAGE

Errors in Using Micrometers

1. Misreading the tool (most common misreading is by .025, incorrectly counting sleeve divisions)
2. Wear on the anvil and spindle faces
3. Miscalibration
4. Damage
5. Heat
6. Pressure too tight
7. Pressure too loose

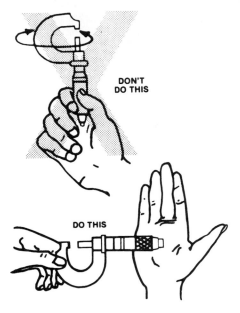

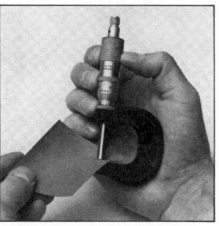

A handy method to clean the contact surfaces is to close the micrometer lightly on a piece of soft paper. It is then withdrawn. The paper will probably leave fuzz and lint on the surfaces. Blow this out with lung power—not the air hose. Never use compressed air to clean any precision instrument. The high velocity forces abrasive particles into the mechanism as well as away from it.

FIGURE 9.45
Opening and closing a micrometer. (Scherr Tumico. Reprinted by permission.)

FIGURE 9.46
Cleaning micrometer contact surfaces. (Scherr Tumico. Reprinted by permission.)

Micrometers Used in Industry

Figure 9.47 shows various micrometers that are used in industry.

FIGURE 9.47a
Micrometer head. (Scherr Tumico. Reprinted by permission.)

(a)

FIGURE 9.47*b*
V-anvil micrometer. (Scherr Tumico. Reprinted by permission.)

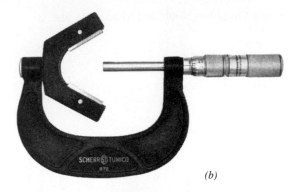

(b)

FIGURE 9.47*c*
Hole location micrometer. (Scherr Tumico. Reprinted by permission.)

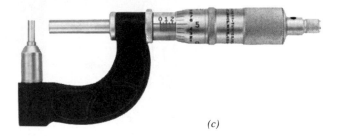

(c)

FIGURE 9.47*d*
Blade micrometer. (Scherr Tumico. Reprinted by permission.)

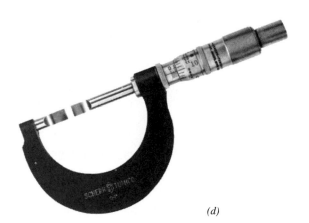

(d)

FIGURE 9.47*e*
Point micrometer. (Scherr Tumico. Reprinted by permission.)

(e)

FIGURE 9.47f
Deep throat micrometer. (Scherr Tumico. Reprinted by permission.)

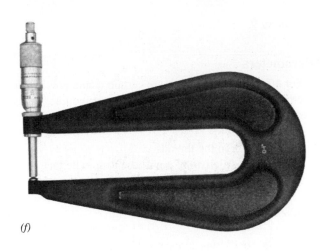

(f)

FIGURE 9.47g
Multimicrometer in use. (M.T.I. Corp. Reprinted by permission.)

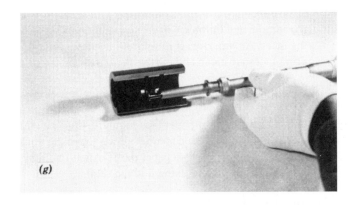

(g)

FIGURE 9.47h
Micrometer set. (L. S. Starrett Co. Reprinted by permission.)

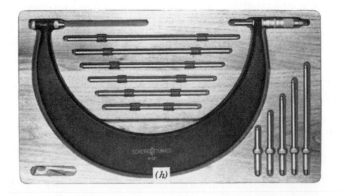

(h)

Calibrating Micrometers

For effective results there are certain techniques and tools to use when calibrating the micrometer. In calibrating the micrometer, you are looking for some specific problems that may cause inaccuracy, such as:

1. Lead error in the threads
2. Out of parallelism of anvil face to spindle face
3. Worn measuring faces
4. Measuring faces not perpendicular to the sleeve and thimble centerlines
5. Zero setting not set at zero

The conditions above must be checked and verified to be acceptable before a micrometer is truly said to be in calibration. *Lead error* can cause a micrometer to reflect good calibration in one area of its range, but have measurement error in another area of its range (or travel). The *parallelism of the anvil face to the spindle face* can cause errors of contact, and the workpiece will not give the same (accurate) reading. *Worn measuring faces* have the same effect as parallelism. *Measuring faces that are not at right angles to the centerline* of the spindle and thimble can cause the micrometer to be inaccurate on sides of the measuring faces. An *incorrect zero setting* can cause many problems. When the micrometer measuring faces are clean and closed, the 0 lines on the thimble and the sleeve should coincide.

Cleaning Micrometer Faces for Calibration and Use.   Before using or calibrating the micrometer, the anvil and spindle faces should be cleaned. A good method for doing this is with a clean sheet of paper (see Figure 9.46). You simply close the micrometer on a clean piece of paper, pull the paper out slowly, open the micrometer, and softly blow on the faces to make sure there are no paper particles left on them. This should be done before calibration and, often, before using it in the shop.

Standards for Calibrating the Micrometer.   Two basic standards are used in calibrating the micrometer: *gage blocks* and *micrometer end standards.* Gage blocks are used more often than end standards because they have the range capability to check for lead error, and they are flat so that complete contact is made with the anvil and spindle faces. End standards are used too, but they are rod-shaped with spherical ends that make only single point contact with the measuring faces. End standards that come with micrometers are generally available only in 1 in. increments. With either standard, the micrometer should read the correct size when the measuring faces are closed and the ratchet stop (if there is one) is used. The ratchet stop, or the friction thimble, has one purpose: to maintain constant pressure that is adequate for measurement.

A good practice is to calibrate the micrometer with the ratchet stop and use it with the same movement of the ratchet stop. In other words, if the ratchet stop clicks twice in calibration, you should click it twice in use.

Calibration Technique.   The basic steps in calibrating the micrometer are outlined below:

Step 1.   Visually inspect the micrometer for damaged parts.

Step 2. Clamp the micrometer to a micrometer stand during calibration. Do not hold it in your hand.

Step 3. Select the standard (gage blocks or end standards).

Step 4. Clean the micrometer faces (anvil and spindle) with the method prescribed earlier.

Step 5. Use a variety of gage blocks to check for lead error. A standard sequence to follow in size order is

.105 in.	.210 in.	.315 in.	.420 in.	.500 in.
.605 in.	.710 in.	.815 in.	.920 in.	1.000 in.

The micrometer should be accurate throughout this range if there is no lead error.

Step 6. Check the zero setting. When the anvil and spindle faces are closed, the zero setting should be correct. If not, the sleeve can be rotated with a key wrench (usually supplied with the micrometer when purchased) (see Figure 9.48).

FIGURE 9.48
Setting the micrometer.
(L. S. Starrett Co. Reprinted by permission.)

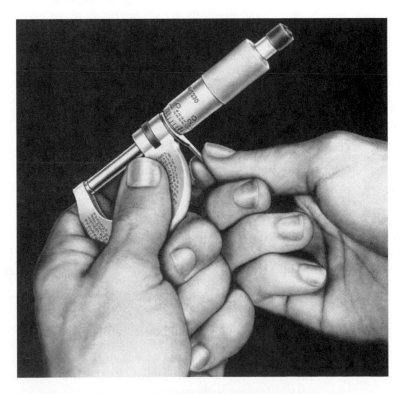

Lining up zeros on thimble
and sleeve

Step 7.   Check the parallelism of the micrometer's measuring faces with an optical parallel. Note, however, that the measuring faces could be parallel to each other, but could be out of square with the centerline (of the spindle or anvil).

Step 8.   Check the flatness of the measuring faces with an optical flat. A perfectly flat measuring surface, when observed through an optical flat, will show a fringe pattern of straight lines that are equally spaced. Other patterns, such as circular patterns, may be seen. These indicate convex or concave conditions.

These calibration techniques should be performed with standards, usually maintained in the metrology lab and traceable to national standards.

Along with the calibration system, which will be discussed later, the user of the micrometer should perform checks on calibration during use. The user can spot check the micrometer by using shop standards or gage blocks. Simply apply the micrometer to the standard (after cleaning the surfaces) and keep a close check on the zero setting during use.

VERNIER INSTRUMENTS

Many measurements today are made with tools that require an ability to read the vernier scale. Examples of these tools are *height gages, vernier calipers,* and *protractors.* In Figure 9.49, notice the parts of the height gage (particularly the bar and the plate). This is where the vernier scale is read.

FIGURE 9.49
Parts of a vernier height gage. (L. S. Starrett Co. Reprinted by permission.)

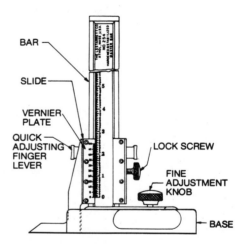

Each large number on the bar is 1 in. The small numbers in between the large numbers represent .100 in. each. The 0 line on the plate is where you begin to read the vernier.

Also on the plate, you will see numbers from 0 to 50 in. increments of 5. The space from 0 to 50 on the plate is equal to one small line on the bar, which is .050 in. Without the 0 to 50 vernier scale on the plate, the closest you could measure something is only to .050 in. accuracy. The starting point, again, is the line 0 on the plate. This line is the first thing you look at when reading the vernier.

FIGURE 9.50
Vernier reading.

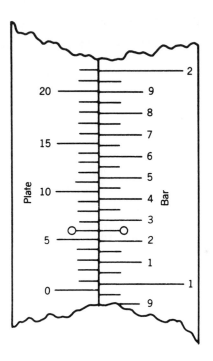

Figure 9.50 is an example of one vernier reading. Following are steps in how to obtain this reading.

Step 1.   Look at the zero line on the plate to see where you start. In this case, it is not quite at 1 in., and the zero line is at .950 in. (plus something).

Step 2.   Remember (or write down) .950.

Step 3.   Now you must go to the vernier scale on the plate to see how much to add to .950 to obtain the correct reading.
The vernier scale has 50 lines on it. However, only one of those 50 lines will line up perfectly with a line on the bar. *This is the line you are looking for.* It is called the *coincidence line.*

Step 4.   When you find the one line that lines up with any line on the bar, you add the amount shown on the plate (at that line) to the .950 you already have. In this case, the line is .006 as shown by the small circles drawn on the scale.

Once again, the zero line showed that your reading starts at .950 plus something. The vernier scale shows .006 line. So, the answer is .956.
Another tool that uses the vernier scale is the vernier caliper. This is a hand-held tool where feel is as important as it is with the micrometer. The vernier caliper measures three different characteristics: *inside dimensions, outside dimensions,* and *depth dimensions.*

The Universal Bevel Protractor

One of the best tools available for measuring angles (not angularity) is the universal bevel protractor (Figure 9.51). Other small protractors are only accurate within 1 degree, but the universal bevel protractor has a vernier scale that gives you accuracy within 5 minutes of a degree. For example, an angle of 20 degrees ± 30 minutes cannot be measured with a simple protractor. It must be measured with the universal bevel protractor because of the vernier capability in minutes.

Remember: There are 360 degrees in a circle and 60 minutes in each 2 degrees.

The top scale of the universal bevel protractor shows you the measurement in degrees, but the vernier scale obtains the accuracy in minutes.

Figure 9.51 shows a universal bevel protractor, which by the use of an attached magnifying glass allows the vernier scale to be read easily.

Reading the Vernier Scale of the Universal Bevel Protractor.   Figure 9.52 shows various applications of the universal bevel protractor. Note, at this point, that the basic principle of reading the protractor vernier scale is the same as reading the vernier caliper.

FIGURE 9.51
Bevel protractor. (M.T.I. Corp. Reprinted by permission.)

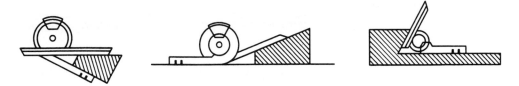

FIGURE 9.52
Bevel protractor applications.

FIGURE 9.53
Bevel protractor reading.

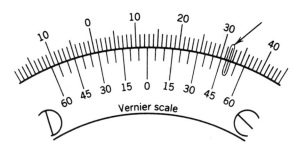

Vernier scale

Figure 9.53 shows one example of how to take a vernier reading.

Step 1. Use the 0 line on the main scale and obtain the reading in degrees. This is 12 degrees (almost 13 degrees) here.

Step 2. Look to the right of the 0 line and see what line (between 0 and 60) lines up precisely with any line on the main scale. In this case it is the 50-minute line.

Note: Read the vernier scale in the same direction from zero as the main scale reading.

Step 3. Therefore, the reading you get is 12 degrees and 50 minutes (12° 50″). The 50-minute line is the only one of all the graduations that lines up.

Remember: The selection of what vernier scale to use is simple. If the numbers on the main scale are getting larger to the right, use the right vernier scale.

TELESCOPING GAGES

Many tools can be used to measure inside diameters. One of these tools is the telescoping gage. A telescoping gage is a tool that transfers an inside diameter to an outside diameter so that it can be measured with a tool such as an outside micrometer. Take care when using the telescoping gage, because it can be very inaccurate if it is not used correctly. These gages generally come in full sets ranging in size, so that you may choose the correct size

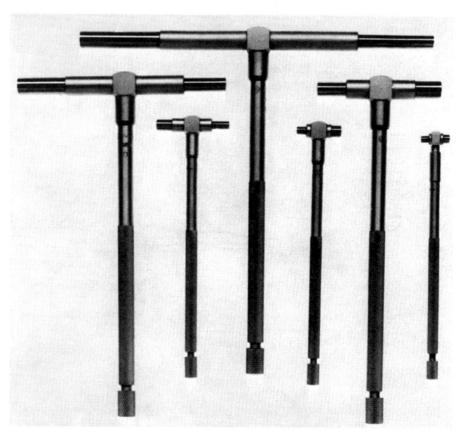

FIGURE 9.54
Telescoping gages. (M.T.I. Corp. Reprinted by permission.)

gage for the particular hole size you are measuring. The usual set ranges from $\frac{5}{16}$ in. to 6 in. capability (see Figure 9.54).

The telescoping gage is a tool with one spring-loaded cylinder and one fixed cylinder. As shown in Figure 9.55, you choose the gage you need, depress the spring-loaded cylinder (so that it will go inside the diameter), and then position it correctly in the diameter and lock it in with the lock provided. Then measure over the ends of the two cylinders with a micrometer.

After locking the gage in (Figure 9.56) to ensure accuracy (due to positioning the gage properly), you next have to "mike" the gage to obtain the actual size of the diameter being measured.

To mike the telescoping gage requires certain steps to ensure a close degree of accuracy (see Figure 9.57).

Step 1.   Fix the fixed cylinder end at the anvil of the micrometer.

Step 2.   Slowly bring the micrometer spindle down on the spring-loaded cylinder end until you feel a slight rub between the telescoping gage and the micrometer spindle.

FIGURE 9.55
Positioning telescoping gages.

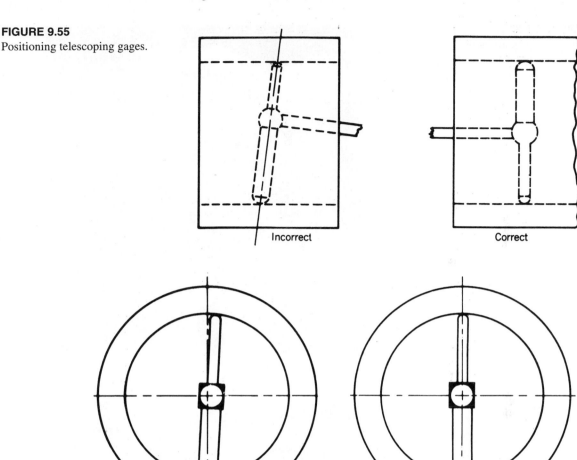

Incorrect   Correct

Incorrect   Correct

FIGURE 9.56
Positioning telescoping gages.

Step 3.   Read the micrometer. The feel of this rubbing when you obtain the actual size comes with experience.

Step 4.   Finally, as shown in Figure 9.58, you need to hold the micrometer properly with one hand because the other hand is occupied in holding the telescoping gage. Never allow the micrometer to be run down far enough to actually hold the gage on its own. Always go by feel.

Caution.   It is important to remember that transfer tools, such as telescoping gages and small hole gages, reduce the reliability of a measurement because of the buildup of possible errors when transferring measurements from tool to tool.

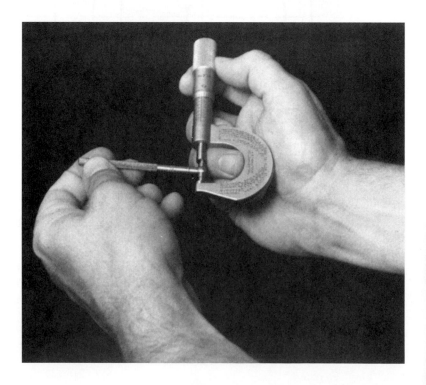

SMALL HOLE GAGES

There are many ways to measure small holes (holes that are .400 in. diameter or less). One way is with *small hole gages*. A small hole gage (Figure 9.59) is a tool with two rounded contact stems, a "spreader," and an adjustment knob. They usually come in sets of four gages. Each gage has a range of approximately .100 in. Small hole gage sets measure hole sizes from about .125 diameter to .400 diameter. These gages are similar to telescoping gages in that they are adjustable and are transfer-type gages. A micrometer (or other tool) must be used in conjunction with this tool to read the measurement. Also, as with telescoping gages, a certain feel is involved for accuracy.

Using Small Hole Gages

To use small hole gages, follow the steps below.

Step 1. Select the gage that has the range for the hole you want to measure.

Step 2. Put the gage in the hole with the adjustment knob loose.

Step 3. Moving the gage up and down for a short distance, tighten the adjustment knob until you feel the contact points rub in the hole (Figure 9.60).

When measuring the small hole gage with a micrometer, make sure that the rounded contact points of the gage are on the same line as the centerline of the anvil and spindle (Figure 9.61). Again, you must have a feel for knowing when to stop tightening the micrometer.

There are other ways to measure small holes such as Go–NoGo plug gages (attribute gages). However, these gages only tell if the hole is in print size or not.

ATTRIBUTE GAGES

Attribute gages are fixed gages that are designed to check a single dimension and/or tolerance limit. There is a difference between *gaging* and *measuring* a dimension. A gage is designed to check the dimension with Go or NoGo (good or bad) results. Measuring instruments measure the dimension with variable (discrete values) results. All gages are designed with gagemaker's tolerance and *wear allowance*. Wear allowance can cause a gage to be produced exactly at the limit of the dimension it was designed to check. In this case, the gage will not fit the part if the part has been made exactly at that limit (an acceptable part). Therefore, it is possible, with all gages, to reject a part that is on the borderline of being acceptable. The rule of thumb for using all gages is to reinspect a borderline suspect part that the gage has rejected using a measuring instrument before making the reject decision.

There are several different kinds of *attribute* (or Go–NoGo) gages. They are designed to tell you only that a dimension is or is not within tolerance limits. Remember that attribute means that a condition does or does not exist. Attribute gages do not tell you the actual size of a dimension, just whether it is good or bad.

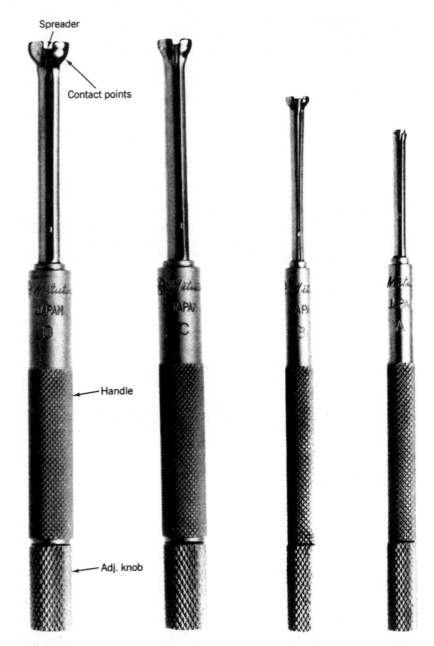

FIGURE 9.59
Small hole gages. (M.T.I. Corp. Reprinted by permission.)

FIGURE 9.60
Small hole gage application. (L. S.
Starrett Co. Reprinted by
permission.)

FIGURE 9.61
Measuring a small hole gage.
(M.T.I. Corp. Reprinted by
permission.)

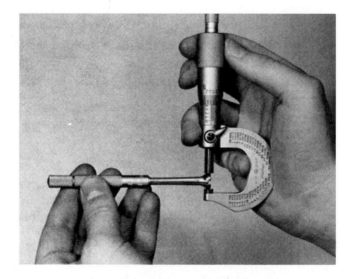

Gages discussed here are the ones most widely used in manufacturing and inspecting components. They include *plug gages, ring gages, flush pin gages,* and *snap gages.* Keep in mind that attribute gages are fixed gages that are designed to check a single dimension.

Plug Gages

These are simply hardened steel "pins" that are accurate in size, with a handle. There are usually two *members,* the Go and the NoGo (see Figure 9.62). They are used to check inside diameters such as drilled holes on an accept or reject basis. There are three basic kinds of plug gages:

Single purpose. A Go or NoGo member only (not both).

Double end. Those that have both a Go and NoGo.

Progressive. Those where the Go and NoGo are on the same side (made from the same piece of steel). The Go is the first part used.

Application of the plug gage is very simple. Start the member at a slight angle into the hole, rotate it to a 90-degree position, and allow it to go into the hole (if it will) under its

FIGURE 9.62
Ring gage (upper). Plug gage
(lower). (Federal Products Corp.
Reprinted by permission.)

own weight. Do not try to force it in. If the Go goes in and the NoGo does not, the hole is
to print. If the Go will not go in, the hole is undersize. If the NoGo will go in, the hole is
oversize. How much undersized or oversized a hole is must be determined with another tool
(such as a small hole gage and micrometer).

Plug Gage Limitations.   There are two main limitations of plug gages.

1. They only tell you that the hole is bad or good, not how bad or how good.
2. They will not detect out of roundness, taper, or barrel-shaped or bell-mouthed holes
   (which can be a problem).

Thread Plug Gages

There are also thread plug gages (Figure 9.63) that will measure (on a collective basis) internal threads such as tapped holes. They work on the same principle; the Go must go and
the NoGo must not. Note, however, that the gage is designed so that the NoGo will at least
start (one full turn, or so) but should not go farther.

Ring Gages

Like plug gages, ring gages (Figure 9.64) are hardened steel with very accurate inside diameters. They are also attribute gages. They come in two basic ways.

FIGURE 9.63
Thread plug gage.

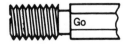

FIGURE 9.64
Two ring gages (Go–NoGo)

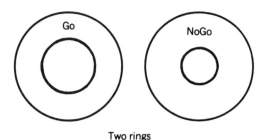

Two rings

1. Two different rings (one a Go and the other a NoGo). (*Note:* NoGo has a slot outside the ring.)
2. One progressive ring (Go and NoGo) with diameters in the same ring (Figure 9.65).

FIGURE 9.65
Progressive ring gage.

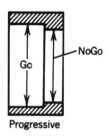

Progressive

Ring gages are used to measure outside diameters on a Go–NoGo basis. Their uses and limitations are about the same as a plug gage, except that they are used on outside diameters instead of inside diameters. When they are a set of two rings, the NoGo ring will have a groove on the outside. This groove is simply there to readily identify it as the NoGo ring.

There are also thread ring gages for Go and NoGo applications on external threads.

Flush Pin Gages

Flush pin gages (sometimes called fingernail gages) are attribute gages that provide additional information similar to that from a variable gage. This is so because the position of the pin (its step) gives you a rough idea of whether the part dimension being checked is near the high limit, low limit, or nominal dimension (Figure 9.66).

Flush pin gages work in two different ways:

1. The tolerance step is on the bar of the tool (Figure 9.66*a*).
2. The tolerance step is on the pin itself (Figure 9.66*b*).

In these figures, the step (on the bar or on the pin) is the exact height of the total tolerance of the dimension being checked.

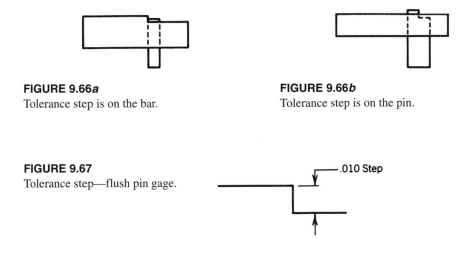

FIGURE 9.66*a*
Tolerance step is on the bar.

FIGURE 9.66*b*
Tolerance step is on the pin.

FIGURE 9.67
Tolerance step—flush pin gage.

Figure 9.67 shows an example of .010 total tolerance. If the part is not within the tolerance limits, the pin will be below the lowest step, or above the highest step. You must look closely at the gage and pin to determine if the part is oversize or undersize.

Figure 9.68 shows examples of determination of *oversize* or *undersize* conditions.

FIGURE 9.68
Determination of
oversize/undersize.

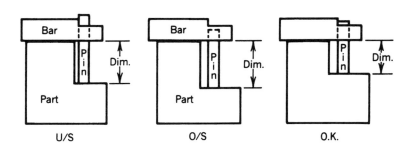

Snap Gages

Snap gages are used quite often in Go–NoGo applications. They are usually shaped like a "C" as far as the frame goes, with adjustable measuring *jaws* (Figure 9.69). They generally have one solid jaw on one side, with adjustable jaws on the other. The first jaw is adjusted to be the Go, while the second jaw is progressively adjusted to be the NoGo. Snap gages are applied to the dimension carefully (not quickly) in a rotating motion. The Go should go on the diameter and the NoGo should not.

Whenever using any of the above attribute gages, you must remember that they will only tell you "good or bad" and nothing else.

FIGURE 9.69
Snap gage (adjustable).

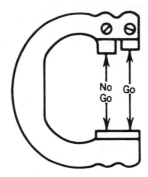

Specific Precautions When Using Attribute Gages

Ring Gages. Ring gages will not detect geometric problems within the feature you are measuring. Examples of these geometric problems are shown in Figure 9.70.

FIGURE 9.70
Geometric problems with fixed gages: (*a*) shaft is out-of-round; (*b*) shaft is tapered.

(*a*)

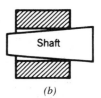

(*b*)

Plug Gages. Plug gages will not detect geometric problems within the feature (hole) you are measuring. Examples of these geometric problems in holes are shown in Figure 9.71.

FIGURE 9.71
Geometric problems with fixed gages: (*a*) hole is hourglass-shaped; (*b*) hole is bell-mouthed; (*c*) hole is tapered; (*d*) hole is barrel-shaped.

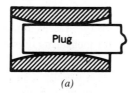

(*a*)

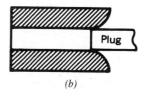

(*b*)

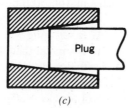

(*c*)

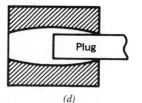

(*d*)

Snap Gages. Snap gages must be applied using the reference surface. Make sure that the piece being measured is straight (not cocked) in the gage (see Figure 9.72).

FIGURE 9.72
Manipulative error.

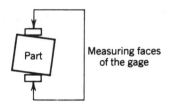

Part | Measuring faces of the gage

At all times when using ring, plug, snap, or any other type of gaging equipment, make sure that you have the correct gage for the dimension, that the part is clean, that the gage is clean, and that you know the limitations of the gage.

Application of Attribute Gages

All gages have a particular application to the part, or application of the part to them. Rules of thumb are:

1. If the gage can be held in one hand and the part in the other, there is no problem.
2. If the gage is heavier than the part, apply the part to the gage.
3. If the gage is lighter than the part, apply the gage to the part.
4. In all cases, use good common sense.
5. Avoid any application that may cause damage to the part or the gage.

RADIUS GAGES

Radius gages (Figure 9.73) are used frequently during manufacturing processes to measure *radii* when the radii are toleranced so that fractional gages are applicable. Radius gage sets come in sizes from $\frac{1}{64}$ to $\frac{1}{2}$ in. in increments of 64ths: for example, $\frac{1}{64}$, $\frac{1}{32}$, $\frac{3}{64}$, and so forth. Each radius gage has five different measuring surfaces to measure various radii as shown in Figure 9.74. Three of the surfaces are for measuring outside radii, and two are for inside radii, as Figure 9.74 shows.

Note: The smallest gage in a set is usually $\frac{1}{64}$ in. (or .016 in. in decimals). Therefore, since the radius is .016, this gage set cannot be used to measure a radius such as .010 or .008. Nor can any radius larger than $\frac{1}{2}$ in. (or .500) be measured, since the largest gage in the set is $\frac{1}{2}$ in.

Radius gages are visual gages. After selecting which one to use, you apply the gage to the part. If the actual radius on the part is smaller than the gage, you will see the conditions shown in Figure 9.75a and c. If the actual radius is larger than the gage, you will see the conditions shown in Figure 9.75b and d.

Radius gages are also a handy tool for Go–NoGo use. You simply choose two gages that are the high and low limit of a radius tolerance and apply them. But if you need to find the actual radius of a part and do not have a radius gage that fits perfectly, you must use other methods.

FIGURE 9.73a
Radius gage set. (M. T. I. Corp.
Reprinted by permission.)

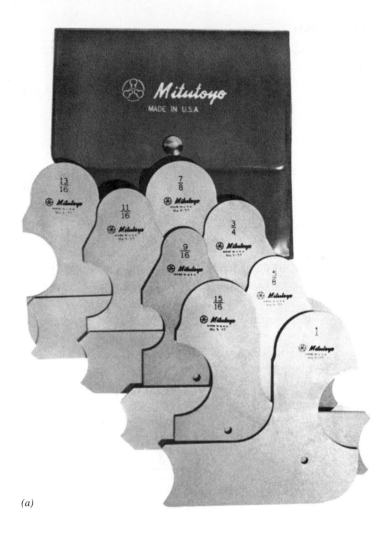

(a)

TAPERED (ADJUSTABLE) PARALLELS

Tapered parallels have a very simple design, yet they are widely used in the production and inspection of a product. They are two pieces that are inclined and joined together by a dovetail slide and a lock screw (see Figure 9.76).

They are joined together in such a way that as the two pieces are moved in opposite directions, the sides spread wider apart but still remain parallel to each other (as shown in Figure 9.77).

Tapered parallels come in various sizes, each with its own specific range. They can be set to a specific width with a micrometer (see Figure 9.78).

After being set, they can be used for machine setups, inspection setups, virtual size gages, Go–NoGo gages, and even measurement of certain size diameters (if the edges have radii). As shown in Figure 9.79, they can be used to measure the width of a slot. To do so,

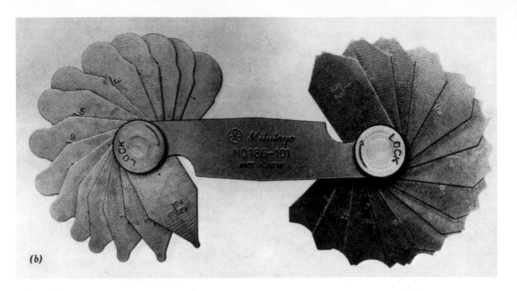

FIGURE 9.73*b*
Radius gage set. (M. T. I. Corp. Reprinted by permission.)

FIGURE 9.74
Radius gage applications.

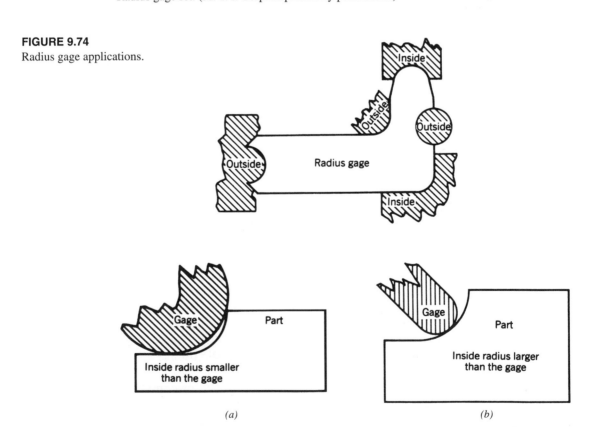

(a)

(b)

FIGURE 9.75*a*
Inside radius is smaller than the gage.

FIGURE 9.75*b*
Inside radius is larger than the gage.

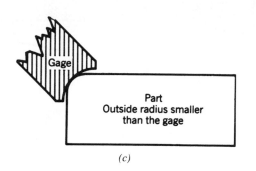

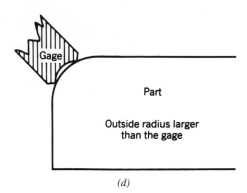

(c)

(d)

FIGURE 9.75c
Outside radius is smaller than the gage.

FIGURE 9.75d
Outside radius is larger than the gage.

FIGURE 9.76
Adjustable parallel.

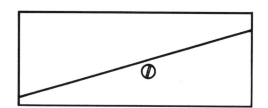

FIGURE 9.77
Parallel expands due to the angle.

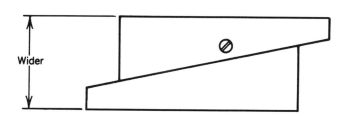

FIGURE 9.78
Measuring an adjustable parallel.

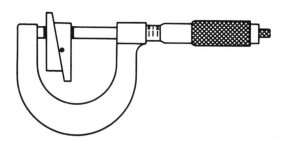

FIGURE 9.79
Adjustable parallel used to transfer
a measurement.

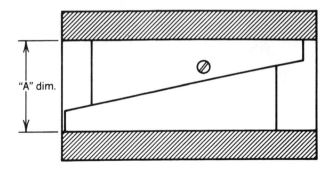

"A" dim.

loosen the lock screw, insert the parallel into the slot, spread it to meet the surfaces of the slot (with correct measuring pressure), lock it in, and remove and measure it with a micrometer or other tool, depending on the tolerance you are working with.

Tapered parallels can also be set and locked in as a Go–NoGo gage for sorting tasks. Another use for them is the same as for fixed parallels: to extend a reference surface, such as the surface plate for inspection. Where tolerances permit, they can be used with a micrometer to replace the need for gage blocks.

Tapered parallels are used for more applications than fixed parallels because they are adjustable and can work through a larger dimensional range than fixed parallels.

CENTERLINE ATTACHMENTS

There are various methods of measuring the *centerline to centerline distance* between holes. One of the quickest ways to do this (and still retain accuracy) is to use centerline measuring attachments. There are several kinds of centerline attachments made today, but the two types covered in this book are the most popular. These two types are the *centerline gage,* shown in Figure 9.80, and the *center master,* shown in Figure 9.81.

FIGURE 9.80
Centerline attachment for a vernier
caliper.

FIGURE 9.81
Center master (used on a height
gage).

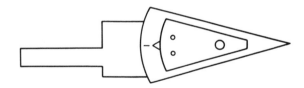

The Centerline Gage

Centerline gages are very simple in their construction. They are made such that when you clamp them onto a vernier caliper, the center of the cone is in line with the measuring face of the caliper (see Figure 9.82).

FIGURE 9.82
Centerline attachments on a caliper. (M.T.I. Corp. Reprinted by permission.)

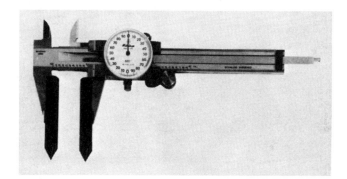

The cone-shaped tip of the gage will automatically find the centerline of any hole that is not larger than the gage itself. Once the gages are properly clamped to the caliper, you can measure the distance between two holes and read the caliper directly.

Figure 9.83 shows the centerline gages in use. Care must be taken when attaching these gages to the caliper. Both must be attached at the same depth unless you desire to measure the distance between two holes that are offset in depth. Also, keep in mind that attachments to any measuring device can contribute to the cumulative error of that device.

FIGURE 9.83
Application for centerline attachments.

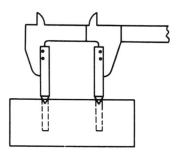

The Center Master

The next type of centerline attachment, called the *center master,* is shown in Figure 9.84. It is used with a height gage and surface plate and attaches directly to the height gage. Using the center master is very easy. The center master has an indicating arrow and a single line on it to show when the blade is in line with the centerline of the hole.

Using a *vernier height gage,* the steps to finding the distance between centerlines of two holes are (see Figure 9.85):

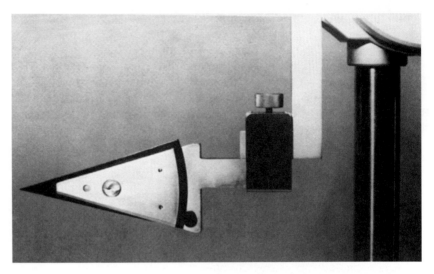

FIGURE 9.84
Center master on a height gage. (M.T.I. Corp. Reprinted by permission.)

FIGURE 9.85
Center master helps measure hole to hole. (M.T.I. Corp. Reprinted by permission.)

Step 1.   Insert the center master into the bottom hole and line it up. Read the vernier scale of the height gage.

Step 2.   Insert the center master into the top hole and line it up. Read the vernier scale again.

Step 3. Subtract the reading at the bottom hole from the reading at the top hole. This is the centerline-to-centerline distance.

MECHANICAL COMPARATORS

Bench comparators are often used for on-site measurements. They can be carried and used anywhere because they have their own built build-in reference surface, indicator stand, and indicator. Some bench comparators use a granite surface plate with a vertical post built into it; others use metal reference surfaces. Some of them have serrated reference surfaces (see Figure 9.86) that will allow dirt or chips to fall between the serrations to avoid errors in the measurements. All bench comparators have a method of vertical adjustment and locking it in. Most have a clamp on the vertical post that, when loosened, allows the indicator to be moved up and down to adjust it when necessary. Some have a rack; turning the knob moves the indicator up and down; then it can be locked into place.

Bench comparators are generally used when quick but accurate measurements of height, thickness, and so on must be made on a variable basis. Examples of this can be seen from time to time on machines for the operator to measure parts during production or at workbenches when parts must be screened, or 100 percent inspected.

All bench comparators require the use of some type of standard to establish the size to be compared. More often than not, gage blocks are used as this standard. For example, the

FIGURE 9.86
Bench comparator with an electronic digital indicator. (M.T.I. Corp. Reprinted by permission.)

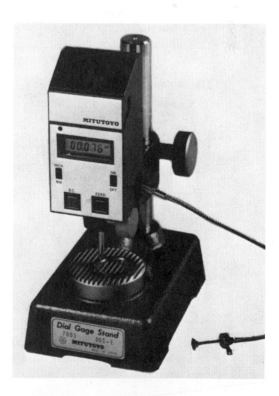

FIGURE 9.87
Diameter.

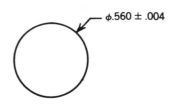

φ.560 ± .004

piece in Figure 9.87 must be measured on a 100 percent basis (each piece must be measured as it is produced). The bench comparator would be set up using a .560-in. stack of gage blocks. The indicator would be set at 0 on the stack, and then the part would be passed under the indicator. Any part that was good would not cause a movement on the indicator farther than ± .004 in. from 0 set.

From time to time, the comparator should be checked with the gage block stack for accuracy. Sometimes a representative master may be used, as shown in Figures 9.88 and 9.89. All the operator has to do is read the device to determine the comparison between the standard and the part.

■ Example 9–6

You know that the comparator was set to 0 with a .560-in. gage block stack. Therefore, any part that passes under the indicator that is actually .560 in. high will cause the indicator to come to 0. If the indicator comes to 0 from a counter-clockwise position, plus movements past 0 would mean that the part was higher than .560 in.

■ Example 9–7

The indicator reads plus .003 in. Therefore, the part is .560 in. + .003 in. = .563 in. actual size.

■ Example 9–8

The indicator reads minus .001 in. Therefore, the part is .560 in. − .001 in. = .559 in. actual size.

Cautions When Using Bench Comparators

1. Ensure clamps are tight.
2. Make sure your gage block stack is dimensionally correct.
3. Make sure the indicator is facing you directly (to avoid parallax error).
4. Make sure the vertical post is not loose.
5. Make sure the indicator has the accuracy you require for the measurements you are making.

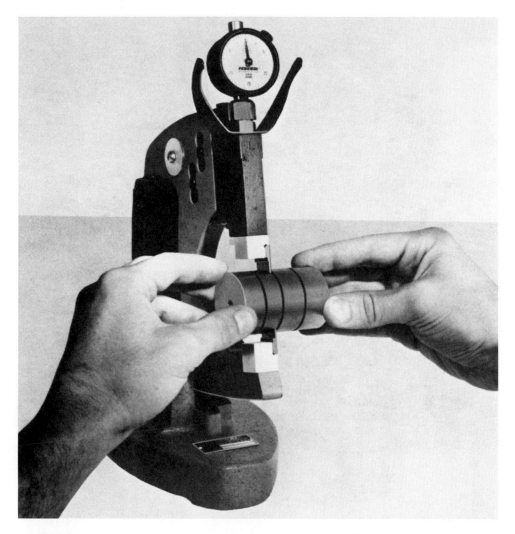

FIGURE 9.88
Dial snap gage with stand. (Federal Products Corp. Reprinted by permission.)

6. Check the indicator calibration with the gage block stack at regular intervals (especially when being used continuously).
7. Make sure parts are clean and deburred before passing them under the indicator.
8. To move the bench comparator, use the base; do not lift it with the upright post.
9. Avoid bumping the indicator rod from the side. It may be necessary to lift the indicator tip up and then put the part under it.
10. Make sure to use the proper tips on the indicator.
11. Do not use the bench comparator in jerking motions, which could cause miscalibration. Use smooth, even motions to make your measurements.

FIGURE 9.89
Bench comparator in use. (Federal Products Corp. Reprinted by permission.)

MEASURING AND GAGING INTERNAL THREADS

Refer to Chapter 6 on internal threads. The nomenclature here changes slightly from that on internal threads in Chapter 6. The *crest* and *root* are looked at in a different manner, as shown in Figure 9.90. The *minor diameter* is actually the original hole (at the crest diameter) that was drilled before tapping. The *major diameter* (after tapping) is now at the root of the thread. (It is at the crest on external threads.) The *pitch diameter,* thread depth, and angle remain the same.

Measurement (in general) includes the pitch diameter, minor diameter, and the depth of the threads in the hole (if required). Sometimes the depth is not a requirement because the hole must be tapped all the way through. The minor diameter can be measured simply with a plug gage. The pitch diameter is usually proved with a *thread plug gage* like the one shown in Figure 9.91. It is a Go–NoGo type of gage.

FIGURE 9.90
Internal threads—nomenclature.

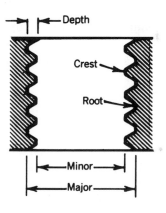

FIGURE 9.91
Thread plug gage

Measuring Depth on Threads

Measurement of the depth of threads can be done in various ways. One of the most accurate ways is the *turn method.* This involves using a thread plug gage's accurate lead to verify the depth of threads (see Figure 9.92).

FIGURE 9.92
Callout for minimum full thread.

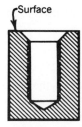

Keep in mind that thread plug gages are accurate within gage tolerances and therefore have an accurate lead. The *lead* is the distance that a thread plug gage will travel in *one full turn* (or the distance between two adjoining threads).

To Inspect Depth

If there are 20 threads per inch, each thread is $\frac{1}{20}$ in. or .050 in. apart. So .050 is the lead. Therefore, if you have an internal thread that is $\frac{1}{4} - 20$ and is supposed to be .600 deep (at the last full thread), and the lead is .050 in., then if the threads were actually .600 in. deep, the plug would take 12 turns to reach the bottom thread.

$$.050 \overline{)\,.600\,}^{\text{12 turns}}$$

■ **Example 9–9**

$$\tfrac{3}{8}\text{ in. } - 16 \times 0.625 \text{ in. deep} = \tfrac{1}{16}\text{ in. } = .0625 \text{ in. lead}$$

and

$$0.625 \text{ deep} \div .0625 \text{ lead} = 10 \text{ turns}$$

Let's say that the previous part is measured by the turn method and you get only nine turns. To find out how deep nine turns is, just multiply the nine turns times the lead. Therefore, $9 \times .050$ in. $= .450$ in., and since the minimum depth is 0.600 in. $\pm$ .010 in., these threads are too shallow.

Remember: When using the turn method, you *must* include the depth of chamfers because the measurement is from surface A to the bottom thread. So the example of 0.450 in. with a .030-in. chamfer is actually 0.480 deep.

MEASURING AND GAGING EXTERNAL THREADS

There are various ways of measuring external threads, depending on the application and the desired accuracy. Usually, the most important dimension of threads of any kind is the pitch diameter.

Various tools can be used to measure threads. Some are explained in the following examples.

■ **Example 9–10**

One of the quickest ways is the use of a pair of Go–NoGo thread ring gages, one of which is shown in Figure 9.93. These gages are set to the tolerance limits of the part. Their application is simply screwing them onto the part. The Go should go on easily without force over

FIGURE 9.93
Thread ring gage.

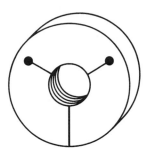

the length of the threads, and the NoGo should start but not go over more than a couple of threads before it drags.

Remember: These rings only tell you whether or not the part is in tolerance. They do not indicate what the actual size of the thread is.

■ **Example 9–11**

Another quick method is with the screw thread micrometer shown in Figures 9.94 and 9.95. These mikes are set up to measure the pitch diameter directly. The only problem with the mikes is that they are designed to measure a range of different pitch values; therefore, an incorrect setup could cause error in measurement. To correct this, the micrometer should be set with a thread plug gage of known pitch diameter before use.

■ **Example 9–12: The Three-Wire Method**

This method of measuring pitch diameter is relatively quick and more accurate than the others (depending on the accuracy of the wires used). As shown in Figure 9.96, you need three wires of the same specified size, setting up two of them on one side and one on the other.

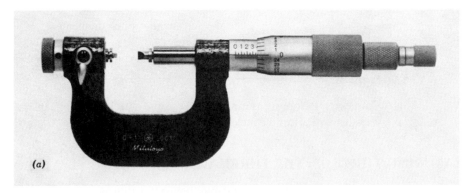

(a)

FIGURE 9.94a
Thread micrometer. (M.T.I. Corp. Reprinted by permission.)

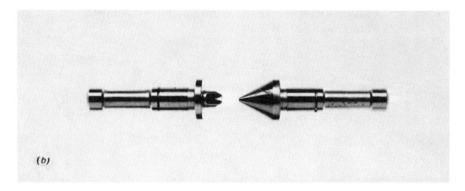

(b)

FIGURE 9.94b
Anvils for the thread micrometer. (M.T.I. Corp. Reprinted by permission.)

FIGURE 9.95
Thread micrometer in use. (L. S.
Starrett Co. Reprinted by
permission.)

There is a formula to determine what wires to use and how to arrive at the pitch diameter. However, most wire sets have a direct reading table that will also give you pitch diameter. Make the setup as shown. Measure the dimension over the wires and use the constant number on the table to arrive at the pitch diameter.

Measurement over the wires:

$$M - \text{constant} = \text{pitch diameter}$$

Without a table of constants, you must use a formula that is described later in this chapter.

Caution.   Before measuring threads, make sure they are clean and not damaged in any way that will affect the measurement.

MEASURING THREADS: THE THREE-WIRE METHOD

One of the most important measurements to be made when measuring threads is the *pitch diameter*. The pitch diameter of external threads can be measured in several ways. There are screw thread micrometers, thread comparators, and many other methods, but the most

FIGURE 9.96

Three-wire method. (L. S. Starrett
Co. Reprinted by permission.)

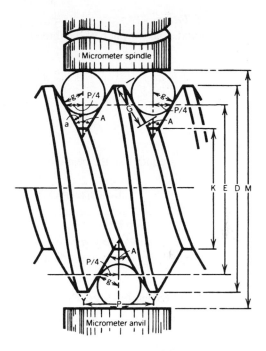

accurate hand-held tools for measuring the pitch diameter of external threads are thread
wires, or the three-wire method. In this method, you use three wires of the same diameter.
Two of these wires are placed on one side of the thread and the other wire on the other side,
as shown in Figure 9.96.

Using the three-wire method, any wire size can be used as long as the wire is small
enough to make contact with the "flanks" of the threads and large enough to also extend
above the crests of the threads.

Best Wire Size

When accuracy is necessary, and to avoid the possibility of error due to the thread angle,
you should use a particular *best wire size* for the measurement. The best wire size is calculated with the following formula:

$$\text{best wire size} = .57735 \, (p)$$

where .57735 is a constant and p is the pitch of the thread you are measuring.

■ **Example 9–13**

You want to measure a $\frac{1}{4}$ in. − 20 thread.

Solution

First, find the pitch of the thread.

$$p = \frac{1 \text{ in.}}{\text{number of threads per inch}}$$

$$= \frac{1}{20} = .050 \text{ in.}$$

Second, find the best wire size using the pitch above.

Best size = .57735(.050) = .0288 in. (or .029-in.-dia. wire)

Remember: Any wire size can be used under the conditions stated above, but try to use the best size to avoid inaccuracy due to varying thread angles.

Three-Wire Method (Step-by-Step Procedure)

To measure threads using the three-wire method, follow these steps:.

Step 1.   Calculate the best wire size. (Make sure you use wires that are accurate enough for the threads you are measuring.)

Step 2.   Place the wires onto the threads (two on one side and one on the side directly opposite the threads).

Step 3.   Using a micrometer (or other instrument), measure the dimension over the wires. (This is called *M*.) Good contact pressure is required.

Step 4.   After finding the measurement *M* (over the wires), use this formula to calculate the actual pitch diameter:

$$\text{pitch diameter } (E) = M + (.86603 \, p) - 3W$$

where E is the pitch diameter
M is the measurement over the wires
p is the pitch of the thread
W is the wire size you used
.86603 is a constant

For example, let's say that M is .280 in., p is .050, and W is .029. So

$$E = M + .86603p - 3W$$
$$= .280 \text{ in.} + .86603(.050) \text{ in.} - 3(.029) \text{ in.}$$
$$= .280 \text{ in.} + .0433 - .087 \text{ in.}$$

$$= .280 \text{ in.} + .0433 - .087 \text{ in.}$$
$$= .3233 - .087$$
$$= .2363 \text{ in.}$$

This is the pitch diameter.

If you find that the pitch diameter is within the tolerance limits on the drawing, the part is acceptable.

Summary

First, find the pitch (p) of the thread.

Next, find the best wire size.

Next, take the measurement over the wires.

Last, inserting the M, p, and W values in the formula, calculate the actual pitch diameter.

BASIC CALIBRATION TECHNIQUES

Calibration is simply defined as comparing an instrument of known accuracy with another instrument.* For the most part, inspectors, machinists, and operators are concerned with calibration at the company level (which is the company's metrology lab) or at the individual level (which is a person calibrating his or her own tools). At either level, it is important that calibration be performed at certain intervals of time. These calibration intervals depend mainly on

1. How often the tool is used
2. How often it is found to be out of calibration

Calibration Intervals

Calibration intervals must be based on both of the above mentioned items, or they will not be effective. The reason for this is that if the calibration intervals are too short, you will be spending too much time calibrating the tool. If the intervals are too far apart, you run the risk of using a tool that is out of calibration. Calibration intervals can (and should) be adjusted based on previous calibration results.

Calibration Systems

In commercial industries, identification systems for calibration vary from company to company. Some companies use the color code system, assigning tools and instruments a color (yellow, green, blue, etc.) that identifies when they are due for calibration. Refer to Chapter 1 for calibration systems.

Other companies use calibration stickers on the tool with a variety of information on them, as covered in Chapter 1.

---

* Traceable to the National Institute for Standards and Technology (NIST).

Companies with large numbers of tools to control use the computer for effectiveness in maintaining calibration intervals. The computer prints out lists of tools that are to be calibrated to meet their particular interval requirements. It can also be programmed to print out those tools that need adjustment at particular intervals.

Calibration Methods

A general rule for calibration is that it should be performed with a standard 10 times as accurate as the tool being calibrated. A micrometer, for instance, would be calibrated with a gage block (for length) and checked for wear on the measuring faces of the spindle and anvil with an optical flat. Both the gage block and optical flat are far more accurate than the micrometer.

Calibration Environment

The calibration environment in most cases is a clean, temperature-controlled, humidity-controlled room where "working standards are kept for the purpose of the calibration of company tools and instruments." When tools are brought in for calibration, they should be cleaned, if necessary, and set aside until they are the same temperature as the room's temperature (usually 68 degrees Fahrenheit). Then calibration can be performed.

Risks Without a Calibration System

In most mass production companies, there are numerous tools and gages to monitor. Without an effective calibration system of some kind, keeping track of calibration intervals for each tool or gage is next to impossible, let alone retaining the ability to fine-tune each interval to the point where you can be assured that calibrated tools and gages are being used.

If operators and machinists use tools or gages that are *not* calibrated, the risk is *producing parts that are not per specifications.* If an inspector uses a tool that is not calibrated, the risks are *accepting a bad part* and *rejecting a good part.* The largest risk that a company or individual takes without a calibration system is that defective products can be shipped to the customer regardless of any other controls it may use.

Production, inspection, and quality control rely heavily on measurements. Measurements can be subject to a variety of possible errors, such as heat, dirt, manipulation, and geometry. There are many errors possible in any given measurement. Calibration of the tool should not be allowed to be one of them. Calibration can be controlled if a company (or individual) wants to control it.

REVIEW QUESTIONS

1. Precision means
   a. hitting the target
   b. getting the true value
   c. getting consistent results
   d. none of the above

2. What is the name of the type of dial that has consecutive numbers around the dial in a clockwise direction?
   a. last word
   b. balanced
   c. amplified
   d. continuous
3. Indicator error (expressed as percent of full scale range) is called
   a. sine error
   b. amplitude error
   c. TIR
   d. cosine error
4. What types of measurements can be made with a caliper?
   a. outside
   b. inside
   c. depth
   d. all of the above
5. What is the common error (amount) made when using a micrometer?
   a. .001" off
   b. .0005" off
   c. .025" off
   d. none of the above
6. Measurement of the pitch diameter of an external thread can be accomplished using
   a. the three-wire method
   b. thread micrometer
   c. ring gage
   d. both a and b
   e. all of the above
7. Using gage blocks, it is a good idea to use _____ _____ on every stack to prevent wear.
   a. careful wringing
   b. a calculator
   c. wear blocks
   d. six blocks
8. What is the discrimination of a universal bevel protractor?
   a. 5 minutes
   b. 30 minutes
   c. 1 degree
   d. none of the above
9. Following the rule of thumb for accurate measurements, what is the discrimination of the gage that should be used to measure a part dimension of a .500–.501?
   a. .001"
   b. .0005"
   c. .0001"
   d. .0002"

10. Measuring a part using a gage block stack is called _____ measurement.
    a. stack
    b. transfer
    c. differential
    d. linear
11. The most important general aspect about measuring pressure is that
    a. it should be predetermined
    b. it should be constant
    c. it should be tight
    d. it should be light
12. What are the different cross-sectional shapes of gage blocks that can be purchased?
    a. round
    b. square
    c. rectangular
    d. all of the above
    e. a and b only
13. How many different combinations of gage block stacks can be made using a typical 81 piece set of blocks?
    a. 5,000
    b. 2,000
    c. 8,100
    d. over 10,000
14. When using wear blocks on gage block stacks one should
    a. keep the same side outward at all times
    b. not forget to factor the wear blocks in the stack dimension
    c. use them on every stack if possible
    d. all of the above
    e. none of the above
15. What specific gage blocks are best to use to stack a 1.4501 dimension?
    a. .1001, .100, .250, 1.000
    b. .1001, .350, 1.000
    c. .450, 1.000
    d. none of the above
16. What is the significance of a yellow face on a dial indicator?
    a. Caution should be taken in measurement.
    b. It is a .0001″ indicator.
    c. It is a metric indicator.
    d. There is no significance of the color of the face.
17. What type of dial indicator face has numbers, for example, that are $+1, +2, +3, +4$ and $-1, -2, -3,$ and -4 from zero?
    a. continuous dial
    b. deviation dial
    c. plus/minus dial
    d. balanced dial

18. Of the following choices, what is the gage discrimination that should be used to inspect a part dimension that has a total tolerance of .002″?
    a. .0001″
    b. .0003″
    c. .00025″
    d. .0005″

19. What is the term used to describe the total travel of an indicator needle during a measurement of runout (for example)?
    a. TIR
    b. FIM
    c. FIR
    d. all of the above
    e. a and c only

20. The two faces of a micrometer that make contact with the part during measurement are the
    a. anvil and frame
    b. anvil and thimble
    c. spindle and anvil
    d. spindle and barrel

21. Which of the following are actual micrometers that can be purchased?
    a. 0–1″ micrometer
    b. 1–2″ micrometer
    c. 2–3″ micrometer
    d. all of the above
    e. a and b only

22. Which of the following micrometers is unique in the way that the observer reads the results?
    a. outside micrometer
    b. hub micrometer
    c. depth micrometer
    d. blade micrometer

23. The _____ line must be found in order to read a vernier scale.
    a. main
    b. coincidence
    c. distance
    d. scale

24. A two-point "transfer" gage that can be used with an outside micrometer to measure inside diameters is called a(n)
    a. snap gage
    b. dial bore gage
    c. telescoping gage
    d. adjustable parallel

25. The three basic types of plug gages are single purpose, double ended, and _____.
    a. dual limit
    b. progressive
    c. attribute
    d. variable

26. One method that can be used to measure the full depth of threads in a tapped hole is the _____ method.
    a. dedendum
    b. turns
    c. full depth direct
    d. three wire

27. The smallest dimension that can be measured using a 2″ micrometer is:
    a. 2.000″
    b. .500″
    c. 1.000″
    d. .750″

28. What is the fewest number of gage blocks needed (using two .100″ wear blocks) to stack up a 1.7591 dimension?
    a. 5
    b. 8
    c. 6
    d. 7

29. Which of the following indicators should be used to measure the TIR of parallelism of a part where the parallelism tolerance is .0005″ TIR?
    a. .0001″
    b. .0002″
    c. .001″
    d. .0005″
    e. .00005″

30. Which of the following indicator tips should not be used on a diameter?
    a. spherical
    b. pointed
    c. flat
    d. none of the above
    e. a and b only

10 *Surface Plate Inspection Methods*

SURFACE PLATES

The primary measurement made when using the surface plate is *height* (Figure 10.1). This is true because all measurements made on the surface plate are made from the plate up. This chapter outlines specific measurements where a surface plate may be used, including coverage on the surface plate itself. Operators, machinists, and inspectors in the mechanical industry use surface plate techniques during the course of their jobs and should understand these basic techniques. Operators and machinists use surface plate techniques mainly to lay out castings and raw stock and to measure parts. Inspectors use them for the same purposes, but mostly for making dimensional measurements. In the following pages we discuss the various basic uses of the surface plate and give helpful hints on the many applications of this valuable tool.

Many reference surfaces on component parts or assemblies are established by *planes*. The ideal plane for dimensional measurement should be perfectly flat, but since nothing is perfect, we must settle for the next best thing. Surface plates provide a true, flat reference surface for dimensional measurement (see Figure 10.2). They are a simulated datum plane.

There are different materials from which surface plates are made. Two of the main materials are cast iron and granite (stone). Cast-iron surface plates have their good and bad points. Two good points are that they are magnetic and usually their configuration allows parts to be clamped down during measurement. A bad point is that if they are damaged, raised material is produced on the surface, which could cause a considerable amount of error in measurements. Another bad point is that since they are iron, they will easily rust.

Granite surface plates are far superior to those of cast iron because:

1. Damaged areas produce no raised material.
2. They are lower in price.
3. They retain their flatness longer than cast iron.
4. They will not rust.

Whichever type of plate is used, it is important to be sure that the plate is level when in use. Dimensions measured on surface plates are taken from the plate up since the plate is the reference surface (see Figure 10.3).

One widely used accessory to the surface plate is a gage block stack (see Figure 10.4). Gage blocks reflect the actual distance from the plate up, so that comparisons can be made to the part being measured with a high degree of accuracy.

As shown in Figure 10.5, a dimension that must be located by three perpendicular planes is not a problem because of another accessory called an *angle plate* (or knee). This is also an approved simulated datum plane.

Some surface plate accessories are shown in Figures 10.6 through 10.11. There are many more accessories that can be obtained to aid in setting up a wide variety of measurements.

FIGURE 10.1
Surface plate equipment (L.S. Starrett Co. Reprinted by permission.)

FIGURE 10.2
Granite surface plate. (L.S. Starrett Co. Reprinted by permission.)

Flat on this surface only

FIGURE 10.3
Reference and measured surfaces.

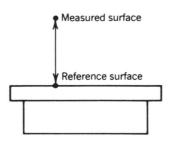

Measured surface

Reference surface

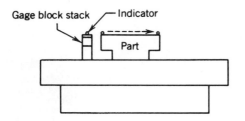

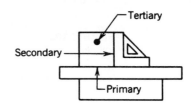

FIGURE 10.4
Transfer measurement on a surface plate using gage blocks.

FIGURE 10.5
Locating primary, secondary, and tertiary datums on a surface plate.

FIGURE 10.6
Angle plate (knee).

FIGURE 10.7
Precision square. (M.T.I. Corp. Reprinted by permission.)

FIGURE 10.8
Matched V-blocks. (M.T.I. Corp. Reprinted by permission.)

269

FIGURE 10.9
Square gage blocks. (M.T.I. Corp.
Reprinted by permission.)

FIGURE 10.10
Fixed parallels. (M.T.I. Corp.
Reprinted by permission.)

FIGURE 10.11
Planer gage. (Brown & Sharp.
Reprinted by permission.)

FIXED PARALLELS

Fixed parallels are precision tools that are often used in production and inspection applications. These applications vary from machine setups to surface plate setups.

Fixed parallels are rectangular in shape and come in various sizes (see Figure 10.12). They are usually purchased in matched pairs. These parallels are made to be square and parallel on all sides to within tooling tolerances.

FIGURE 10.12
Fixed parallel extends the surface
plate.

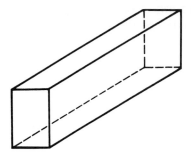

Their main purpose is to extend a reference surface where it is necessary. This reference surface can be either on a machine that is producing the part or on the surface plate during the inspection of the part.

Fixed parallels can also be used in some sorting tasks. For example, some washers must be sorted to their *virtual thickness*. Here, matched parallels could be used. They are separated by feeler stock (or gage blocks if applicable) to the maximum virtual thickness allowed on the washers. Next, they are clamped securely, and the feeler stock is removed. Now, the washers can be rolled through them. Any washer that is beyond the virtual size allowed will not go through.

In Figure 10.13, we see another example of the uses of fixed parallels. This part must be located on datum C to inspect it for parallelism, but datum C cannot be located directly on the surface plate because there is a diameter in the way.

FIGURE 10.13
Parallelism—interrupted datum.

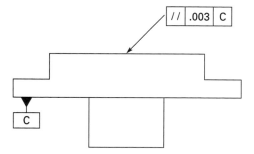

As shown in Figure 10.14, with parallels, locating datum C for inspection is no problem. The parallels extend the surface plate so that the diameter that was once in the way is now clear from the plate.

In production and inspection fixed parallels have a wide variety of uses, most of which are learned with experience.

FIGURE 10.14
Part located using fixed parallels.

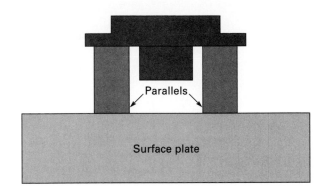

CARE OF SURFACE PLATES

Listed below are some hints that are helpful in taking good care of a surface plate.

1. Do not abuse surface plates; they are a precision tool.
2. Do not allow unnecessary objects to be placed on the surface plate.
3. Do not store tools on the surface plate. Keep on the plate only what you are using at any given time.
4. Clean the surface plate before and after use.
5. Keep the surface plate covered when it is not in use.

The proper method of putting items on, or removing items from, a surface plate is shown in Figure 10.15.

To avoid damage to the surface plate, the methods described above should be used. By putting items on the plate in this manner, you avoid the possibility of damage to the plate because you ensure that the surface of the item and the surface of the plate meet evenly. You will also avoid damage when removing items by always sliding the item from the plate.

FIGURE 10.15
(*a*) Proper method of putting parts on a surface plate. (*b*) Proper method of removing parts from a surface plate.

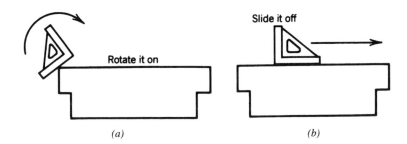

SURFACE GAGES

Surface gages (Figure 10.16) are used in many measurement applications, but since they do not have a direct-reading scale, they are used in transfer (or comparison) measurement. In Figure 10.17, you see the surface gage transferring a dimension established by a stack of

FIGURE 10.16
Parts of a surface gage. (L.S.
Starrett Co. Reprinted by
permission.)

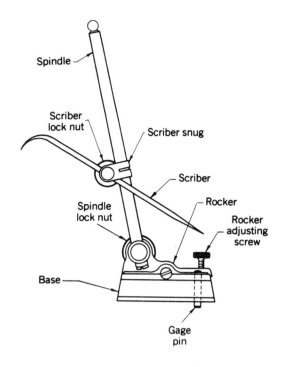

FIGURE 10.17
Surface gage and indicator in use.
(L.S. Starrett Co. Reprinted by
permission.)

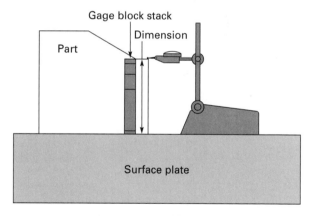

gage blocks to the part in order to measure it. This is simply a comparison of a known standard to the part.

Remember: All methods of measurements are comparisons to known standards. Gage blocks are a good example of a standard traceable to the National Institute of Standards and Technology (NIST).

Uses of the Surface Gage

Surface gages have many uses, including the transfer measurement shown above, scribing parallel lines (using the two pins provided), and inspecting runout, concentricity, and perpendicularity. These are just a few uses. Many more are learned by obtaining experience with the tool.

Possible Errors in Using the Surface Gage

Although the surface gage is a worthwhile tool for use in production and inspection, remember these few points to avoid errors. The causes of errors include (1) loose clamps, (2) a too-long indicator support rod, (3) an upright post at an angle to the base, and (4) completely loose fine adjustment bolts (Figure 10.18).

The Cantilever Effect

An example of a cantilever is a beam that is only supported at one end. Because of this single support and length, it is very unstable and can bend with weight on the other end. How much it bends depends on how much weight is on it and how far the weight is from the supported end. The farther the weight is away from the supported end, the more the beam will bend (see Figure 10.19). This deflection, resulting from long support rods, can cause measurement errors.

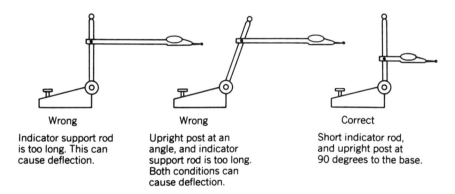

Wrong	Wrong	Correct
Indicator support rod is too long. This can cause deflection.	Upright post at an angle, and indicator support rod is too long. Both conditions can cause deflection.	Short indicator rod, and upright post at 90 degrees to the base.

FIGURE 10.18
Possible errors using a surface gage. (L.S. Starrett Co. Reprinted by permission.)

FIGURE 10.19
Cantilever effect. (L.S. Starrett Co. Reprinted by permission.)

10 pounds here will bend it a little

10 pounds here will bend it a lot

POSSIBLE SOURCES OF ERRORS IN MEASUREMENT

Below are some possible sources of errors when using hand tools and the surface plate.

Errors Using Hand Tools

This list is repeated here for emphasis and because hand tools are used in conjunction with surface plate measurements.

1. *Miscalibration.* Hand tools should be calibrated regularly (according to use).
2. *Dirt.* Parts should be clean before measurement is made.
3. *Wear.* Measuring faces of hand tools should be checked regularly for wear.
4. *Heat.* Hand-held measuring instruments should not be held for an extensive period of time.
5. *Burrs.* Parts should be deburred before measurement.
6. *Training.* The person using the tool must completely understand how to use it (reading, manipulation, etc.).
7. *Manipulation.* Hand-held instruments must be applied properly to the workpiece.
8. *Geometry.* The workpiece may be out of round or tapered or have other geometrical shapes.

Errors in Surface Plate Measurements

1. *Gage blocks.* Gage blocks are not properly wrung together, or gage blocks do not include the proper size blocks.
2. *Clamps.* Clamps on height gages, surface gages, and other tools must be tight enough to avoid movement.
3. *Wear.* Surface plates may have excessive wear in high-use areas.
4. *Vibration.* The bench holding the surface plate must be firm and virtually vibration free.
5. *Parallelism.* The workpiece is not parallel, and without "sweep measuring" the workpiece, this would not be discovered.
6. *Dirt.* Surface plates should be kept clean at all times on the working surface.
7. *Cosine error.* Indicator tips should be kept parallel to the workpiece surface being measured.
8. *Support rod length.* Support rods for indicators should not be too long (cantilever effect).
9. *Tip wear.* The indicator tip is worn flat and does not provide spherical point contact.
10. *Reading.* Errors in reading the vernier scales of height gages occur. (Sometimes they are due to parallax or a person's poor eyesight.)
11. *Parallax.* Reading errors due to parallax can be minimized by holding one's eye perpendicular to the surface being read (for example, the indicator face).

MAKING LINEAR MEASUREMENTS USING THE TRANSFER METHOD OR THE DIRECT READING METHOD

Linear (or length) measurements are most often made by using the transfer method. Simply set up the dimension with gage blocks and transfer the dimension from the gage block stack to the part.

Example of Using the Transfer Method

When Figure 10.20 appears as part of the drawing, it means that the tab must be 1.800 in. ± 0.0005 in. from datum surface A (no matter what size the tab is). Figure 10.21 shows how to inspect this part using the transfer method. Follow these steps:

Step 1. Set up gage blocks to 1.800 in.

Step 2. Set 0 on the indicator (on the blocks).

Step 3. Carefully slide the surface gage to the tab surface and take a reading.

Step 4. The plus or minus amount shown on the indicator tells you how much over or under 1.800 in. the actual dimension is.

Step 5. In this example the indicator should not show more than plus 0.005 in. or minus 0.005 in. (otherwise, the part is not to print).

Measuring the Same Dimension Using the Direct Reading Method

This method of linear measurement is also often used. Here you use the surface plate as a starting point (since surface A is where the dimension begins, and surface A is resting on the plate).

FIGURE 10.20
Height requirement.

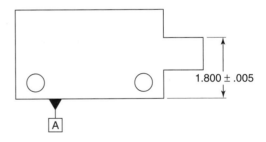

1.800 ± .005

FIGURE 10.21
Transfer measurement with gage blocks.

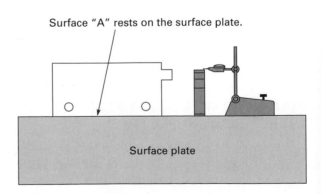

Surface "A" rests on the surface plate.

Surface plate

FIGURE 10.22
Dual-reading technique: zero on the
surface plate.

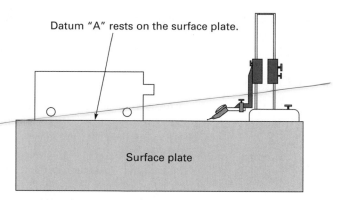

The first step is to rest the tip of the indicator on the surface plate and (using the fine adjustment knob) set 0 on the indicator (see Figure 10.22). Next, read the vernier scale on the height gage and write down the reading. This reading will be used later. For this example, let's say that the reading was 1.200 in.

The next step is to raise the indicator making it rest on the tab surface (as shown in Figure 10.23). Take care in doing this and do not bump the indicator against anything. If the indicator does get bumped, start over. Now, using the fine adjustment knob, set 0 again on the indicator. At this point, take a second reading on the vernier scale and write it down. Let's say that this reading was 3.002 in.

Now, the actual dimension of the part is the difference between the highest and lowest readings you have taken. This is true because you set the indicator to 0 on the plate and 0 on the tab surface. Hence

$$
\begin{array}{r}
3.002 \\
-1.200 \\
\hline
1.802
\end{array}
$$

Since the tolerance is 1.800 in. ± .005 in., the part is to print.

FIGURE 10.23
Dual-reading technique: zero on the
part surface.

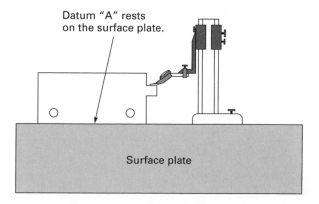

There are other ways to make the type of measurements shown above, such as transferring measurements from height masters or using digital height gages for direct reading.

CENTERLINE MEASUREMENT OF FEATURES

Many measurements require that the centerline of a feature (a feature being a hole, slot, tab, etc.), must be a specified distance from a datum or reference surface. To obtain an accurate measurement, the centerline must first be found. For example, the centerline of a hole depends on the actual size of the hole. Linear measurements to the centerline of a feature cannot be accurately made unless the feature size is first measured. Below are a few basic steps in making measurements to centerlines of features.

Surface Plate Inspection Methods

Step 1.   Locate the datum surface first on the surface plate.

Step 2.   Measure the actual size of the feature. (In this example, it is a hole.)

Step 3.   Write down the actual feature size.

Step 4.   Measure the distance from the reference surface to your choice of either side of the feature.

Step 5.   Write down this distance.

Step 6.   Now, depending on which side of the feature you chose to measure, you simply add or subtract one-half the actual hole size from your measurement.

An Example of Centerline Measurement of a Hole

The drawing in Figure 10.24 means that the actual centerline of the .410 − .400 in. hole must be within 1.500 ± .010 in. from datum A no matter what the actual hole size is.

FIGURE 10.24
Coordinate dimension.

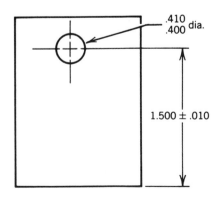

.410
.400 dia.

1.500 ± .010

FIGURE 10.25
Coordinates measured using transfer technique.

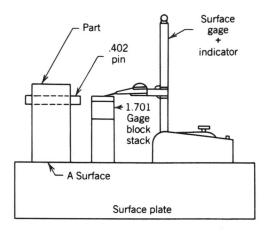

Therefore, you can see the importance of the need to measure the hole first, because it has .010 in. tolerance on it and could be any actual size within that tolerance. You can never assume the hole is a specific size.

Let's say we measured the actual hole size here and that it was .402 in. The preferred way to measure this part, in most places, is to get a gage pin that fits the hole and insert it into the hole. After doing this, you set the part on the surface place with datum A resting on the plate, since it is the locating surface (see Figure 10.25). In this case we choose to measure to the top of the hole. Calculate the distance to the top of the pin assuming that the hole is exactly 1.500 in. from surface A to its centerline. This is simply 1.500 in. plus one-half the pin size, or 1.500 in. + .201 in. = 1.701 in. Next, you set up a gage block stack of exactly 1.701 in. and zero the indicator on this gage block stack. Now, the zero on your indicator is set at 1.701 from the surface plate. Transfer this measurement by sliding your surface gage to where the indicator tip is at the very top of the pin. If the part is within blueprint tolerance, your indicator will not show a reading lower than minus .010 in. or higher than plus .010 in.

Remember: Never allow the indicator to be bumped after you have set zero. If you bump the indicator, start over.

Make certain you contact the very top of the pin during the measurement.

PARALLELISM MEASUREMENT

Parallelism is an important functional tolerance of *orientation* on machined parts. Perfect parallelism is when two surfaces are exactly the same distance apart throughout their length. But since nothing is perfect, we use tolerances to define the amount of acceptable *nonparallelism* of machined surfaces to each other.

The symbol for parallelism is //, signifying two planes the same distance apart (Figure 10.26).

FIGURE 10.26
Parallelism callout.

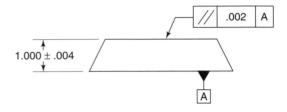

■ Example 10–1

When Figure 10.26 appears in the drawing, we know that Datum A is the reference plane. If located on datum A, the other surface must lie within the .002 tolerance zone, as shown in Figure 10.27, and still be within the 1.000 ± 0.004 dimension at the same time. Simply locate surface A on a surface plate as in Figure 10.28. Set an indicator at one end of the part on 0, then move the indicator across the surface at random to the other end while watching the total movement of the indicator. This is the total indicator reading (TIR). The TIR shown in Figure 10.29 is .003; therefore, the part is *not* to print. It must be .002 TIR to be within the specified limits.

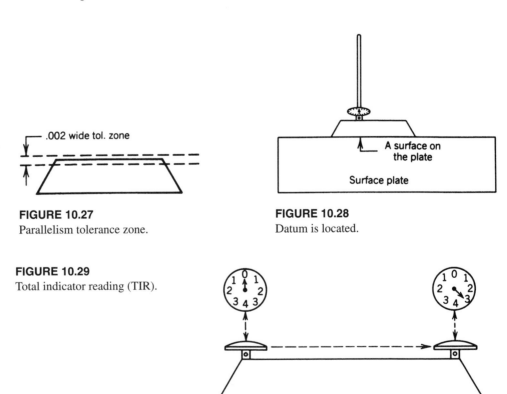

FIGURE 10.27
Parallelism tolerance zone.

FIGURE 10.28
Datum is located.

FIGURE 10.29
Total indicator reading (TIR).

■   **Example 10–2**

The meaning of Figure 10.30 is the same as in Example 10–1. The top surface must be parallel to surface A (the datum) within 0.001 TIR. This example, however, is difficult because surface A is not easy to locate.

   To inspect the part, you must use something to locate surface A, and at the same time, keep surface B from touching the plate. This is often done with a special fixture or a set of parallels; sometimes three pins of equal height will do the job. As shown in Figure 10.31, the ideal tool is a fixture. Note that surface B is clear from the plate.

   After this "special setup" to the hidden reference (or datum) surface, the part is inspected in the same way as in Example 10–1. If the TIR is more than .001, the part is not to print.

FIGURE 10.30
Parallelism to a hidden datum.

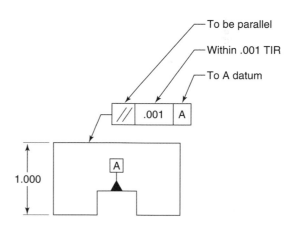

FIGURE 10.31
Fixed parallel locates the hidden datum.

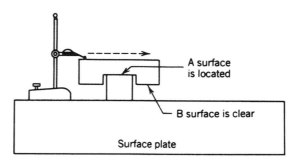

FLATNESS MEASUREMENT

Flatness is another important functional tolerance of *form* on some machined parts. For example, if you had an assembly that required two parts to be bolted or clamped tightly together to form an airtight seal, you would want the two mating surfaces to be flat enough to do so. Therefore, the drawing would indicate a flatness tolerance for the two surfaces. Flatness is also used to "qualify" datum surfaces. Flatness has been described as the geometric characteristic of a reference plane.

FIGURE 10.32
Flatness callout.

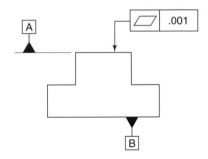

FIGURE 10.33
Measuring flatness using jack
screws to set up the optimum plane.

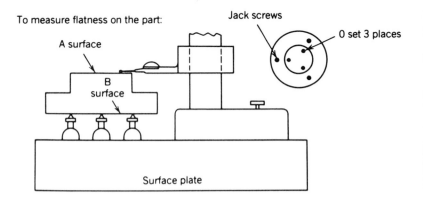

The symbol for flatness is ⟋⎯⎯⟋, showing a surface (also, indicating *one* surface)
(Figure 10.32).

1. Flatness only relates to itself; there are no reference surfaces from which to measure it.
2. Flatness measurement is always total indicator reading (TIR) unless otherwise specified,
   and TIR means the total movement of the indicator.

When you see Figure 10.32 as part of a drawing, it means that the surface indicated by
the arrow must be flat within .001 TIR.

Remember: Flatness is to itself. You cannot use surface B for reference. If you did reference surface B, you would be measuring parallelism.

To Measure Flatness on the Part

To measure flatness on a surface plate, set the part on three jackscrews (Figure 10.33). (Put
surface B on the jack screws.) Then put the indicator on surface A and set 0 at three points
as shown. Setting zero involves adjustments of the jack screws. This is getting surface A parallel to the surface plate so that the flatness can be measured. After the part is zeroed in to
the surface plate, sweep the indicator randomly over the surface, keeping watch on the TIR.

Jack screws are tools used for setting up the part to be measured. They consist of a base
and a threaded locator so that you can set a part on them and adjust the height simply by
screwing them up or down.

CONCENTRICITY MEASUREMENT

There are times when cylindrical features must be controlled with respect to their location to each other. The two basic reasons for this are *fit* and *function.* Concentricity is a tolerance of location related to cylindrical features.

The symbol for concentricity is ◎. This symbol stands for two cylinders that share the same axis (the true definition of concentricity). Since it is impossible for any two cylinders to share exactly the same axis, the word *eccentricity* can be used to describe features that are not concentric to each other.

It must be noted at this point that concentricity is not supposed to be considered on a full indicator movement (FIM) basis. Concentricity is simply the direct linear distance between two different axes. These axes are the datum axis and the axis of the measured feature. Therefore, a concentricity tolerance that is stated on the drawing as FIM is not correct.

The problem in measuring concentricity with one indicator is that you cannot determine the axis of the feature from one side. You must use two indicators and differential measurement to find the *exact* distance between the axes. This is where concentricity and runout differ from each other. Concentricity is an axis-to-axis relationship, and runout is an axis-to-surface relationship. Concentricity is measured with at least two indicators and differential measurement. Runout is measured with one indicator and direct measurement (TIR).

Example of Concentricity

The drawing in Figure 10.34 means that the *axis,* of .500 in. diameter, must be concentric to *datum axis* A within a cylindrical zone of .003 in.

To Inspect the Part.   First, the datum axis must be established by locating the datum feature and rotating the part. This is normally done with a variety of tools, such as V-blocks, rotary table, precision spindle, centers (if the datums are centers), and others.

For this example we will use a rotary table. Datum diameter A is located in the jaws of the rotary table as shown in Figure 10.35. This will establish the datum axis in rotation. Next, two indicators are placed on the measured diameter (the .500-in. diameter) 180 degrees apart. Both indicators are brought to zero on the .500-in. diameter. This step is shown in Figure 10.36.

In Figure 10.37 the part is rotated for maximum readings on the indicators. The .500-in. diameter axis shows a .0015-in. movement in relation to the datum A axis.

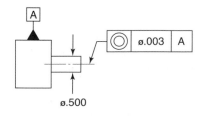

FIGURE 10.34
Concentricity callout.

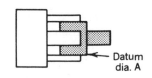

FIGURE 10.35
Simulated datum: three-jaw chuck.

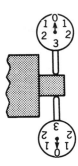

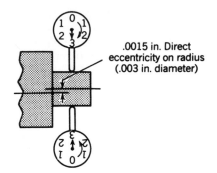

.0015 in. Direct
eccentricity on radius
(.003 in. diameter)

FIGURE 10.36
Indicators set up for a differential
measurement (should be a minimum of two
indicators).

FIGURE 10.37
A .0015-in. direct eccentricity (or .003-in.
diametral zone).

Differential Measurement to Establish Actual Eccentricities

Let's say that at one point during the rotation (shown in Figure 10.37), the indicator on top read +.003 in. and the opposing indicator on the bottom read −.003 in. In this case the actual movement of the axis of the measured feature is .003 in. because both indicators show the same amount of movement. However, in another reading you see that the top indicator shows a +.004 reading and the bottom indicator shows a −.002 reading. The eccentricity is still .003 because the interpolation between the two readings is .003.*

The two indicators will find the true axis position related to the datum axis regardless of problems in the diameter. These problems range from taper to out of roundness and other geometrical conditions that can be misleading to a single indicator reading.

Always be sure of what is required in the measurement. Is it concentricity or runout? They both have a different purpose, symbol, and measurement technique. The odd thing about them is that they are commonly confused with each other.

Note: The specification says that the use of concentricity should be carefully considered. In fact, runout can replace concentricity in several cases.

The inspection of concentricity with two indicators is the minimum. More opposing indicators on one part would be better.

RUNOUT TOLERANCES

Two kinds of runout tolerances are used on machined parts. These are shown with their symbols in Figure 10.38.

Runout tolerances apply to features in rotation and are measured with respect to a datum feature. The datum feature is usually located with a tool that will allow it to rotate true (establishing a datum axis of rotation); then the measured feature surface is compared to this axis.

---

* +.004 in. minus −002 in. = spread or differential of .006 in., and one-half of this total (.003 in.) represents axis eccentricity.

FIGURE 10.38
Runout symbols.

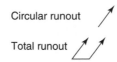

When measuring runout tolerances, an indicator is usually used. This could be a mechanical or electronic indicator, depending on the application. Measuring runout is a matter of understanding what the datum is, locating it properly, and using the indicator on the measured surface in the correct manner. An example of a runout requirement is shown in Figure 10.39.

FIGURE 10.39
Circular runout callout.

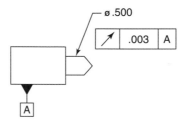

The example in Figure 10.39 means that the .500-in. diameter must not run out over .003 in. TIR. This .003-in. TIR is measured at individual points (if circular runout) or across the entire diameter in one sweep (if total runout).

To Inspect the Part

Runout measurements can be made with several different inspection devices. Some of these are

1. A rotary table
2. A V-block
3. Centers (if the datum is the center of the part)

Note: These devices are "simulated datums."
A rotary table will locate the datum feature and allow it to rotate true so that comparisons can be made. A V-block will do the same thing for a runout on a diameter.

Steps for Circular Runout Measurement

Step 1.   Locate the datum feature in the V-block as shown in Figure 10.40.

Step 2.   Place the indicator tip on the measured diameter at various points along the diameter while rotating the part.

Step 3.   For each point measured (in circular runout) the TIR should be .003 in. or less. Each point measured is evaluated separately from the others.

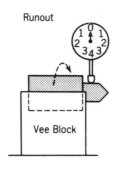

FIGURE 10.40
Measuring circular runout.

FIGURE 10.41
End surface total runout.

Steps for Total Runout Measurement

Step 1.   Locate the datum feature in the V-block as shown in Figure 10.40.

Step 2.   Place the indicator tip on the measured diameter at one end, rotate the part, and move the indicator tip along the entire length of the diameter using a uniform sweeping motion.

During the movement of the indicator, watch for the extreme minus and plus readings. These extreme readings will establish the TIR for total runout.

End Surface Runout

At times, the end surface of a shaft will have a runout tolerance applied to it. This is shown in Figure 10.41. To measure end surface runout, you need a tool that will not allow the shaft to move axially during the measurement. In this case, a V-block will not serve the purpose. A rotary table is suited for this task.

To measure end surface runout, follow the steps outlined below.

Step 1.   Locate the datum diameter in a tool (such as a rotary table) that will not allow the part to move axially.

Step 2.   Place the indicator in the center of the end surface with plus and minus travel available.

Step 3.   Rotate the part and move the indicator tip slowly toward the outer edge of the shaft. Watch for the TIR during this movement. See Figure 10.42 for an illustration of the measurement.

FIGURE 10.42
Measuring end surface total runout.

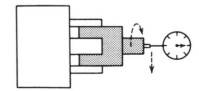

Runout tolerances are very easily measured when you understand the datums and the tolerance zones. Runout tolerances can be very helpful in controlling other geometric characteristics, such as

- Roundness
- Taper
- Cylindricity
- Concentricity
- Straightness

In all cases, the datum that is specified must be located properly, and a variable instrument (such as an indicator) must be used. Runout tolerances are regardless of feature size, so there can be no bonus tolerances applied.

PERPENDICULARITY (SQUARENESS) MEASUREMENT

Perpendicularity is another tolerance of orientation that is required for many component parts. However, some industries use the terms *normality* and *squareness,* which mean the same thing. Perpendicularity is always related to a datum plane or feature. For example, the statement "This surface to be perp. to A dia. within .002 TIR" means that you *must* use A diameter to locate the part for the measurement. The symbol for perpendicularity is ⊥. This symbolizes two surfaces at a 90-degree angle to each other.

Let's review some definitions.

Feature.   Any portion of a part that can be used as a basis for a datum.

Datum.   Points, lines, planes, or other geometric shapes that are assumed to be exact for use as a reference for measurement.

In Figure 10.43, the datum feature is B.

Remember: Datum features (or surfaces) are the surfaces you must locate the part on, not measure the part. The measured surface in Figure 10.43 is the short leg.

In Figure 10.44 the surface can be anywhere in the .002-in.-wide tolerance zone. The vertical line shows what perfect perpendicularity would be. The dashed line shows the width of the zone, which is (.002 in.). The view in Figure 10.44 shows the worst possible condition *within the tolerance zone.*

FIGURE 10.43
Perpendicularity callout.

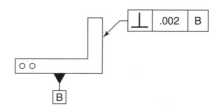

FIGURE 10.44
Perpendicularity tolerance zone.

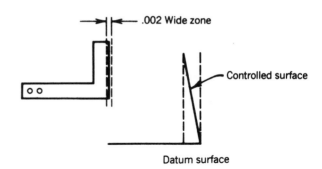

To Inspect the Part

Contact datum B completely on a simulated datum plane (e.g., surface plate); then measure the perpendicularity by using one of the tools shown in Figure 10.45, depending on the tolerance allowed. The machinist's square (Figure 10.45a) and feeler stock can be used when tolerances permit. The cylindrical square (Figure 10.45 b) is easy to use. You simply line up any side of the square that matches the measured surface and follow the *topmost* dashed curve to read the out-of-squareness directly. The cylindrical square is accurate to .0002 in. The master sequences gage (not shown) has a very accurate vertical member that allows you to measure perpendicularity within .0001 in.

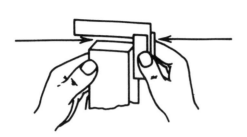

FIGURE 10.45a
Precision square in use. (L.S. Starrett Co. Reprinted by permission.)

FIGURE 10.45b
Cylindrical square. (Brown & Sharp. Reprinted by permission.)

COAXIALITY MEASUREMENT (USING POSITION TOLERANCE)

Coaxiality is a positional tolerance used to control the alignment of two or more holes shown on a common axis when rotation is not involved. In the case below, a bolt must go through two holes at the final assembly. The largest size of these holes can be a .412-in. diameter. If the two holes are not in their true position and/or not at the low limit of size, the bolt will not go through. Therefore, the coaxiality of the two holes is important.

Let's say that in this case, the two holes were drilled at different times in the process. This tells us that their alignment must be inspected in-process to ensure coaxiality.

Coaxiality is the same thing as concentricity because they are both concerned with two features being on the same axis to a certain tolerance. However, using positional tolerance allows you to use MMC and get bonus tolerance.

Example of Coaxiality Measurement

The drawing (Figure 10.46) shows two holes with coordinate basic dimensions locating the position (with basic dimensions). The tolerance zone (where the axis of the holes must lie) is a cylindrical zone .010 in. in diameter with the true position in the center of the zone.

Remember: Perfect true position here is 2.200 in. and 1.100 in. *exactly.* Note that the holes can vary .005 in. from the position in any direction.

According to Figure 10.46, the actual centerline of each hole must lie inside the cylindrical zone shown in Figure 10.47.

The view in Figure 10.48 shows the worst possible condition that the part is allowed to be in (within the tolerance). If one hole centerline is in the extreme of the diametral zone and the other hole centerline is at the opposite extreme *and* they are both at the smallest diameter allowed (.402 in.), it will cause the condition shown at the left of the figure and contribute to the interference of the bolt going through both holes.

To Inspect the Part

Step 1.   First inspect the position of each hole from each of the datums A and B. If they are to size (in diameter) and are in the true position zone, there is no problem.

FIGURE 10.46
Position tolerance: coaxial.

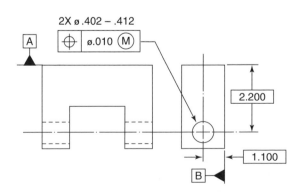

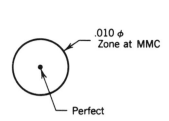

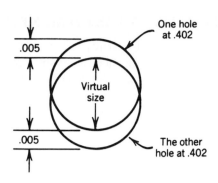

FIGURE 10.47
Coaxial tolerance zone.

FIGURE 10.48
Possible location error (virtual size) at MMC.

Step 2. To inspect quickly to see if the mating part (the bolt) will go into these parts, simply check the virtual size of the two holes. This can be done by calculating the virtual size. Now a functional gage can be made. This gage will be a virtual size pin.

Virtual size = MMC (of hole) − true position tolerance.

So

.402 in. − .010 in. = .392 in.

Now, in Figure 10.49, you have a .392-in. pin that will go through any good part. Remember, this can be done only if the position tolerance is specified at the MMC condition.

FIGURE 10.49
Functional gage for coaxiality at MMC.

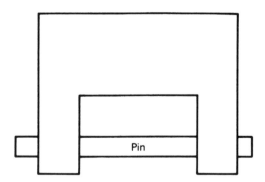

ANGULARITY MEASUREMENT: ANGULARITY VERSUS SIMPLE ANGLES

Angularity is another important tolerance of orientation, which is used quite often when the angle of a certain surface must be tightly controlled. The symbol for angularity is ∠. This symbolizes a surface at an angle to another surface. There is a big difference between angularity toleranced in linear units and a simple angle having a tolerance in degrees and minutes.

The differences between them are shown below. Example 10–3 is a simple angle callout, but Example 10–4 is an angularity callout. Note that they are both always related to a reference surface (or datum).

■ Example 10–3: Simple Angle Callout

Figure 10.50 shows a simple angle callout. This means that the angle of this part must be 30 degrees plus or minus 1 degree. Therefore, the angle could be anywhere from 29 to 31 degrees and still be to print (see Figure 10.51).

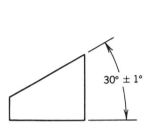

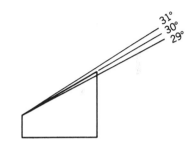

FIGURE 10.50
Simple angle callout.

FIGURE 10.51
Simple angle tolerance zone.

FIGURE 10.52
Bevel protractor measures simple angles.

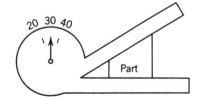

This part can be inspected with a protractor as shown in Figure 10.52. Protractors can show angles in degrees and minutes.

■ Example 10–4: Angularity Callout

Shown in Figure 10.53 is an angularity tolerance of .015 in. The symbol for angularity is shown in what is called a *feature control frame.* The main difference between this angularity and a simple angle callout is the fact that angularity is toleranced in linear dimensions and simple angles are toleranced in angular dimensions.

Angularity within .015 in. gives you a .015-in.-wide tolerance zone. The surface can lie anywhere within the tolerance zone as long as it is in the zone throughout the entire feature. The view in Figure 10.54 shows that a perfect 30-degree angle cuts through the middle of the zone and the zone is equally distributed around the perfect 30-degree plane.

FIGURE 10–53
Angularity callout.

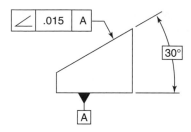

FIGURE 10.54
Angularity tolerance zone.

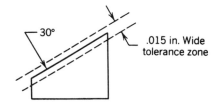

This part requires the use of a sine bar (sine plate) to inspect the 30-degree angle.

The view in Figure 10.55*a* shows the inspection of the part with a sine bar. A sine bar is a simple tool that works using right-triangle trigonometry. You can set up any angle with this tool using gage blocks. An actual sine bar is shown in Figure 10.55*b,* and a sine plate is shown in Figure 10.55*c.* For compound angles, a compound sine plate (Figure 10.55*d*) could be used.

The example in Figure 10.53 shows a part that has a 30-degree "basic" angle from surface A and must be within an angularity tolerance of .015, as shown. To inspect this part, you could use a sine bar set up at 30 degrees. For example purposes only, the length of the sine bar here is 10 in.

Remember: The sine bar forms the hypotenuse of the angle you are working with (see Chapter 14). So, the angle is 30 degrees. The hypotenuse is 10 in. Now solve for the opposite side (sine 30° = .5).

FIGURE 10.55*a*
Angularity measured using a sine bar.

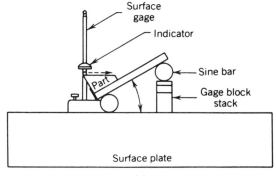

(a)

FIGURE 10.55*b*
Sine bar on a surface plate. (Brown & Sharp. Reprinted by permission.)

(b)

FIGURE 10.55*c*
Sine plate. (Grinding Technology, Inc. Reprinted by permission.)

(c)

FIGURE 10.55*d*
Compound sine plate. (Grinding Technology, Inc. Reprinted by permission.)

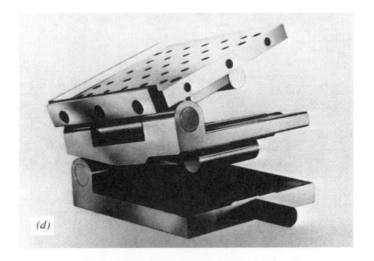

(d)

Formula: Sine 30° × hyp. = opp.; or

$$.5 \times 10 = 5.0 \text{ in.}$$

This tells you that you must put 5 in. of gage blocks under the sine bar to set up a 30-degree angle.

The picture in Figure 10.55*a* shows how the part is inspected. As you can see, the angled surface of the part becomes parallel to the surface plate, so you can simply run an indicator across it to obtain the *total indicator reading.* If it is less than .015 in., the part is to print. There are other devices used to measure angularity such as angular gage blocks, rotary tables, and indexing heads.

Remember: The length of the sine bar used must be known when you are using the formula. If the bar were only a 5-in. length, you would have to put only 2.5 in. of gage blocks under it to attain a 30-degree angle.

Understanding the right triangle concepts and knowing that the sine bar forms the hypotenuse of the triangle will allow you to use the sine bar effectively to measure angles toleranced by angularity.

Note: A further study of measuring angles will describe alignment datums that may be necessary. If the part is rotated on the sine bar, you can get any reading you want. Secondary alignment datums are not mentioned in any specification but should be considered.

SYMMETRY

Symmetry is another location tolerance. It is defined as a condition in which a feature (or features) is symmetrically disposed around the center plane of a datum feature. The symbol for symmetry is ⎓ . This symbolizes two like lines that are both exactly the same distance from the line in the center, so they are symmetrical.

Figure 10.56 shows a part with a slot in it. This slot, per the drawing, must be symmetrical within .010 in. to datum feature B. In short, the actual centerplane of the slot must lie within a .010-in.-wide tolerance zone that is equally disposed around the centerplane of the datum feature B, as shown in Figure 10.57.

FIGURE 10.56
Symmetry callout (use position symbol).

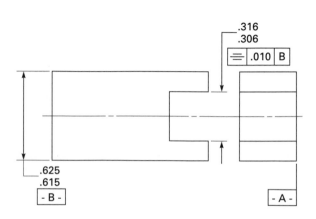

FIGURE 10.57
Symmetry tolerance zone (RFS).

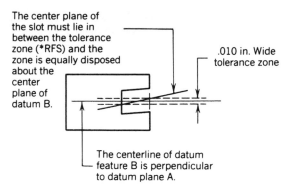

The center plane of the slot must lie in between the tolerance zone (*RFS) and the zone is equally disposed about the center plane of datum B.

.010 in. Wide tolerance zone

The centerline of datum feature B is perpendicular to datum plane A.

*RFS—means regardless of feature size.

To Inspect the Part

Let's say that we measure the B dimension and it is .620 in. We measure the width of the slot and it is .310 in. Therefore, the thickness per side should be .620 in. minus .310 in. and the answer divided by 2. So each side would be .155 in. thick (as shown in Figure 10.58) if the slot and datum B were perfectly symmetrical.

Note: It is important to understand at this point that if the centerplane of the slot were to move by .001 in., the thickness on one side would be .154 in. and the other side would be .156 in. thick in Figure 10.58.

FIGURE 10.58
Example of perfect symmetry.

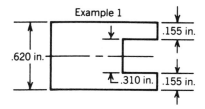

Example 1

.620 in.

.310 in.

.155 in.

.155 in.

To Measure the Part

You simply measure the thickness of each side, subtract the smallest thickness from the largest, and divide your answer by 2. The answer you get will be the actual distance that the centerplane of the slot has moved away from the centerplane of the datum, and it is allowed to move only .005 in. in either direction with a .010 total tolerance zone.

Let's say that the part in Figure 10.59 measures as shown. One side is .160 in. thick and the other side is .150 in. thick. So .160 in. minus .150 in. equals .010, and that divided by 2 equals .005 in. that the centerplane has moved. Here the part is acceptable because the centerpiece is allowed to move up to .005 in., but it is at the worst allowable condition.

FIGURE 10.59
Borderline acceptable part.

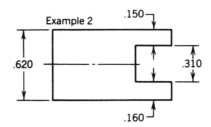

STRAIGHTNESS (OF AN AXIS)

Straightness is another important tolerance of "form" when it is related to mating parts. The symbol for straightness is ⸻, symbolizing a straight line (or axis). For example, Figure 10.60 means that the pin can range in diameter from .615 in. to .605 in. and can be straight up to .015 in. when it is at MMC. (MMC is .615 in. in this case.)

FIGURE 10.60
Straightness of an axis at MMC.

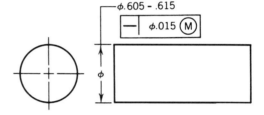

The sketch in Figure 10.61 shows the worst condition that the part can be and still be acceptable (its virtual condition).

Remember: The virtual size equals the MMC plus the tolerance. So

.615 in. + .015 in. = .630 in.

Note: The drawing says that the part is to be straight within .015 in. at MMC. Therefore, if the actual part is smaller than MMC, the straightness tolerance (per ANSI) increases by the same amount that the diameter decreases. For example, if the pin diameter were ac-

FIGURE 10.61
Diameter versus virtual size.

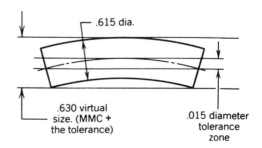

tually .610 in., the diameter would be .005 in. smaller than the MMC. So the straightness tolerance would increase to .020 in. automatically (.015 in. + .005 in.).

To Inspect the Part

This part could be inspected on the surface plate, or a functional gage could be made that has a .630 inside diameter.

Note: The diameter of the gage must be longer than the pin itself. This gage is simulating the virtual condition (see Figure 10.62).

FIGURE 10.62
Part is at MMC size and perfectly straight.

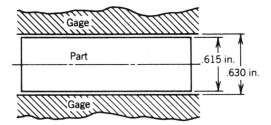

As you can see in Figure 10.62, a straight pin will fall through the gage, because its virtual size is simply its largest size. Also, in Figure 10.63, you can see that the pin at MMC and .015 in. out of straight will still fall through. In Figure 10.64, observe that the pin is at its lowest allowable size and out of straight by .025 in., but will still go through because the gage allows for the bonus tolerance you get when the part is under MMC.

FIGURE 10.63
Part is at MMC size and bent 0.015 in.

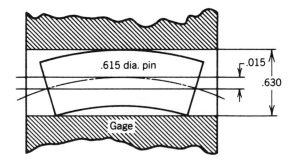

Hence, inspecting the part this way simply involves measuring the diameter of the pin and attempting to slip it through the gage. If the diameter is within its size limits and will go through this gage, this part is straight within tolerances. Always measure the diameter!

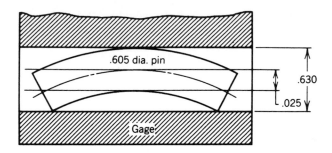

STRAIGHTNESS MEASUREMENT: STRAIGHTNESS OF SURFACE ELEMENTS

Another control of straightness is that which is applied to the elements of the surface only (instead of its axis). When straightness is applied to surface elements, neither MMC nor RFS applies because a surface element has no size.

When straightness is applied to surface elements, the feature control frame will have a leader line pointing to the surface, as shown in Figure 10.65. In the figure the tolerance zone is two perfectly parallel imaginary lines .002 in. apart. The pin (during measurement) is rotated such that all elements of its surface are passed between these lines and no portion can extend beyond the tolerance zone, as shown in Figure 10.66.

Measuring straightness of surface elements can be done in two basic ways, with the optical comparator and the surface plate setup.

Comparator

With a comparator, the surface of the pin is projected and an overlay of the tolerance zone is used on the screen. The pin is then rotated 360 degrees and each surface element must lie between the two tolerance zone lines.

Surface Plate Setup

Two jack screws equipped with V-anvils can set up the straightness plane in one element. This setup is shown in Figure 10.67.

Once you establish the straightness plane (in each element), you simply sweep the surface and it should be within the stated tolerance across the surface. This measurement

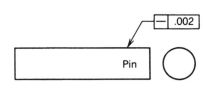

FIGURE 10.65
Straightness of surface elements callout.

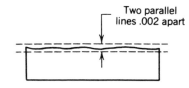

FIGURE 10.66
Tolerance zone for the straightness of surface elements.

FIGURE 10.67
Straightness of surface elements measured with jack screws to set up the optimum line element.

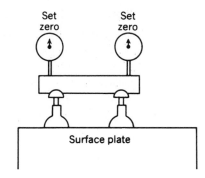

would then be seen using TIR (FIM) of the indicator after establishing 0. Rotate the pin to another element and repeat the procedure. Obviously, use of the optical comparator is a much faster method.

Remember: Straightness of surface elements demands variable measurement. Because MMC cannot apply, no functional gage can be used.

USING PROFILE TOLERANCES FOR COPLANARITY

There are times when it is important that two or more interrupted surfaces be in the same plane, or *coplanar.* Coplanarity can be accomplished easily using *profile of a surface tolerances.* Coplanarity is generally applicable when two or more surfaces are interrupted by a feature or features such as slots, holes, pins, tabs, and so on.

■ **Example 10–5**

In this example we have a part that has two surfaces interrupted by a slot. It is important that the two surfaces be coplanar. So we apply profile of a surface tolerance to the part as shown in Figure 10.68.

The tolerance zone is two parallel planes that are .005 in. apart; both surfaces must lie between these two planes, as shown in Figure 10.69.

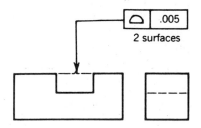

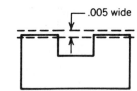

FIGURE 10.68
Coplanarity control using profile of a surface.

FIGURE 10.69
Tolerance zone for the part.

■ **Example 10–6**

In this example a part has several slots cut into it and two important surfaces that establish the datum plane for measurement. This part is shown in Figure 10.70.

In this case, the tolerance zone is two parallel planes that are .005 in. apart. The tolerance zone is equally distributed around the datum plane established by datums A and B simultaneously. This is shown in Figure 10.71.

FIGURE 10.70
Coplanarity: multiple datums.

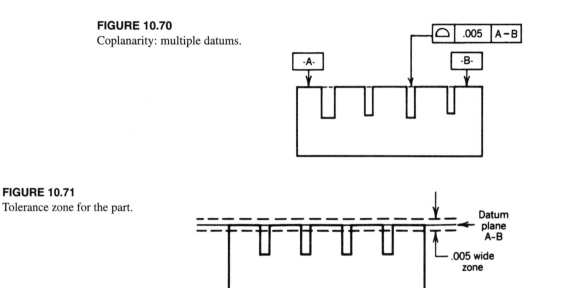

FIGURE 10.71
Tolerance zone for the part.

Measurement

In Example 10–5 we need to establish one plane and compare the other plane to it. This can be done by direct location of one of the surfaces on the surface plate and sweeping the other surface with an indicator.

In Example 10–6 we must locate datums A and B and compare the other surfaces to them. This can be accomplished by locating datums A and B on a set of parallels (upside down), then sweeping the other surfaces with an indicator.

Coplanarity is well defined when using profile of a surface tolerancing, and the datums and measured surfaces are very clear.

REVIEW QUESTIONS

1. Which of the following methods is used to inspect flatness?
   a. three-wire method
   b. jack screw method
   c. three parallel planes
   d. two datums method

2. Which of the following geometric controls requires one composite TIR value for measurement?
   a. concentricity
   b. circular runout
   c. axial position
   d. total runout

3. Which of the following geometric controls can be used to control coaxial features?
   a. concentricity
   b. position
   c. runout
   d. all of the above
   e. a and c only

4. Which of the following geometric tolerances can be inspected using a sine bar?
   a. angularity
   b. perpendicularity
   c. parallelism
   d. flatness

5. The primary measurement that is made on a surface plate is
   a. outside diameter
   b. parallelism
   c. flatness
   d. height
   e. all of the above

6. Which of the following geometric controls is used to control coplanar surfaces?
   a. position
   b. profile of a surface
   c. profile of a line
   d. parallelism

7. Which of the following geometric controls replaced the symmetry symbol in the ANSI Y14.5M–1982 standard?
   a. concentricity
   b. parallelism
   c. position RFS
   d. profile

8. Which of the following accessories to a surface plate can be used to contact two datum surfaces of a part that are interrupted by a raised surface between them?
   a. matched V-blocks
   b. fixed precision parallels
   c. right angle plates
   d. precision square

9. Which of the surface plate accessories can be used to contact multiple datum diameters?
   a. matched V-blocks
   b. fixed precision parallels
   c. right angle plates
   d. precision square

10. What is the proper way to remove items from the surface plate?
   a. Lift them off carefully.
   b. Use both hands.
   c. Slide them off.
   d. Get help.
11. Parallelism must be inspected with respect to at least one _____.
   a. point
   b. plane
   c. line element
   d. datum
12. Primary datum surfaces are effectively contacted by mounting them on the surface plate because
   a. it is a stable surface
   b. the three or more highest points are contacted
   c. it has no raised material
   d. none of the above
13. Concentricity measurement involves
   a. direct readings
   b. TIR
   c. FIM
   d. differential measurement
   e. FIR
14. Which of the following tolerance zone shapes apply to concentricity?
   a. two parallel lines
   b. a cylinder
   c. two parallel planes
   d. axial belt
15. Which of the following geometric errors are automatically controlled by total runout tolerance?
   a. concentricity
   b. roundness
   c. taper
   d. all of the above
16. Using a gage block stack on the surface plate to compare the height of a part to the height of the stack is called _____ measurement.
   a. transfer
   b. differential
   c. inferential
   d. circumferential
17. Which of the following geometric controls involves a 90 degree angle?
   a. angularity
   b. profile
   c. parallelism
   d. perpendicularity
   e. none of the above

18. Which of the following types of tolerances has a vertex?
   a. angularity
   b. perpendicularity
   c. simple angles
   d. none of the above
19. What type of tolerance zone applies to straightness of an axis tolerance?
   a. two parallel planes
   b. cylinder
   c. two parallel lines
   d. two dots
20. What tolerance zone applies to straightness of surface elements?
   a. two parallel planes
   b. cylinder
   c. two parallel lines
   d. two dots
21. What is the height of a gage block stack that should be placed under a 10 in. sine bar to obtain a 12-degree angle? (Round answer to the nearest .0001 in.) _____
22. What is the height of a gage block stack that should be used to set up a 30-degree angle with a 5 in. sine bar? _____
23. What angle would result if a 5.000 in. gage block stack were placed under a 10 in. sine plate? _____
24. (Yes or no) Is it possible to have alignment problems that cause error in the TIR value when measuring angularity with a sine bar? _____
25. What is the height of a gage block stack that should be placed under a 3 in. sine bar to obtain a 30-degree angle? _____

11 Special Measuring Equipment and Techniques

This chapter covers various types of special equipment and techniques for measuring product characteristics that are different and more sophisticated, beyond the ability of the hand tools covered in Chapter 9 or the surface plate methods covered in Chapter 10. The special measuring and test equipment covered in this chapter include electronic equipment, pneumatic (air) gaging equipment, coordinate measuring machines, hardness testing equipment, roundness measuring machines, optical flats, and so on.

DIGITAL MEASURING EQUIPMENT

There is a variety of measuring and test equipment with electronic digital displays of measured values. Some have the ability to automatically transmit values obtained during the measurement into computers (via certain types of interface devices). The ability to transmit data into computers can enable users to perform certain functions more efficiently, such as:

- Regulating the process through in-process (on-line) gages
- Automatically preparing statistical control charts
- Producing inspection reports and printouts
- Performing other types of special analyses of product or process data

The use of this type of equipment also improves the timeliness of obtaining and using data for the aforementioned purposes. Data from measurements taken on a process variable, for example, are obtained in real time and can be fed through interface devices to a computer for making real-time process decisions.

Electronic digital equipment ranges from small hand tools to special measuring machines and testing equipment covered later in this chapter. Typical hand tools such as micrometers, calipers, and dial indicating gages (not shown here) are all available with electronic digital display from a variety of gaging manufacturers. Surface plate accessories such as height gages and/or accompanying dial indicators are also available. The more complex measuring instruments and measuring machines are equipped with electronic digital displays, and most have the capability of transmitting the data to computers.

Discrimination

Discrimination of electronic digital devices provides added value to the use of the tool. Where older vernier (or dial) calipers discriminated to .001 in., the electronic caliper discriminates to .0005 in. Most of these electronic gages also give the operator the option to switch from English to metric discrimination (such as inches to millimeters) by pushing a button. Some of the electronic indicators are also capable of producing a TIR (Total Indicator Reading) value automatically displayed for measurements that require TIR results.

Benefits of Electronic Instruments

The benefits of using electronic digital display instruments include:

- They are easier to read.
- They can transmit data for real-time use (such as automatic preparation of SPC charts; see Chapter 13).
- They have better (finer) discrimination.
- They can be used with a computer to produce automatic inspection reports.

Pitfalls of Electronic Instruments

There are some pitfalls to using electronic instruments:

- Battery replacement is necessary.
- They are more delicate than vernier or dial instruments.
- Repair costs are generally higher.
- Most electronic digital instruments require special interface devices to transmit data.
- Cables or transmitters needed to "hook up" the tool tend to confine the use of the tool to a specific location.

Although examples of hand tools with electronic digital display were not available at the time of writing of this edition, refer to the topics of air gages, roundness measuring machines, and coordinate measuring machines covered later in this chapter for examples of electronic digital displays.

The remaining pages in this chapter are devoted to discussing special measurement equipment and techniques for dimensional measurement. The techniques covered are not all-inclusive, nor are they the only way to measure the characteristics discussed, but they are popular methods for measurement. Once again, the term *special measuring equipment,* as it is used in this book, means equipment that is used for making dimensional measurements that are beyond the capability of the simple hand tools or surface plate methods covered in Chapters 9 and 10.

FLATNESS MEASUREMENT (USING OPTICAL FLATS)

There are times when flatness must be measured in millionths of an inch. Some examples that require this degree of accuracy are:

1. The measuring faces of gage blocks
2. The anvil and spindle faces of a micrometer
3. Other parts with lapped surfaces

FIGURE 11.1
Measuring flatness with optical flats. (M.T.I. Corp. Reprinted by permission.)

When you are working with surfaces made of material that reflects light and you need flatness measured in millionths, an optical flat may be used. Optical flats (as shown in Figure 11.1) come in various diameters from 1 to 12 in. and various thicknesses from $\frac{1}{4}$ to 1 in. They are generally made from Pyrex glass or fused quartz.

Optical flats can be purchased in various degrees of accuracy, from a reference grade that is flat about 1 millionth of an inch, to a commercial grade that is flat to about 8 millionths of an inch. You can also get optical parallels, which have two measuring faces that must be parallel to each other.

Methods

There are two methods for measuring flatness with optical flats: the *air wedge method* and the *contact method*. In the air wedge method, the flat is held at a very small angle to the surface of the workpiece. In the contact method, the flat is in full contact with the surface being measured.

Visual Measurement

Optical flats measure flatness visually. The observer must be able to interpret interference bands (or fringes) to determine the flatness of a surface. For example, a surface that is perfectly flat will reflect straight, evenly spaced fringes through the optical flat as shown in Figure 11.2.

If the bands are curved and not evenly spaced, this indicates that the surface is not flat. Bands should be viewed from a recommended distance of 10 times the diameter of the flat being used and they should be viewed perpendicular to the flat itself. The lighting used should be a monochromatic light (or a light with one wavelength), such as a light utilizing helium gas. Some optical flats are coated to block out other light sources.

FIGURE 11.2
Straight and equally spaced bands
appear if the surface is flat. (M.T.I.
Corp. Reprinted by permission.)

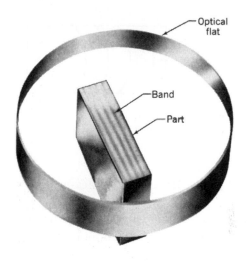

Reading the Bands

When the bands are observed in the recommended manner (perpendicular to the flat), the distance between each band is 11.57 millionths of an inch. In decimal form, this would be .00001157 in. The surface in Figure 11.3 represents a surface that is out of flat by one band, as shown by the dashed line that just touches the next band. This surface is out of flat by 11.57 millionths and the surface is convex.

Another example is if a part were to show a three-band flatness error. The amount of flatness error is then equal to

$$3 \times .00001157 \text{ in.} = .000035 \text{ in. (or 35 millionths)}$$

Figure 11.4 shows a surface that is two bands out of flatness (which means about 23 millionths).

Is the Surface Concave or Convex? When reading the bands, you must be able to tell if the surface is concave or convex. In the air wedge method, if the bands curve around the point of contact, as shown in Figure 11.5, the surface is convex. If the bands curve around the air wedge itself, as shown in Figure 11.6, the surface is concave.

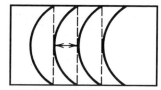

FIGURE 11.3
A one-band error equals 11.7 millionths if
viewed at a 90-degree angle to the flat.

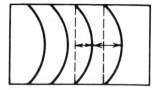

FIGURE 11.4
A two-band error equals 23.4 millionths.

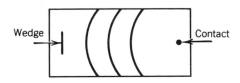

FIGURE 11.5
Wedge method: surface is convex.

FIGURE 11.6
Wedge method: surface is concave.

Basic Steps in Measuring Flatness with an Optical Flat

Step 1. Make sure that the surface of the workpiece has no burrs or nicks.

Step 2. Remove any dust that may be on the workpiece or the flat.

Step 3. Put a clean sheet of paper over the surface to be measured.

Step 4. Place the optical flat on the paper. (This is to protect the flat from possible damage when placing it on the workpiece.)

Step 5. Holding the optical flat in place, carefully slide the paper out from between the flat and the workpiece.

Step 6. Look for interference bands (fringes). If you do not see any, repeat steps 1 through 6.

Step 7. Observe the bands perpendicular to the flat and from a distance of 10 times the diameter of the flat being used.

What If Bands Do Not Appear?

If bands do not appear, there may still be dust or nicks on the mating surfaces that are preventing close proximity of the surfaces. Try repeating steps 1 through 6. There may be too much of a wedge (gap) between the workpiece and the flat. Press down on the flat in a uniform manner, and the bands may appear. There may still be moisture or an oil film that is causing the flat to "wring" to the surface. There may also be too great an angle between the workpiece surface and the flat. Applying pressure at different points around the outer edge of the flat may cause bands to appear.

The Distance Between the Bands

Sometimes it is difficult to see the exact distance between the fringes. The thin imaginary line to do this could be simulated by attaching a piece of thread or fine wire to the monochromatic light source and lining up the thread or wire with the bands.

Irregular Surfaces: The Contact Method

There are times when surfaces (which are both convex and concave and have random peaks) are irregular. In such cases the contact method is recommended, and you should not attempt to maintain a wedge. The same steps should be followed as mentioned earlier except that you press down evenly around the flat to allow it to come in full contact with the surface being measured. The flat will make contact at the high points of the surface. These high points will show up as a round band. The flatness error in this method is measured by counting each band (including one side of the high spots) and dividing the total number of bands by 2 (see Figure 11.7).

Cautions When Using Optical Flats

1. The amount of 11.57 millionths (stated earlier) applies only when you view the bands perpendicular to the flat. If you view them at an angle, the amount is different. Refer to measurement handbooks to view bands at an angle.
2. All surfaces of the part and the flat must be clean.
3. Make sure that the flat and the part have reached temperature equilibrium (the same temperature) before making the measurement.
4. Take good care of the optical flat. Avoid damaging it.
5. Use the proper light source (monochromatic light).
6. If bands do not appear, do not wring the optical flat to the surface. It may be that the surface is not flat enough for bands to appear.

When Surfaces Are Both Concave and Convex

When surfaces are both concave and convex, a fringe pattern like the one in Figure 11.8 will appear. The fringe pattern in Figure 11.8 is, of course, dependent on where the point of contact and wedge are located.

When they are used properly, optical flats can be an accurate method of measuring flatness. The most frequent source of errors made when using optical flats is related to the technique used and the interpretation of the bands.

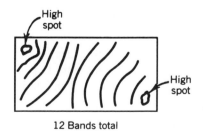

FIGURE 11.7
Contact method: 12 bands total.

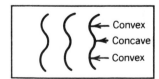

FIGURE 11.8
Concave–convex.

Optical Parallels Used in Calibration

Optical parallels come in very handy when calibrating measuring tools such as the micrometer. Errors in measuring with the micrometer can occur if the anvil and spindle faces are worn such that they are not parallel to each other. Optical parallels can measure this directly. Remember, however, that the anvil and spindle faces could be parallel to each other and still not be perpendicular to the axis of measurement.

MEASURING SURFACE FINISH

Most component parts today require particular surface textures, depending on how they are to be used. In the old days, this surface roughness was commonly referred to as "coarse" or "fine." Today, the surface texture of a component part is more stringently controlled in many applications and requires more specific classifications than simply coarse or fine.

Before we discuss surface finish measurement, some definitions must be considered. These definitions are listed below.

Roughness. Closely spaced irregularities on the surface.

Waviness. Widely spaced irregularities on the surface.

Lay. The direction of the predominant surface pattern. (Surface finish measurements should be made in a direction that is perpendicular to the lay of the surface pattern.)

Flaws. A defect in the surface that is not part of the surface roughness or lay. This may be a deep tool mark, a void on the surface, or other flaws. The effect of flaws should not be included in roughness average measurements.

Roughness width cutoff. A dimension that distinguishes surface roughness from surface waviness for a particular distance. This cutoff length must be preset prior to using the profilometer. When the cutoff length is not specified on the drawing, you would, in most cases, use a .030-in. cutoff length.

Arithmetic average (AA). The average distance between the peaks and valleys of surface roughness.

Cutoff. The electrical response characteristic of the instrument that is selected to limit the spacing of the surface irregularities to be included in the roughness measurement (.030 in. if none specified).

Surface Finish and Lay Symbols

Certain symbols are used to describe the direction of lay for the surface. These are used in conjunction with the surface finish symbol shown in Figure 11.9.

The symbols appear under the surface finish symbol and designate the lay of the surface. The symbols are shown in Figure 11.10.

FIGURE 11.9
Basic surface finish symbol.

FIGURE 11.10
Lay symbols.

//   which means a parallel lay.

⊥   which means a perpendicular lay.

C   which means a circular lay.

R   which means a radial lay.

M   which means a multi-directional lay.

X   which means an angular lay in both directions.

FIGURE 11.11
Examples of lay directions.

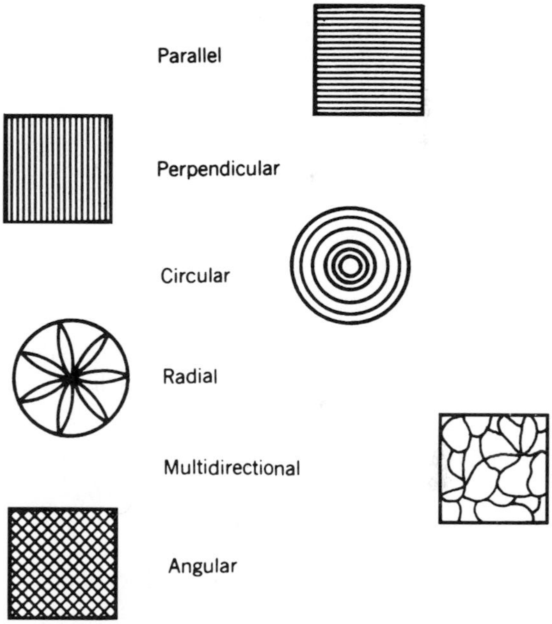

Parallel

Perpendicular

Circular

Radial

Multidirectional

Angular

FIGURE 11.12
Other surface finish symbols.

Surface may be produced by any method.

Material removal by machining is required.

Material removal is prohibited.

Examples of each lay (per the symbols in Figure 11.10) are shown in Figure 11.11. Figure 11.12 shows other symbols.

Always remember that roughness measurement is usually performed in a direction that is at a 90-degree angle (perpendicular) to the direction of the lay.

The Difference Between Roughness and Waviness

As stated earlier, roughness is considered closely spaced, while waviness represents widely spaced irregularities. Figure 11.13 helps show the difference between roughness and waviness.

FIGURE 11.13
Waviness versus roughness.

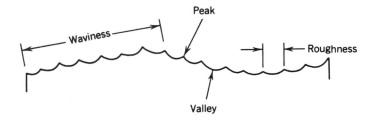

Peaks and Valleys.   Some other terms also require definitions. They are not unlike the definitions used when discussing the land around us. In the mountains, a peak is considered to be the highest point and a valley is considered to be the lowest point. It is the same for surface texture.

Surface texture takes the shape of several peaks and valleys. These peaks and valleys, when magnified, look somewhat like threads from a side view (as is shown in Figure 11.13). The measurement of surface roughness is generally the average height of these combined peaks and valleys.

Surface Finish Measurement

In this chapter we do not intend to describe the many methods in which surface finish can be measured. Two methods, however, enjoy wide use in the shop: a *fingernail comparator* and a *profilometer*.

Fingernail Comparators.   In those situations where surface finish is widely toleranced, fingernail comparators (Figure 11.14) are usually used. These fingernail comparators come in a variety of different styles. One style provides several coins with each coin representing a different surface roughness. Other types are a sheet of metal or plastic that has a variety of roughness samples that you can compare to the part. Each sample is identified with a finish designation, for example, AA32.

They are commonly called fingernail comparators because you run your fingernail across the comparator, then across the surface of the part. Run your fingernail across the comparator (perpendicular to the lay) and you will feel the roughness. Then run your fingernail across the surface of the part (also perpendicular to the lay) and decide which surface is rougher. If the comparator feels rougher than the part, it is likely that the part is acceptable.

■ Example 11–1

The drawing calls for an AA32 (arithmetic average) finish. You scratch the comparator in the area marked AA32 with your fingernail, then scratch the surface of the part. The part surface seems smoother to the fingernail than the comparator was. This means that the part surface is better than an AA32 finish.

Fingernail comparators are widely used (when applicable) because they are simple and inexpensive.

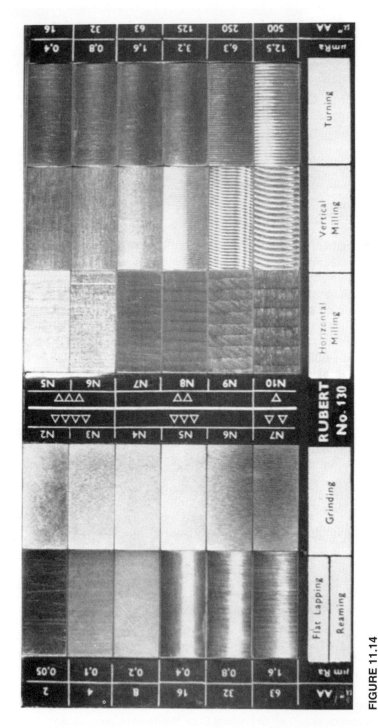

FIGURE 11.14
Surface roughness comparison standard (fingernail comparator). (Photo by James Pluma.)

FIGURE 11.15
Profilometer in use. (Federal
Products Corp. Reprinted by
permission.)

Electronic Instruments for Surface Finish Measurements.   There are several different
types of electronic instruments used to measure surface finish. The one that is used primarily is the profilometer (Figure 11.15). This instrument measures surface finish on a *variable* measurement basis; in other words, the actual average surface of a part can be measured. There are several things to consider when using the profilometer.

1. *Treat the stylus with care.* The stylus of the profilometer in Figure 11.16 is the part that
   actually follows the surface. It is a diamond-tipped, conical-shaped part that is small
   enough to get into the peaks and valleys of the surface.

FIGURE 11.16
Profile tracer stylus. (Federal Products Corp. Reprinted by permission.)

2. *Set the instrument to read your roughness value.* Most profilometers have a switch on the dial to convert the dial to read whatever range of roughness you are measuring, as shown in Figure 11.15. There are generally two scales. The top scale goes from 0 to 30 and the bottom scale goes from 0 to 10. Depending on the setting, the multiplier 30 can mean 300, or 10 can mean 100 finish.

■ Example 11–2

When the selector is set on 10, you can read the bottom scale directly. When the selector is set on 100, you would read the bottom scale as 0, 10, 20, 30, 40, 50, 60, 70, 80, 90, 100 (in other words, add a zero to the numbers shown, which are 0, 1, 2, 3, 4, 5, 6, 7, 8, 9, 10).

When the selector is set on 30, read the top scale directly. It is 0, 10, 20, 30. When the selector is set on 300, read the top scale by adding 0 to all of the numbers. So the top scale, 0, 10, 20, 30, now becomes 0, 100, 200, 300.

By selecting the proper position for the application, you can use the same scale for larger readings.

■ Example 11–3

You want to measure a surface finish of AA250. Set the selector at 300 and read the top scale (which is 0, 10, 20, 30). If the top scale reads 20 (and you are set at 300), it means 200.

---

3. *Make sure that you have set the prescribed cutoff length of the instrument.* Using the cutoff selector (Figure 11.15), you can preset the prescribed cutoff for the measurement. Usually, this cutoff is 0.030 in. but can be other lengths, such as 0.010 in., 0.100 in., and so forth.
4. *Make sure that the stylus is tracing perpendicular to the lay.* The stylus should be tracing in a direction that is perpendicular to the lay of the pattern. In other words, it should cross over the lay (see Figure 11.17).
5. *Make sure that the stylus is set to a stroke that will not cause it to fall off the surface being measured* (Figure 11.18). A poorly set stroke length can damage a stylus by causing it to fall off the surface being measured. There is usually a knob on the instrument that sets the stroke distance the stylus will travel during the measurement. It moves back and forth along this distance. Locate this knob and set the stroke according to your requirements and the part you are measuring. For accuracy, the stroke length must be at least five times the cutoff length.
6. *Never allow the stylus to drop onto the part; set it on gently.* This is to avoid damaging the stylus.
7. *The arm holding the stylus should be as parallel as possible to the surface being measured.* This ensures that the stylus is in good contact with the surface.
8. *Locate the workpiece securely* (Figure 11.19). The workpiece should be located so that it will not move around during the measurement. On cylindrical pieces, it is good practice to use a V-block or at least support the part on both sides so that it will not move.

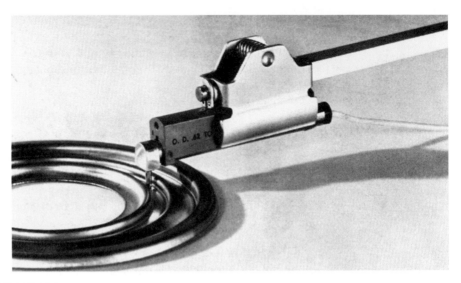

FIGURE 11.17
Tracing perpendicular to lay. (Federal Products Corp. Reprinted by permission.)

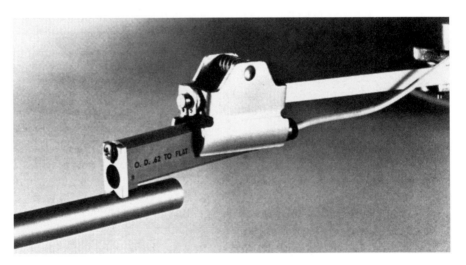

FIGURE 11.18
Dual skid tracer. (Federal Products Corp. Reprinted by permission.)

9. *Use a stylus with the proper type of skids* (Figure 11.20). The *skid* on a profilometer is that part just under the stylus that references (or rides along the surface) so that the stylus can follow the peaks and valleys of the surface. There are basically two types of skids:

a. *Single skid.* This is basically for flat surfaces, where there is no probable side-to-side movement.

FIGURE 11.19
Checking surface finish on an inside diameter. (Federal Products Corp. Reprinted by permission.)

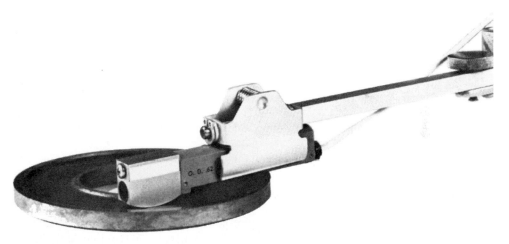

FIGURE 11.20
Double skid tracer in use. (Federal Products Corp. Reprinted by permission.

    b. *Double skid.* This is where there are two skids surrounding the stylus. It is used mainly on cylindrical parts to assist the stylus in avoiding side-to-side movement. The dual-skidded tracer is recommended in all cases.

10. *Calibrate the profilometer periodically* (Figure 11.15). There is usually a calibration patch supplied with the profilometer that is used to check calibration of the instrument. This block has a known surface finish, and the profilometer should read very close to this finish when the stylus is moved across the patch (within tolerances that are usually stated on the patch). The statistical average of a number of measurements on the calibration patch is recommended.

Basic Steps to Measuring with a Profilometer

Step 1. Turn the profilometer on and let it warm up.

Step 2. Set the prescribed range for the measurement.

Step 3. Set the prescribed cutoff length (if none is specified, you use 0.030 in.).

Step 4. Set the stroke length of the stylus according to the part you are measuring. This should be five times the cutoff length.

Step 5. Set the unit of measurement. This is usually AA (which means *arithmetic average*).

Step 6. Position (and secure if necessary) the part being measured. Remember, cylindrical parts should not be allowed to roll.

Step 7. Make sure that you have the correct stylus for the finish being measured.

Step 8. Set the arm (holding the stylus) parallel to the surface being measured.

Step 9. Make sure the stylus is in the *back* position of movement. Holding up the stylus by hand, turn the machine on and stop it when the stylus is back all the way.

Step 10. Gently rest the stylus on the part.

Step 11. Turn the machine on and allow the stylus to trace.

Step 12. Read the dial (at the proper scale) for the surface finish. The average roughness is the mean reading about which the needle tends to dwell, or fluctuate, under small amplitudes.

Surface texture is of prime concern in many industrial applications today. The important thing is to know your particular application and equipment to obtain the desired results.

PNEUMATIC COMPARATORS (AIR GAGES)

Pneumatic comparators (commonly called air gages) are often used for tight-tolerance measurements on machined and other precision parts. A pneumatic gage uses an air supply of constant pressure, an indicator, and specific-size nozzles in its operation. There are several kinds of pneumatic gages on the market, and they all have certain common limitations, such as:

1. Gage heads that accompany air gages have limited measuring ranges.
2. The air gage must have a master (such as a master setting ring for measuring diameters).
3. Air gaging requires a specific level of surface finish to be effective. Rough surface finishes (or porous conditions) limit the effectiveness of the gage.
4. The air supply must be clean and dry for the gage to function properly.

Air gaging (Figure 11.21) has several advantages:

1. The training involved in using an air gage is minimal.
2. Variables data can be obtained on several product dimensions quickly and accurately.
3. Air gaging, in the proper environment, can achieve accuracies as small as .000010 (ten-millionths) inch.

FIGURE 11.21
Pneumatic comparator. (Courtesy of Federal Products Corporation)

4. Important part surfaces can be gaged without contacting or damaging the surface and with little wear on the gage member.

5. The air gage tends to blow loose contaminants from the surface to avoid measurement errors associated with dirt, coolant, and other contaminants.

6. The gage can be rotated (or traversed) when checking dimensions so that geometric error (such as out of roundness, taper, and barrel shape) can be detected.

7. Measurement pressure is constant, as opposed to manual gages, which depend on operator skill.

8. One or more air nozzles can be arranged in a variety of locations for effective gaging. An example of this is three air nozzles 120° apart in an air gage for diameters. This configuration is optimal for detecting trilobe diameters. There are air gages today that have been placed into granite surface plates for comparative measurements of flatness.

Air gaging heads can be obtained in many styles (such as those in Figures 11.22 and 11.23). The common use for air gage heads are plug gaging (for inside diameters), ring gaging, snap gaging (for outside diameters), and other variations, such as the flatness checking instrument mentioned previously. Several companies have used air gaging heads in multiple-dimension inspection fixtures for fast and accurate inspections.

Each gaging head can have several air jets, depending on the application. Heads with one jet are often used for TIR measurements such as runout, flatness, straightness, and position (or location). They are also used for linear measurements such as length or depth. Heads with two jets are often used for diameters (and geometric errors in diameters such as roundness, taper, barrel shape, etc.) and thickness measurements. Based on the limitations and advantages previously stated, pneumatic comparators (air gages) have a wide variety of applications in precision dimensional measurement.

FIGURE 11.22
Pneumatic comparator heads.
(Courtesy of Federal Products
Corporation.)

FIGURE 11.23
Pneumatic comparator head.
(Courtesy of Federal Products
Corporation.)

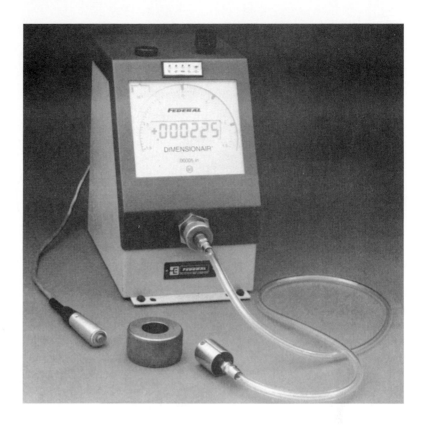

CIRCULARITY (ROUNDNESS) MEASUREMENT

Roundness is a characteristic that is not easily measured. Features assume a variety of out-of-roundness errors while the part is being made, depending on the type of process and several other variables. These different "shapes" are caused by lobes or high areas around the circumference of the part. The lobes can be symmetrical (180 degrees apart) or nonsymmetrical (equally spaced, but not opposite each other). There are three types of lobing of out-of-round parts: *odd, even,* and *random* lobes, as shown in Figure 11.24.

Diametrical Measurement of Roundness

A part with *nonsymmetrical* (odd) lobes can be very deceiving when measured with a tool such as a micrometer, vernier caliper, snap gage, or V-block. The reason is that no matter in which direction you measure the part, it can look perfectly round to a diameter measuring tool, when actually it is not. This is shown in Figure 11.25 with 3-, 5-, and 7-lobe examples. Note that in these odd-number lobes, the diameter measurement is the same in all directions.

However, parts with *symmetrical* (or even-numbered) lobes can be measured with diametrical tools because they are not deceiving to the tool. An example of symmetrical lobes is the oval part shown in Figure 11.26.

FIGURE 11.24
(*a*) Three-lobe; (*b*) 5-lobe;
(*c*) 7-lobe; (*d*) random lobe;
(*e*) round; (*f*) oval.

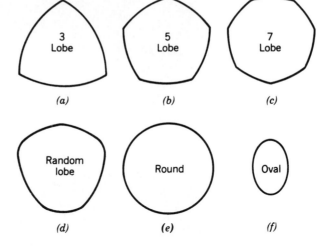

FIGURE 11.25
Measurements are the same in all
directions with odd lobes.

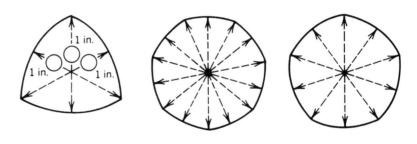

FIGURE 11.26
Oval part in V-block.

Note: V-blocks are effective in measuring roundness if they are the correct angle:

$$V \text{ angle} = 180° - \left[\frac{360°}{n} \right]$$

where *n* is the number of equally spaced lobes on the part (see Figure 11.27).

FIGURE 11.27
V-block used for roundness
measurement.

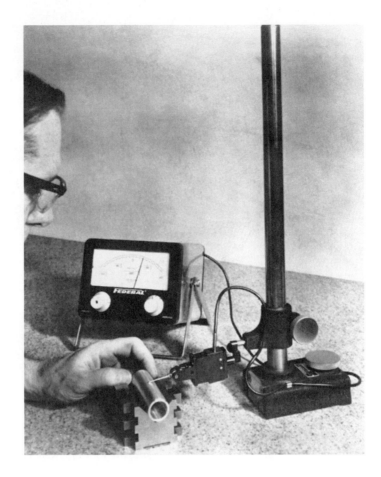

The Virtual Size

Another interesting fact is that you cannot measure odd-lobed parts for size with a diameter measuring tool (other than a ring gage). Because of the spacing and height of the lobes,
a part can be virtually larger (if it is a shaft) or smaller (if it is a hole). About the worst virtual size to be found is the 3-lobe part. It can be virtually much larger than the tool measures it to be.

One method for measuring the virtual size of an out-of-round shaft is a *ring gage* (use
a plug gage for a hole). When matching a shaft to a ring gage, the virtual size can be detected (see Figure 11.28).

Radial Measurement of Roundness: The Preferred Method

A good method for measurement of out of roundness, regardless of the number of lobes, is
radial measurement. An example of radial measurement is the precision spindle (see Figure 11.29). This device either has a rotating base with a fixed indicator or a fixed base with

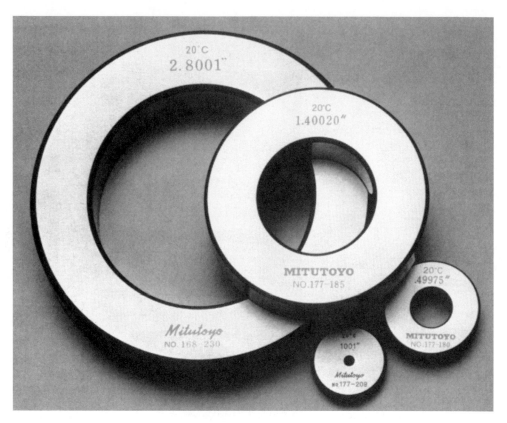

FIGURE 11.28
Ring gages measure the effective size. (M.T.I. Corp. Reprinted by permission.)

a rotating indicator. Either way, the device rotates "true" and allows you to compare the actual part to two concentric circles that are established by the true rotation of the spindle. Or, simply said, the spindle rotates in such a way that if you put a truly round part in it, you will get no movement on the indicator. The indicator (or probe) will detect the amount, height, and spacing of the lobes, giving you a true picture of the roundness of the part.

The interpretation of roundness measurement according to geometrical and positional tolerancing, is radial, with the tolerance zone being two concentric circles. This requires radial measurement of odd-lobed parts and accepts diametrical measurement of even-lobed parts. However, for all purposes where roundness is very important to the function or fit of component parts, radial measurement is preferred.

CYLINDRICITY

Cylindricity is a *form* tolerance that may be considered as an extension of circularity (roundness). The symbol for cylindricity is a circle with tangent lines.

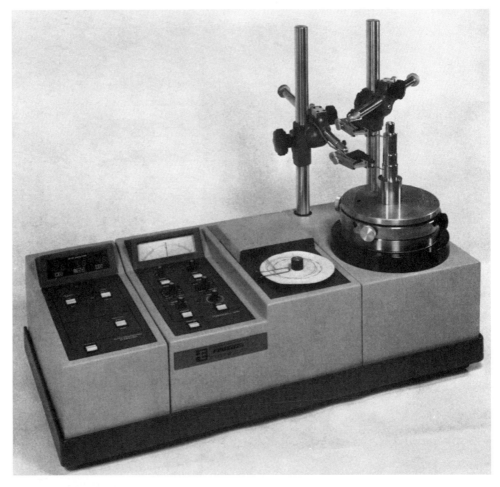

FIGURE 11.29
Formscan 3000 uses a precision air bearing spindle. (Federal Products Corp. Reprinted by permission.)

The only difference between *cylindricity* and *circularity* (roundness) is that cylindricity controls the roundness of a feature throughout its entire length. The tolerance zone for cylindricity is two concentric imaginary cylinders, as shown in Figure 11.30.

The surfaces of the shaft at all elements must lie between the two concentric cylinders .001 in. apart. Cylindricity controls the feature throughout its entire length. When you apply cylindricity, it controls other characteristics at the same time, as shown in Figure 11.31.

Measurement of cylindricity is best accomplished using radial techniques such as those described in the earlier section on true circularity (roundness) measurement. The precision spindle (refer to Figure 11.29) may be used in the same way, except that a precision vertical post must be used so that a vertical sweep may be made on the part. This is depicted in Figure 11.32.

FIGURE 11.30
Cylindricity callout and tolerance
zone.

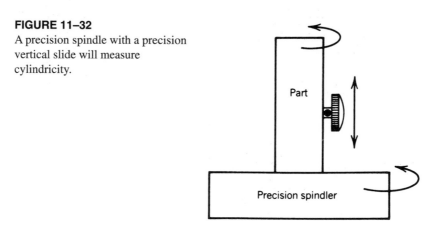

φ.500 ± .003

| ⌀ | .001 |

.001   Apart

Straightness
(a)

Taper
(b)

Roundness
(c)

FIGURE 11.31
Cylindricity controls *(a)* straightness; *(b)* taper; *(c)* roundness.

FIGURE 11–32
A precision spindle with a precision
vertical slide will measure
cylindricity.

Part

Precision spindler

Cylindricity is very difficult to measure; it requires sophisticated equipment for accuracy. In cylindricity, you are making measurements that are affected by the collective geometry of the part. Roundness of circular elements, straightness, taper, surface flaws, and surface texture are all combined to add to the difficulty of the measurement.

FREE STATE VARIATION

There are certain components in industry where free state variation is applicable. *Free state variation* refers to that variation seen in flexible parts when they are released by the holding device of the machine. For example, when a thin wall aluminum casting is located in a chuck on the lathe and the casting is located in the jaws of the chuck, the diameter that is turned will be round; however, when the casting is released from the pressure of the hold-ing device (the chuck jaws) it returns to its original form, causing the diameter that was ma-chined to be out of round. Due to this variation, some geometric tolerances applied to flex-ible parts can exceed the size tolerance of the feature being machined. Note that General Rule 1 does not apply to free state parts.

As an example, a diameter has a size tolerance of 1.500 − 1.495 and a roundness tol-erance of .015 free state.

When tolerances are shown on a drawing in the free state the circled F modifier will be shown and the term "avg. dia." will be shown next to the size tolerance, as shown in Figure 11.33.

FIGURE 11.33
Free state callout.

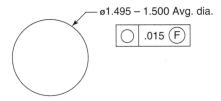

To measure the diameter in Figure 11.33, an inspection is required. Measure the diam-eter in four different directions, find the average of the four measurements, and check that the average diameter is within the size tolerance, as shown in Figure 11.34.

FIGURE 11.34
At least four measurements should be taken.

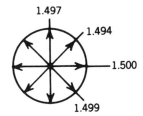

■ **Example 11–4**

Four measurements are taken in four different directions. The average of the four meas-urements is

$$
\begin{array}{r}
1.497 \\
1.494 \\
1.500 \\
+\ 1.499 \\
\hline
5.990
\end{array}
\qquad
\begin{array}{r}
1.4975 = \text{average diameter} \\
4\overline{)5.990}
\end{array}
$$

Note: The average diameter (1.4975) is within the size limits.

■ **Example 11–5**

Four measurements are taken on another part (Figure 11.35). The average of these four measurements is

$$
\begin{array}{r}
1.499 \\
1.503 \\
1.506 \\
+\ 1.497 \\
\hline
6.005
\end{array}
\qquad
\begin{array}{r}
1.5012 = \text{average diameter} \\
4\overline{)6.005}
\end{array}
$$

FIGURE 11.35
Measurements on another part.

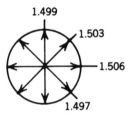

1.499
1.503
1.506
1.497

Note: On this part the average diameter is not within size limits and should be rejected.

The average diameter rule cannot be applied to any part unless "avg. dia." and "free state" are shown on the drawing. If these notations are not shown, size tolerances must be met. Also note that General Rule 1 does not apply to parts subject to free state variation in the unrestrained condition.

OPTICAL COMPARATORS (PROFILE TOLERANCES)

Optical comparators (Figure 11.36) are used throughout the industry for making comparison measurements. The standards used in these measurements vary and may include:

Built-in micrometer heads. To allow you to measure actual movements from a reference point to the measured point.

Graduated screens. To make direct comparisons using the screen itself (such as angular measurements with a screen graduated in degrees and minutes).

Simple overlays. To scale to make comparisons for different shapes.

Pantograph mechanisms. For use in "blind" areas.

The optical comparator projects the profile of a part on the screen after it has been magnified in size.

FIGURE 11.36
Optical comparator. (M.T.I. Corp.
Reprinted by permission.)

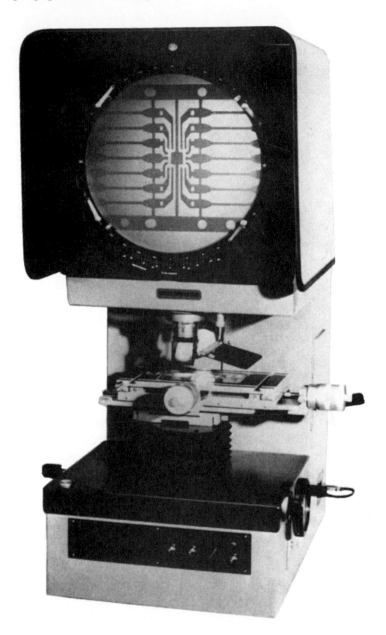

■ **Example 11–6**

Using a 10× comparator, if a part is placed in front of the light source, the image on the screen will be 10 times the size of the actual part. If the part has a .250-in. diameter, the image on the screen will appear as a 2.50-in. diameter.

The comparator table (or staging area) is capable of movement from left to right, up and down, toward or away from you. The movement toward or away from you allows you to focus the part for a sharp image on the screen.

Crosslines

There are usually crosslines built into the comparator screen for the horizontal and vertical referencing (see Figure 11.37). Direct measurements can be made starting at one of these crosslines and using the micrometer heads to measure actual movement from one surface to another.

FIGURE 11–37
Crosslines on the comparator screen. (M.T.I. Corp. Reprinted by permission.)

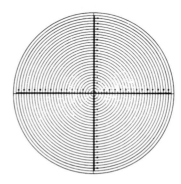

Overlays

Various overlays can be used on a comparator. These overlays are simply a clear material, such as Mylar, that can be clamped onto the comparator screen (usually having their own crosslines for referencing them to the comparator's crosslines).

Most overlays are specially made to compare a particular shape on the product, such as a contour. The overlay will generally have the perfect shape drawn with tolerance lines on each side of it or sometimes will show just the tolerance lines (see Figure 11.38).

FIGURE 11.38
Overlay description with tolerance lines.

Angular Measurement

Most comparators have screens with a built-in scale of 360 degrees, allowing for measurement within minutes of a degree using a vernier scale. Crosslines divide the circle into its quadrants (Figure 11.39).

With a part on the staging area, the screen can be turned to match the angle of the part; then the angle can easily be determined.

FIGURE 11.39
Angles and images. (M.T.I. Corp.
Reprinted by permission.)

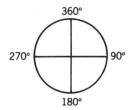

Linear Measurement

Using the crosslines as reference, linear measurements can be made by moving the staging area (using its micrometer heads, or other attachments). By starting at one edge of the part and referencing it on a crossline, then zeroing the micrometer head, the part can be moved to the measured surface, and the micrometer head reading will show the movement.

An advantage of linear measurement with a comparator is that no contact is made with the part. Therefore, contact pressure and wear on measuring surfaces is not a problem. The comparator is also handy when the reference points are in "space" (not physical points).

Contour Measurement

The comparator cannot see inside a part. If there is an internal contour or shape, there must be some sort of attachment to compare this shape. This attachment is often in the form of a *pantograph* that allows you to physically trace internal shapes. The pantograph system has two parts that move together. The stylus follows the shape of the part being measured, and its counterpart shows this movement on the screen. When the stylus moves, the counterpart moves exactly the same way so that you can work easily with internal shapes on a comparator.

Parts of a Comparator

In this book we do not elaborate on the parts of the comparator, but you should have an idea of what makes it work. (Refer to Figure 11.40.)

FIGURE 11.40
Parts of a comparator. (M.T.I. Corp.
Reprinted by permission.)

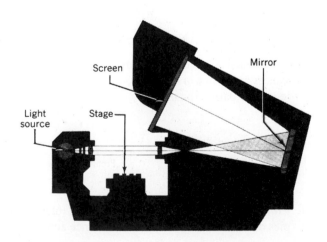

Light source. The part is placed in front of the light source so that its image can be viewed on the screen.

Lens system. The lens system magnifies and transmits the image to the screen.

Screen. The screen must be large enough to show the magnified image of the part.

Holding devices. There must be a staging area for the part to be held in place. The kind of holding device depends on the parts that are to be measured. This can vary from simple fixtures, to centers, or other clamping devices.

Projection of a Surface

Some comparators are equipped with the ability to project a surface (sort of a mirror image) onto the screen (Figure 11.41). This requires a strong light source and the ability of the material to reflect light.

FIGURE 11.41
Projection of a surface. (M.T.I. Corp. Reprinted by permission.)

Surface image

Comparators are all a little different. The important thing is to familiarize yourself with how the staging area works, which overlays you must use, what the magnification power of the comparator is, and how to operate the pantograph (if there is one).

PROFILE TOLERANCES PER ANSI Y14.5M–1982

Profile tolerances are used to control the shape of irregular parts. The symbols for profile tolerances are shown in Figure 11.42. Profile tolerances can be used in many ways. We will review some basic examples.

FIGURE 11.42
Profile symbols.

Profile of a line   Profile of a surface

1. *Profile all around.* This is where a profile tolerance is applied all around a part. Examples of the tolerance callout and tolerance zone are shown in Figure 11.43.

FIGURE 11.43
Profile (all around) and tolerance
zone.

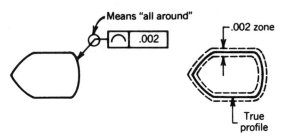

2. *Profile (bilateral tolerance zone).* When a profile tolerance is applied with a leader line pointing directly to a surface, the tolerance zone is the specified width equally disposed around the true profile, as shown in Figure 11.44.

FIGURE 11.44
Profile of a surface.

3. *Profile tolerance (unilateral zone).* If a profile tolerance is shown using a *phantom line* and two arrows, it means that the tolerance band is totally distributed either inside the true profile or outside it, as shown in Figure 11.45.

FIGURE 11.45
Profile tolerance: unilateral zones.

4. *Profile tolerance (using datums).* When a profile tolerance is used in reference to a datum plane, it controls shape, location, and size simultaneously. An example of this is shown in Figure 11.46.

FIGURE 11.46
Profile tolerance: datums specified.

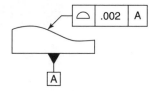

Measuring Profile Tolerances

Most profile tolerances are typically measured on an optical comparator (Figure 11.47) with the assistance of overlay charts for comparison measurement and translation devices such as tracer units. There are times when direct measurement, such as a sweep gage establishing the true profile, can be employed. The gage follows the template, and the indicator shows high and low areas in comparison to the true profile.

FIGURE 11.47
Optical comparator. (M.T.I. Corp. Reprinted by permission).

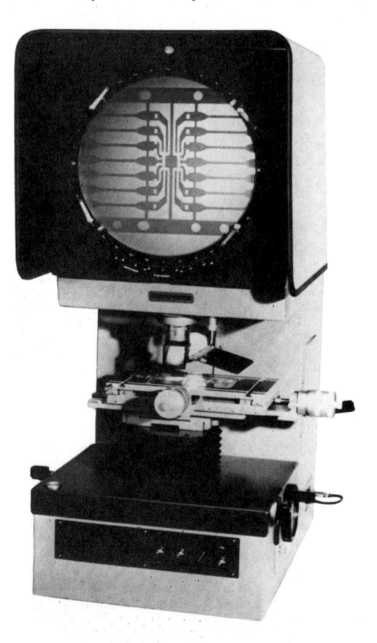

Regardless of the technique or tool used, the important thing is to be able to establish a true profile from which to compare the surface and some means of measuring the difference. Where it is important for a surface to be in the correct shape and position, profile tolerances should be established from datums.

Profile of a Line

Profile of a line simply means that the profile tolerance must be checked for various elements of the surface, and each line element should be within the stated tolerance. An inspector or machinist would establish the true profile and then measure one line element at a time completely across the surface.

Remember: Each line element must be within the tolerance zone, but they have no relationship with each other collectively.

Profile of a Surface

The profile of a surface simply means that the profile tolerance applies throughout the surface. Several line elements are measured collectively, and all must be within profile tolerance.

COORDINATE MEASURING MACHINES

Coordinate measuring machines were originally designed to keep up with the production rate of numerically controlled machines. It can be very time consuming and costly for an inspector to go into a detailed layout of most products that are produced on a numerically controlled machine. These products are usually highly complex, with many features (holes and the like) that must be inspected for position with reference to datums, planes, or surfaces.

An example of this might be a part with an 18-hole bolt circle that is referenced from three datum planes perpendicular to each other. The numerically controlled machine can produce this part very quickly, but for the inspector to lay out the first piece is another matter. Traditionally, the inspector must establish the three datum planes on the surface plate with right-angle plates and clamps. Each of the 18 holes in the part has two coordinates. The inspector must measure manually 18 coordinates in one direction, then rotate the part 90 degrees and measure 18 more coordinate distances. Performed with a vernier height gage, this job can take a long time to perform, and there are *many* chances for error.

The coordinate measuring machine (CMM) can do this job in a few short minutes, and the production machine can be under way to produce additional parts. This is the difference between coordinate measuring machines and traditional methods. The inspection is more accurate (less chance for errors) and takes a lot less time to perform. When considering how important production time is, the CMM is a tool that pays for itself in a very short period of time and soon starts improving profits because of the time and money saved.

How CMMs Work

The basic principle of a coordinate measuring machine is simple. The CMM has three basic directions of movement. These are called X, Y, and Z axes. When standing in front of the machine, the *X* axis is the movement from left to right, the *Y* axis is the movement toward and away from you, and the *Z* axis is the movement up and down. Some have a *W* axis, which is rotation.

CMMs come in various makes and models. They also come with different measuring devices and capabilities. Some are very basic with vernier scales in the *X* and *Y* directions only. Some have dials in the *X, Y,* and *Z* directions. Some have electronic digital display that can be "zeroed" anywhere, and some have computers attached that can record your measurements and even draw you a picture of the part. All CMMs still work on the same basic principle of movement. An example of this movement is shown in Figure 11.48.

FIGURE 11.48
X, Y, and *Z* coordinate directions.

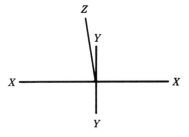

Various probes are used with CMMs, depending on the application. They range from the basic conical style (which will find the center of a hole automatically), to electronic probes that make a beeping noise when contact is made with the part. One type of coordinate measuring machine is shown in Figure 11.49.

Functional Measurements with CMMs

One problem with using a coordinate measuring machine is the fact that many users misunderstand its ability. CMMs are fast and accurate, but one cannot forget the basics. Several examples could be given with reference to the standard for geometric tolerancing. This standard defines particular rules that must be followed if inspections on the product are to be functional.

Locating Functional Datums.   The CMM (with either type of probe) cannot properly locate a secondary or tertiary datum plane. The functional datums must be contacted at the highest points on the surface. The probes of the CMM can contact points, but not the highest points. The only functional datum that can be contacted by the CMM is the primary datum plane (this is done by the staging area on the machine). The secondary and tertiary datum planes must be contacted by using surface plate accessories such as right-angle plates, parallels, and so forth. One only needs to "beep" the right angle plates (two places for secondary, and one place for tertiary) and then place the part against them. (Refer to pages 122–123 on contacting datums.)

Locating Datum Target Points.   Datum target points are easier to locate on a CMM. However, some accessories are still needed. Datum target points are specified points on a part (see datum targets). The only problem is the primary datum (which must be contacted at three specified points). The staging area of the CMM cannot directly contact these three

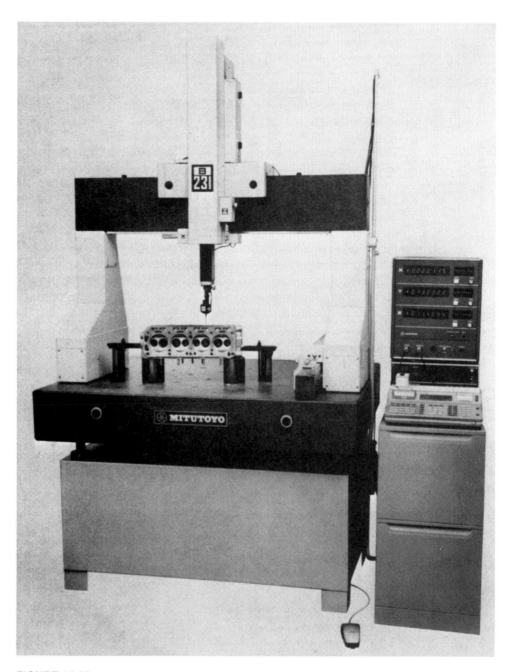

FIGURE 11.49
Coordinate measuring machine. (M.T.I. Corp. Reprinted by permission.)

points. Often, three pins (spherical on the end) are used for the primary datum target points. Then the machine's probe can be used to contact the secondary (two specified points) and the tertiary (one specified point).

Locating Datum Target Areas.   Datum target areas are specified areas that must be contacted with pins that are flat on the end. The CMM probes cannot directly contact an area, so pins must still be used to assist the CMM in the appropriate contacting of datums.

Measuring Hole Positions.   Hole locations are measured by finding out if the axis of the hole lies within its tolerance zone. To measure the location (position) of a hole, therefore, you must find its axis. The beep probe of the CMM does not find an axis. What the beep probe does find is the position of one point along the axis of the hole. It does not tell you about the actual location of the entire axis. The fixed (hard) conical probe of the CMM only tells you the position of the part of the axis at the end of a hole. Furthermore, remember that the position tolerance on a hole applies at both ends, so that measurements must be taken twice (the coordinates at one end, then the other).

Coordinate measuring machines are useful equipment that perform very well. Many labor-hours can be saved without loss of accuracy when they are used. It is essential, however, that they be used properly. Drawing requirements must be functionally verified, or the effort could prove to be a waste of time.

CMM Operations

CMMs are operated in three basic ways: manual, computer assisted, and computer driven. The older manual CMMs, although much better than the alternative of open setup layouts on the surface plate, are not as efficient as computer-assisted or computer-driven CMMs because the operator must read and remember starting points and linear distances, write them down, and then analyze the results. This is a problem specifically in layout and measurement of geometric tolerances (such as position and concentricity). Reporting from the manual CMM must be made on paper after the inspection has been completed, and the proper analysis of position tolerances (see Chapter 8) is difficult. Computer-assisted CMMs are more efficient, in that a computer and a variety of software is added to the CMM for on-line reporting, memorization of zero points, and linear distances. Computer-assisted CMMs are also more efficient in analyzing geometric tolerances, such as position, concentricity, and perpendicularity (to name a few). Software can be added to the computer, which is designed to assist the operator in evaluation of these types of tolerances and reports directly from the machine. Computer-driven CMMs are the highest level of efficiency, due to the fact that they can be programmed (one time) to measure a product, and once programmed, can measure subsequent products without help from the operator. The operator can be doing other important things while the measurements are being made automatically.

CMMs Compared to Functional Gages

CMMs are also more versatile than functional gages in measuring products. Functional gages are usually designed to measure one function at a time. The cost of functional gages is limiting, and they must be maintained to avoid wear and damage. CMMs may have advantages over functional gages because (1) functional gages are designed only to gage the virtual condition (worse case) of products, (2) functional gages cannot be used to gage geometric tolerances specified at LMC or RFS condition (refer to Chapter 7 for LMC/RFS definitions), (3) functional gages are designed with wear allowances and gage tolerances such that they will reject a borderline good product, (4) functional gages usually check only one functional aspect of the product, and (5) functional gages are a Go–NoGo relationship that does not measure the product. By comparison, CMMs can (1) measure all conditions of the product; (2) measure all tolerance modifiers, including LMC and RFS; (3) avoid rejections of borderline good product because they measure actual conditions and are not designed to wear allowances; (4) measure all functional tolerances of the product, not just one; and (5) measure the product and give variable information to the operator for process adjustments and process control purposes.

CMMs Improve Multiple Setups

Another important advantage about CMM use is the avoidance of multiple setups as compared to open surface plate techniques. In many applications the product must be reoriented on various datum structures to inspect various functional characteristics. This need for multiple setups adds time and potential error to the results obtained. CMMs can often measure multiple characteristics in one setup without moving the part.

CMMs and Fixturing

It is important to identify and use the appropriate fixturing (or standard surface plate accessories) as indicated earlier in this chapter. Electronic probes or the fixed probes used on CMMs are limited in the information they provide (especially for establishing functional datum planes). The appropriate fixturing or accessories allow proper functional setup of the part to be measured. The type of datums on the part should be considered so that the correct datum plane is generated from the contact points. A primary datum target on a surface, for example, is established by contacting a minimum of three specific locations on that surface. Often, the probe on the CMM can contact these points, and there are few problems. On the other hand, a primary functional datum plane (for a surface) is established by contacting at least three of the *highest points* on that surface. Fixturing will be needed which will make contact with the highest points so that a plane can be derived from the fixture. Another example of functional inspection on the CMM is the use of gage pins in holes when measuring position, perpendicularity, and other geometric tolerances applied to holes. As with open surface plate setups, there is still no substitute for a snug-fitting gage pin to establish the functional axis of the hole for the purpose of axial measurement. Using the variable and fixed probes of the CMM for axial location can be a problem. The variable probe simulates a dial indicator point of contact. It is often used (when there is no fixturing) in a manner that touches the inside of a hole in a triangle of points (sometimes each point is a little deeper into the hole). This practice is attempting to establish the axis.

Part of the problem here is that holes are out of round, tapered, and have other forms of geometric error that can give false signals of the functional orientation and location of an axis. The practical problem, however, is much easier to understand. A probe touching a hole at specific points is not as functional as a full-form cylinder interfacing with the entire hole (when one considers the fact that holes are usually produced to put a bolt or a pin through them). The form of a gage pin is representative of the bolt or pin that will be used in assembly. The same thing holds true for the fixed (conical) pins used on CMMs. When used, these pins only find a point of the axis at the end of the hole. They do not identify the effect of the third dimension (depth), nor do they fill the hole as well as a gage pin. The importance of fixturing on CMMs is of growing concern to users. The key to the answer is function. Functional measurement of products is important enough to take that extra measure and make the CMM results true to the way the product works. A considerable amount of time and money is saved using CMMs compared to open setups. A little more effort should be made to use accessories or fixturing that make the results functional. Refer to examples shown in Chapter 8.

TORQUE MEASUREMENT

Torque measurement is a vital part of today's industrial measurement, especially when the product is held together by nuts and bolts (fasteners). The wrong torque in an assembly can result in the assembly failing due to a number of problems. Two of these problems are: (1) parts are not assembled securely enough for the unit to function properly; and (2) threads are stripped because torque is too high, causing the unit to fail.

Torque is simply described as a force producing rotation about an axis. This force could be in pounds, grams, ounces, or other units. Torque is based on the law of leverage. The formula for torque is

$$torque = force \times distance$$

The distance could be feet, inches, meters, or other units. As shown in Figure 11.50, a distance D and a force F produce torque T.

FIGURE 11.50
Description of torque.

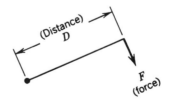

■ **Example 11–7**

A force of 1 lb is applied at a distance of 1 ft. So.

$$
\begin{aligned}
torque &= force \times distance \\
&= 1\ lb \times 1\ ft \\
&= (1 \times 1)\ lb\text{-}ft \\
&= 1\ lb\text{-}ft
\end{aligned}
$$

Note: This is usually referred to as foot-pounds (ft-lb).

■ **Example 11–8**

A force of 12 lb applied at a distance of 10 in. from the center of a bolt will torque the bolt at 120 lb-in. (pound-inches, usually referred to as inch-pounds).

Measuring Torque

Torque on fasteners is measured by a torque wrench. There are many types of torque wrenches. The two types most commonly used are the *flexible beam* and the *rigid frame*. The flexible beam type, shown in Figure 11.51, works on the basic principle that a beam of a certain size, geometry, and material will bend a certain distance at an applied force and return to the original position after the force has been removed. The flexible beam has a pointer that shows the torque applied. This beam can be round or flat.

FIGURE 11.51
Flexible beam torque wrench.

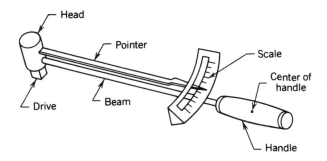

The rigid frame torque wrench works with a torsion bar connected to the frame, as shown in Figure 11.52, and a dial indicator attached so that the torque applied can be read directly. This indicator works using the rack on the lever arm.

FIGURE 11.52
Rigid frame torque wrench.

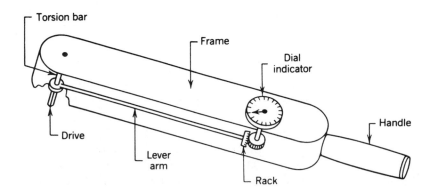

Another type of torque wrench (not shown) is the type for which you preset the desired torque; when you reach the desired torque on the bolt, the wrench will either make a distinct clicking sound or you will simply feel the wrench "give" in the handle.

Cautions in Using Torque Wrenches

The following points should be kept in mind while using torque wrenches.

1. To use the wrench accurately, hold the handle in the center at all times.
2. Apply the force slowly and smoothly; do not jerk the wrench.
3. Hold the wrench steady for a short time after the desired torque has been reached (this is generally considered to be good practice).
4. Handle torque wrenches carefully, especially the pointer.
5. Use torque wrenches for applications only within 80 percent of their range.
6. Beware of false applications of torque. *Example:* The bolt is too long, and it bottoms out before the bolt head reaches the surface for tightening. *Or:* the threads in the hole are too shallow and the bolt bottoms out before tightening.
7. Keep torque wrenches calibrated periodically with a torque wrench measuring instrument that is a known standard.
8. If it is necessary to extend a torque wrench, ensure that you compensate for the change in distance. Also ensure that the extension is in line with the centerline of the wrench handle to avoid cosine error.
9. Avoid rechecking the torque on fasteners that have already been torqued. This can over-torque the fastener. It is best to use a torque wrench the first time.

HARDNESS TESTING

Industrial materials must often be inspected for hardness, which is a material's resistance to penetration. There are many methods of testing material hardness today, but most widely used are the *Brinell and Rockwell hardness tests.* Other tests include Vickers, Knoop, and Scleroscope. In this book we discuss only the Brinell and Rockwell tests. In the following discussion we outline, in step-by-step form, how to perform Brinell and Rockwell hardness tests and describe some particularly important characteristics, parts, and cautions for each type of test.

Brinell Hardness Testing

Hardness testing is done throughout the industry using several different methods. One of these methods is the Brinell hardness test. Hardness is a vital characteristic of some products because of their function, machinability, strength, or other necessary properties. It can be defined as *the ability of a material to resist penetration.*

The Brinell hardness test is very simple. It is a matter of applying a constant load of 500 to 3000 kilograms (kg) (depending on the material) to a 10-millimeter (mm)-diameter steel ball on a flat surface of the part (see Figure 11.53).

FIGURE 11.53
Brinell hardness test.

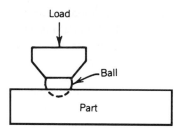

Note: A 500-kg load is usually used for soft materials such as aluminum or copper. A 3000-kg load is usually used for harder materials such as steel or cast iron.

Once the load has been applied for a certain amount of time (15 seconds for hard materials and 30 seconds for soft materials), it is then removed and the diameter of the indentation (dent) is measured in millimeters (mm). This diameter should be the average of two measurements.

Next, the Brinell hardness number must be found. Tables are available to convert this number directly, or you can use the following formula:

$$\text{BHN} = \frac{L}{\dfrac{\pi D}{2}\,[D - (D^2 - d^2)^{1/2}]}$$

where L is the load (in kg)
 D is the diameter of the steel ball (in mm)
 d is the diameter of the "dent" (in mm)
 π is 3.1416 (a constant)

■ **Example 11.9**

The load is 3000 kg, and the diameter of dent is 4.0 mm.

$$\text{BHN} = \frac{3000}{\dfrac{3.1416(10)}{2}\,[10 - (10^2 - 4^2)^{1/2}]}$$

$$= \frac{3000}{15.708(0.835)} = \frac{3000}{13.116} = 229\ \text{BHN} \quad \textit{Answer}$$

When measuring Brinell hardness, note that the surface preparation is important. The surface should be very smooth so that the diameter of the indentation can be clearly defined and measured. It may be necessary to grind the surface (such as a casting surface) before testing.

In some cases, Brinell testing can be considered a "destructive" test because it leaves an indentation that may be a problem if the surface finish in that area is important. If so, test the part in another area.

Most often, a special microscope is used to measure the indentation accurately. This microscope has a built-in millimeter scale and a light for easy reading. Be careful not to make Brinell indentations too close to the edge of a part, because there is no support for the load at an edge, and the indentation will shift and go out of round, giving you a poor reading.

There are many kinds of Brinell hardness testers. Some are manually operated, some are automatic, and some are portable. Know your particular tester and how it operates prior to performing any test with it.

Rockwell Hardness Testing

One of the most widely used tests for hardness of materials in industry is the Rockwell hardness test. There are many reasons for its wide use: (1) it is a simple test to perform; (2) hardness is directly read on a dial; (3) it does not require much skill; (4) it leaves a very small indentation on the part (unlike Brinell tests); and (5) it has a wide range of applications, from soft aluminums to carbides. The Rockwell test also requires that the surface being tested be "prepared" by removing surface roughness or scale. However, some parts need no preparation.

There are two types of "indenter" used in Rockwell testing: the *brale* (which is diamond-cone shaped), and the *ball* indenter (with diameters of $\frac{1}{16}$ in., $\frac{1}{8}$ in., $\frac{1}{4}$ in., and $\frac{1}{2}$ in.) (see Figures 11.54 and 11.55).

FIGURE 11.54
Brale indenter.

FIGURE 11.55
Ball indenter.

The Rockwell test uses two loads (or forces): a *minor* load to seat the indenter into the material and then a *major* load. The actual hardness is measured by the additional depth of penetration caused by the major load.

Calibrating the Rockwell Tester.   This is done with an accurate test block of known hardness. You should take at least five different readings on the test block, and the average of all five readings should match the hardness specified on the test block.

Types of Rockwell Tests.   There are two types of Rockwell tests: *regular* and *superficial*. Each type of test may require the use of a different tester.

1. *Regular Rockwell test.* Uses a minor load of 10 kg and various major loads of 60 kg, 100 kg, or 150 kg.

2. *Superficial Rockwell test.* Uses a minor load of 3 kg and major loads of 15 kg, 30 kg, or 45 kg.

Precautions Prior to Testing

1. Check the indenter chosen for damage or wear.
2. The surface to be tested should be flat and smooth.
3. The anvil used should be clean and undamaged.
4. If the part is round, grind a flat area for the test.
5. Do not make the test too near the edge of the part or too near a previous indentation.

Steps in Performing a Rockwell Test (Regular or Superficial).   *Note:* The minor loads are built into the machine.

FIGURE 11.56
Rockwell hardness tester.

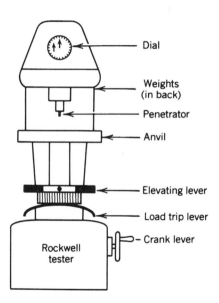

Refer to Figure 11.56 when following these steps:

Step 1.   Prepare the surface if necessary.

Step 2.   Choose the scale to be used (if not already specified) (see tables in Figure 11.57).

Step 3.   Choose the major load to be used. This depends on the scale you chose (see Figure 11.57 tables). The load is often shown in the specification. Weights to set the major load are in back of the machine.

Table of Scales Versus Metals

Scale	Used for:
B	Copper, soft steel, aluminum, malleable iron
C	Steel, hard cast iron, malleable irons, titanium, deep case hardened steels, other metals harder than RB100
A	Carbides, thin steel, shallow case hardened steel
D	Thin steel, medium case hardened steel, malleable iron
E	Cast iron, aluminum, magnesium alloys, bearing materials
F	Annealed copper alloys, thin soft sheet metals
G	Bronze, beryllium, copper, malleable irons
H	Aluminum, zinc, lead
K, L, M, P, R, S, V	Soft bearing materials, other very soft or thin materials

(a)

Normal Rockwell Hardness Scales, Loads, and Indenters

Scale	Indenter	Diameter	Major Load	Dial Color
B	Ball	$\frac{1}{16}$	100 kg	Red
C	Brale	—	150 kg	Black
A	Brale	—	60 kg	Black
D	Brale	—	100 kg	Black
E	Ball	$\frac{1}{8}$	100 kg	Red
F	Ball	$\frac{1}{16}$	60 kg	Red
G	Ball	$\frac{1}{16}$	150 kg	Red
H	Ball	$\frac{1}{8}$	60 kg	Red
K	Ball	$\frac{1}{8}$	150 kg	Red
L	Ball	$\frac{1}{4}$	60 kg	Red
M	Ball	$\frac{1}{4}$	100 kg	Red
P	Ball	$\frac{1}{4}$	150 kg	Red
R	Ball	$\frac{1}{2}$	60 kg	Red
S	Ball	$\frac{1}{2}$	100 kg	Red
V	Ball	$\frac{1}{2}$	150 kg	Red

(b)

FIGURE 11.57
(*a*) Table of scales versus metals. (*b*) Table of Rockwell scales, loads, and indentors.

FIGURE 11.57
(*c*) Superficial Rockwell scales, loads, and indenters.

Superficial Rockwell Scales, Loads, and Indenters

Scale	Indenter	Diameter	Major Load
15N	N Brale	—	15 kg
30N	N Brale	—	30 kg
45N	N Brale	—	45 kg
15T	Ball	$\frac{1}{16}$	15 kg
30T	Ball	$\frac{1}{16}$	30 kg
45T	Ball	$\frac{1}{16}$	45 kg
15W	Ball	$\frac{1}{8}$	15 kg
30W	Ball	$\frac{1}{8}$	30 kg
45W	Ball	$\frac{1}{8}$	45 kg
15X	Ball	$\frac{1}{4}$	15 kg
30X	Ball	$\frac{1}{4}$	30 kg
45X	Ball	$\frac{1}{4}$	45 kg
15Y	Ball	$\frac{1}{2}$	15 kg
30Y	Ball	$\frac{1}{2}$	30 kg
45Y	Ball	$\frac{1}{2}$	45 kg

(*c*)

Step 4. Choose the penetrator to be used. This is done according to the type of test and the scale used. [For example, regular *C* scale uses the brale indenter penetrator (see Figure 11–57 tables).]

Step 5. Choose the proper anvil to use for the part being tested. See Figure 11–58 for various anvils.

Note: The surface being tested must be perpendicular to the centerline of the anvil.

Step 6. Make sure that the *crank lever* is locked in the up position.

Step 7. Carefully install the indenter and anvil you have chosen.

Plane
anvil

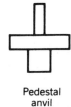

Pedestal
anvil

V-anvil

Special
anvil
(for pipes)

FIGURE 11.58
Various anvils.

Step 8. Place the part on the anvil and position it for the test. *Avoid bumping the indenter. If you do, inspect the indenter visually under a minimum of 10× magnification for damage.*

Step 9. Apply the minor load by carefully turning the *elevating lever* clockwise until the part makes contact with the indenter and the small and larger pointers on the dial are pointing straight up.

Step 10. Turn the *knurled adjustment ring* until the set arrow (on the dial face) is in line with the large pointer.

Step 11. Apply the major load by pressing down on the *load trip lever* (the crank lever will now move forward).

Step 12. When the *crank lever* has stopped moving forward, carefully pull it back to its locked upright position.

Step 13. Take the reading on the proper scale color while the minor load is still applied.

> If *brale* indenter, read the black scale.
> If *ball* indenter, read the red scale.

Step 14. Remove the minor load by turning the *elevating lever* counterclockwise to release the part from the indenter.

Step 15. Remove the part (*you must still avoid bumping the indenter*).

A minimum of three readings should be taken on the part, and the average reading for the hardness should be used.

REVIEW QUESTIONS

1. Which of the following devices can be used to measure flatness in millionths?
   a. sine bar
   b. jack screws
   c. three equal pins
   d. optical flats
2. When measuring surface finish with a profilometer, what cutoff value is understood if not specified on the drawing?
   a. .015
   b. .030
   c. .020
   d. .025
3. Surface finish should always be measured in a direction that is _____ to the lay.
   a. angular
   b. perpendicular

    c. in line with

    d. adjacent

4. Which of the following should be avoided when measuring surface finish?

    a. arithmetic averages

    b. cutoff values

    c. flaws

    d. finish

5. What kind of lay is required when this symbol is shown (C)?

    a. parallel

    b. radial

    c. circumference

    d. circular

6. What set of tools can be used to optically measure the parallelism error between the anvil face and spindle face of a micrometer?

    a. optical flats

    b. precision parallels

    c. optical parallels

    d. none of the above

7. There are two methods for using optical flats to measure flatness. One is the contact method and the other is the _____ method.

    a. angle

    b. wedge

    c. noncontact

    d. clearance

8. An air gage head that would be ideal to measure the roundness of an inside diameter that has three lobes is:

    a. three nozzles (or jets)

    b. two nozzles (or jets)

    c. five nozzles

    d. none of the above

9. Problems in roundness measurement include

    a. it cannot be gaged

    b. no knowledge of lobes

    c. two-point measuring devices do not accurately measure roundness

    d. all of the above

10. What angle of V-block is necessary to use the V-block to measure roundness of a part that has five equally spaced lobes?

    a. 90 degrees

    b. 108 degrees

    c. 60 degrees

    d. 120 degrees

11. The tolerance zone for cylindricity tolerance is

    a. a cylindrical zone

    b. two parallel planes

   c. two parallel lines
   d. two concentric cylinders

12. Cylindricity is a control that automatically controls
   a. straightness
   b. taper
   c. roundness
   d. all of the above

13. When "free state variation" applies to the size of a diameter, the _____ diameter must be within the stated size limits.
   a. absolute
   b. minimum
   c. average
   d. maximum

14. Using optical comparators at 10×, what size would the nominal radial line on an overlay have to be to measure a 0.375 in. radius on a part?
   a. 3.75 in.
   b. 37.5 in.
   c. 0.0375 in.
   d. none of the above

15. A profile tolerance, unless otherwise specified, is automatically
   a. unilateral
   b. maximum limit
   c. bilateral
   d. minimum limit
   e. none of the above

16. What type of line would be used on a profile tolerance to change the tolerance zone to unilateral?
   a. hidden
   b. border
   c. section
   d. phantom

17. If a circle is drawn around the leader arrow from a profile tolerance, this means that the profile tolerance applies
   a. all over
   b. all around
   c. circular
   d. round zone

18. A machine that was designed to support numerically controlled equipment and can be used to quickly and accurately inspect complex parts is the
   a. precision spindle
   b. pneumatic gage
   c. CMM
   d. profilometer

19. Which of the following are benefits of using a coordinate measuring machine instead of a functional gage to measure, for example, the position of holes in a pattern?
   a. It can measure all modifiers (MMC, RFS, LMC).
   b. It can measure more than one part feature.
   c. It can give variable results.
   d. all of the above
   e. a and b only

20. Convert 24-in. pounds torque to foot pounds.
   a. 2 ft-lb
   b. 3 ft-lb
   c. 12 ft-lb
   d. none of the above

21. A torque wrench that gives variable results is a _____ _____ wrench.
   a. rigid frame
   b. flexible beam
   c. variable beam
   d. variable frame

22. Hardness is the ability of a material to resist _____.
   a. pulling apart
   b. stretching
   c. compression
   d. penetration

23. The tolerance zone for circularity (roundness) is
   a. two parallel lines
   b. two cylinders
   c. two concentric circles
   d. two parallel planes

24. A mechanical device used to measure surface finish is called a _____ comparator.
   a. finish
   b. fingernail
   c. roughness
   d. cutoff

25. When dealing with lobes or roundness error, the effect on the actual size of a shaft that is out of round is
   a. the effective size is smaller than it appears
   b. the effective size is smaller than the measured size
   c. the effective size is larger than it appears
   d. none of the above

Lot-by-lot acceptance sampling plans are often used to judge whether a particular lot submitted for inspection is acceptable or not, depending on the quality levels agreed upon by the producer and the consumer. These plans are often used on completed lots at one stage of processing or another to judge the quality of the lot based on statistical samples taken at random. There are a variety of lot-by-lot sampling plans from which to choose, whether one is looking at attributes or variables of the output of the process (refer to the glossary for definitions). Attribute sampling plans covered in this chapter include ANSI/ASQC Z1.4 (formerly Mil-Std-105E) and Dodge–Romig sampling plans. Variables sampling plans covered in this chapter are ANSI/ASQC Z1.9 (formerly Mil-Std-414).

The basis for lot-by-lot acceptance sampling is a percentage called an acceptable quality level (AQL). An AQL (or the maximum percent defectives in a given lot that are considered acceptable as a process average) must be agreed upon by the producer and the consumer before lot-by-lot acceptance sampling can be used. Although acceptance sampling can be, and has been, more effective than 100% inspection at judging lot-by-lot quality, neither method is generally acceptable today as a process control tool. One of the main reasons for this is that neither is effective at *preventing* defects. They are both designed to *find* defects in the process.

Acceptance sampling is often used in conjunction with process control to judge the results of the process and to feed back information to the processors regarding defective output at various stages of operations. The sampling tables covered in the rest of this chapter are designed based on probabilities and, when used properly, will help reduce inspection and still detect poor lot-by-lot quality levels. It is recommended that lot-by-lot acceptance sampling should not be the only method for attempting to control outgoing quality.

Before reviewing acceptance sampling plans, we will first review probability laws and applicable probability distributions.

BASIC PROBABILITY

Statistical quality control applications involve knowledge of the *probability* of some event. Probability is the basis for lot-by-lot sampling plans and statistical process control applications (for attributes and variables). This section will review basic probability laws. Probability is expressed in numbers between 0 and 1. Zero (or zero percent) means that there is no chance of the event occurring, and one (or 100%) means it will definitely occur.

Probability Laws

Some probability problems deal with the additive law and some with the multiplicative law. The **additive law** means that the probabilities of each event must be added. The **mul-

tiplicative law means that the probabilities of each event must be multiplied. The following examples will help to explain these laws. One method to decide which law is applicable is to look at the probability statement (or question) itself. If it contains the word "and," it will generally refer to the multiplicative law. The word "or" generally refers to the additive law.

Mutually Exclusive or Not Mutually Exclusive Events

Certain events are **mutually exclusive** if either one or the other can occur in the same trial, but not both. Events are **not mutually exclusive** if both events can occur at the same time. To solve some probability problems, one must know if the events are mutually exclusive or not mutually exclusive. For example, in the flip of a coin, there are only two possible events (heads or tails) and both cannot occur at the same time. These events, then, are mutually exclusive events. An example of events that are not mutually exclusive is the probability of rolling a 3 or a 5 on a single roll of two dice. Either the 3 or the 5 or both can appear.

Dependent or Independent Events

Events are said to be **dependent** if one event depends on the occurrence of another event. If not, then the events are **independent.** For example, a jar has 10 marbles in all. Three of them are red and seven are white. What is the probability of drawing two red marbles in a row (drawing them one at a time and not replacing them in the jar)? The probability of finding two red marbles in a row makes these events dependent events. The probability of finding the second red marble is greater or less depending upon whether a red marble is found in the first draw. The solution to this problem is in the example of dependent events, multiplicative law.

Single (Simple) Events

The probability of a single event occurring is related to the chance that that event will occur out of all possible events.

■ Example 12–1

What is the probability of rolling a 4 on a single roll of one die?

Solution
There are six numbers in all on one die, so the probability of rolling a 4 is:

$$P_4 = \frac{1}{6} = .166 = 16.6\%$$

■ Example 12–2

What is the probability of drawing a king from a 52-card deck?

Solution
There are 52 cards in all, and 4 of them are kings. The probability of drawing a king is:

$$P_{KING} = \frac{4}{52} = .077 = 7.7\%$$

Complementary Events

Complementary events are those events whose probability of occurrence add to 1. If there are two complementary events, and the probability of occurrence of one event is .30 (30%), the probability of the other event occurring is .70 (70%).

■ Example 12–3

If the probability of finding one defect is .01 (or 1%), what is the probability of *not* finding a defect?

Solution
The probability of not finding a defect, in this case, is $1 - .01 = .99$ (or 99%).

Dependent Events (Multiplicative Law)

Dependent events are where the occurrence of one event depends on the occurrence of another event. Look for the key word or phrase in the probability statement in Example 12–4 that calls for the multiplicative law.

■ Example 12–4

A jar has 10 marbles in all. Three of them are red and seven are white. In two *consecutive* draws of one marble each, what is the probability of drawing a red *and* then another red marble? *Note:* The marbles are not replaced in the jar.

Solution
Drawing two red marbles in a row makes this a dependent event. The key word *and* takes us to the multiplicative law. The probability of drawing two red marbles in a row is equal to the probability of drawing the first red marble times the probability of drawing another red marble on the second draw given that one was drawn on the first draw. Since the marbles will not be replaced once drawn, the probability of the first red marble is 3/10, but the probability of the second red marble being drawn is 2/9 (two red remaining and nine total remaining).

$$P_{RED,RED} = \frac{3}{10} \times \frac{2}{9} = .066 = 6.6\%$$

In this case, the probability of drawing two red marbles in a row (without replacement) is .066 (or (6.6%).

Independent Events (Multiplicative Law)

A good example of this type of probability problem is in reliability (see the glossary on reliability). An electronic system has resistors that are connected in series with each other. The output can be adversely affected if either of the resistors fail, but the failure of either resistor does not affect the performance of the other.

■ Example 12–5

The probability of success of resistor A is 0.90, and the probability of success of resistor B is 0.95. If they are connected in series, what is the probability of success of the system? *Note:* The system success depends on the success of both resistor A *and* resistor B.

Solution

The probability of system success is the product (multiplication) of the probabilities of resistor success.

$$P_{SYSTEM} = P_A \times P_B$$
$$= .90 \times .95$$
$$= .86$$

In this case, the system success (reliability) is lower than the components because the system can fail if either resistor fails.

Mutually Exclusive Events (Additive Law)

If there are two or more events and any none can occur at the same time, the events are *mutually exclusive.* If the probability statement asks about the probability of one event *or* the other, then it is the *additive* law. Note that the probability of both events occurring is zero.

■ Example 12–6

On a single draw of 1 card, what is the probability of drawing a king or a queen?

Solution

The probability of drawing a king or a queen on 1 draw of 1 card is the additive law:

$$P_{KING\ or\ QUEEN} = P_{KING} + P_{QUEEN} = \frac{4}{52} + \frac{4}{52} = .154\ (\text{or } 15.4\%)$$

■ Example 12–7

On one roll of one die, what is the probability of rolling a 2 or a 6?

Solution

The probability of rolling a 2 or a 6 on one roll of one die is:

$$P_{2\ or\ 6} = P_2 + P_6 = \frac{1}{6} + \frac{1}{6} = .333 = 33.3\%$$

Not Mutually Exclusive Events (Additive Law)

If there are two or more events and more than one can occur at the same time, the are *not mutually exclusive* events. If the probability question asks for the occurrence of either one event *or* the other even though both can occur, it is the *additive* law. In Example 12–8, we are interested in the probability that either event will occur.

■ **Example 12–8**

The success of a system depends on either subsystem A or subsystem B starting. The probability that subsystem A will start is .8 and the probability that subsystem B will start is .7. What is the probability of success of the system? Note that both can start as well.

Solution

The probability of success of this system is the additive law, but it does not require both subsystems to start. Therefore, the probability of success of the system is the addition of the probabilities of each subsystem minus the probability of both subsystems starting.

$$P_{SUCCESS} = P_A \times P_B - (P_{A\&B})$$
$$= .8 + .7 + (.8 \times .7)$$
$$= .94$$

Summary

These basic laws of probability (and specific probability distributors) are the foundation for acceptance sampling plans and statistical process control. They all have to do with the laws of chance. Understanding these laws helps us to understand statistical quality control methods.

SAMPLE STATISTICS VERSUS POPULATION PARAMETERS

Statistical quality control methods involve taking a sample (n) from a population and making an inference about the population based on the results of the sample. Results taken from samples are called **statistics;** results taken from populations are called **parameters.** Therefore, statistics are used to make inferences about parameters.

Table 12–1 shows the relationship between samples statistics and population parameters.

TABLE 12–1
Sample Statistics and Population Parameters

Measures	Sample Statistic	Population Parameter	Remarks
Central tendency (average)	$\overline{X}$	μ	$\overline{X}$ is a point estimate of μ
Dispersion (standard deviation)	s (see note)	σ (see note)	s is a point estimate of σ

Note: Different statistical calculators use different buttons to represent the standard deviation of a sample versus the standard deviation of a population. In general, these buttons are:

Sample standard deviation: σ_{n-1} or s or s_x

Population standard deviation: σ_n, or σ', or σ

PROBABILITY DISTRIBUTIONS FOR ATTRIBUTES

There are a variety of probability distributions that provide the foundation for statistical quality control methods. These distributions are categorized as discrete or continuous. The most widely used attribute distributions in statistical quality control are binomial, Poisson, and hypergeometric, which are covered in this section.

It is important to note, at this point, that the following coverage on probability distributions and the examples calculate the probability of *exactly x* number of defects or defectives in a sample. There are tables for these distributions that show the probabilities associated with finding *x or fewer* defectives in a sample. For this reason, the calculated examples will be different from the values in those tables.

Discrete versus Continuous Distributions

The following distributions for attributes (binomial, Poisson, and hypergeometric) are all **discrete distributions.** A discrete distribution is a distribution that deals with discrete data (such as things we count). For example, if 5 samples were taken from each of several successive lots, in each sample, we may find *x* number of defects or defectives. Since this is discrete data, there can only be six results in each sample of 5 parts (0, 1, 2, 3, 4, or 5 defectives). There cannot be any fractional defectives in the samples, therefore the data is discrete. The **normal distribution,** which applies to variables data (see Chapter 13), is a **continuous distribution** that applies to measured data on some continuous scale. It is a continuous distribution because the results (or data) can be in many forms including fractional data (e.g., 1.501 diameter).

The Binomial Distribution

The **binomial distribution** is one of the distributions for statistical sampling plans for attributes and statistical process control (SPC) methods for attributes such as the *p* chart (covered in Chapter 13). The binomial distribution is a *discrete* distribution for sampling *with replace-*

ment. It generally applies to processes where np ≥ 5; samples are small compared to lot size (e.g., sample sizes are less than 10% of lot size); and items will be replaced in the lot.

Before going further, it is important to study **factorials.** Factorials are used in all of the attribute distribution equations.

A Review of Factorials

The factorial of any number is the product of multiplying all numbers (starting from 1) to that number. For example, the factorial of *n* is:

$$n! = 1 \cdot 2 \cdot 3 \cdot 4 \ldots \cdot n$$

The following are examples of factorials:

$$2! = 1 \times 2 = \mathbf{2}$$

$$3! = 1 \times 2 \times 3 = \mathbf{6}$$

$$4! = 1 \times 2 \times 3 \times 4 = \mathbf{24}$$

$$8! = 1 \times 2 \times 3 \times 4 \times 5 \times 6 \times 7 \times 8 = \mathbf{40{,}320}$$

For the following problems, use a scientific calculator (with scientific notation) because the factorials of certain numbers result in very large numbers.

Calculating the Binomial Distribution

The equation for calculating the probabilities under the binomial distribution is as follows:

$$P_X = \frac{n!}{x!(n - x)!} \cdot p^x \cdot (1 - p)^{n-x}$$

where: P_x = the probability of *x* occurrences
  n = the sample size
  p = the constant proportion defective

■ **Example 12–9**

A lot from a process is being inspected. The process average (*p*) is .10 proportion defective (or 10% defective). The sample size (*n*) is 50. What is the probability of finding exactly *x* = 2 defectives in the sample? Use a scientific calculator.

$$P_X = \frac{n!}{x!(n - x)!} \cdot p^x \cdot (1 - p)^{n-x}$$

$$= \frac{50!}{2!(50 - 2)!} \cdot .10^2 \cdot (1 - .10)^{50-2}$$

$$= 1225 \cdot .01 \cdot .00636$$

$$= .0779 \text{ (or } 7.79\%)$$

Therefore, there is a 7.79% chance of finding 2 defectives in the sample.

The Poisson Distribution

The **Poisson distribution** is also the basis for statistical sampling plans for attributes, and statistical process control (SPC) methods for attributes such as the *c* chart (covered in Chapter 13). The Poisson distribution is a *discrete* distribution for sampling *with replacement*. The Poisson distribution applies to sampling from a *continuous time or space* (such as a 300-foot roll of steel). It can also be used, at times, to approximate the binomial distribution. It generally applies to processes where np < 5; samples are large compared to lot size (e.g., sample sizes are greater than 10% of lot size); and items will be replaced in the lot.

The equation for calculating the probabilities under the Poisson distribution is as follows:

$$P_x = \frac{(np)^x \cdot e^{-np}}{x!}$$

where: P_x = the probability of x occurrences
n = the sample size
$!$ = the factorial of a value (e.g., $3! = 1 \times 2 \times 3 = 6$)
p = the constant proportion defective
e = constant value 2.71828

■ **Example 12–10**

A lot from a process is being inspected. The process average (p) is .05 proportion defective (or 5% defective). The sample size (n) is 100. What is the probability of finding exactly $x = 3$ defectives in the sample? Use a scientific calculator.

Solution

$$P_x = \frac{(np)^x \cdot e^{-np}}{x!}$$

$$= \frac{(100 \times .05)^3 \cdot 2.71828^{-5}}{3!}$$

$$= \frac{125 \cdot .006738}{6}$$

$$= .1404 \text{ (or } 14.04\%)$$

In this case, there is a 14.04% chance of finding exactly 3 defectives in a sample of 100 parts.

The Hypergeometric Distribution

The main differences between the hypergeometric distribution and the binomial and Poisson distributions are:

- Sampling is without replacement.
- The hypergeometric distribution applies to noncontinuous processes.
- The hypergeometric distribution applies to small lots and small samples.
- The hypergeometric distribution applies when the number of defects (not defectives) occurring in a process is known.

The equation for calculating the probabilities under a hypergeometric distribution is somewhat more complex than the others. It involves *combinations.*

$$P_x = \frac{C_X^d \cdot C_{n-x}^{N-d}}{C_n^N}$$

where: $d =$ is the occurrences in the population
$P_x =$ the probability of finding x defects
$N =$ the population (lot size)
$n =$ the sample size

First, a study of combinations is necessary to work with the hypergeometric distribution equation. Note that factorials are also used in the equation.

Combinations.   Combinations have to do with the different ways a sample of n things can be taken x number of them at a time. For example, how many combinations are there for 5 different things taken 2 at a time? The equation for combinations is:

$$C_x^n = \frac{n!}{x!(n-x)!}$$

$$= \frac{5!}{2!(5-2)!}$$

$$= \frac{120}{12}$$

$$= 10 \text{ combinations}$$

■ Example 12–11: The Hypergeometric Distribution

A noncontinuous process produces products that are delivered in small lots, and inspection samples taken are also small. A lot of $N = 22$ parts has been presented for inspection from a process where there are $d = 3$ occurrences in the population.

A sample of $n = 5$ parts will be taken to inspect the lot. What is the probability of finding exactly $x = 1$ defect in the sample?

$$P_x = \frac{C_X^d \cdot C_{n-x}^{N-d}}{C_n^N}$$

$$= \frac{C_1^3 \cdot C_{5-1}^{22-3}}{C_5^{22}}$$

$$= \frac{3 \cdot 3,876}{26,334}$$

$$= .442 \text{ (or 44.2\%)}$$

Summary

These attribute distributions, as stated earlier, are the foundation for certain statistical sampling plans and statistical process control methods in which attribute data are collected. For an overview of these distributions, see Table 12–2.

RELATIONSHIPS BETWEEN SAMPLING AND 100% INSPECTION

When deciding whether to inspect items 100% or use the various sampling schemes, you must first understand the main advantages and disadvantages of both. Listed below is a comparison of sampling inspection to 100% inspection.

100% Inspection

Definition: An inspection of every part produced.

Costs: 100% inspection is *costly* due to the time involved to inspect every part.

Effectiveness: Due to boredom, fatigue, and other human problems, 100% inspection is not 100 percent effective. In fact, it is usually only 70 to 90 percent effective.

Application: In certain situations (such as destructive tests) 100% inspection cannot be done.

Sampling Inspection

Definition: Inspecting a *random* sample of those parts produced.

Costs: Sampling inspection takes less time and is, therefore, less expensive.

Effectiveness: With sampling inspection, there are also risks involved, but these risks are calculated.

Application: Sampling inspection is a must for destructive tests.

100% Inspection

Handling: A higher probability of damaging a part because you must handle every part.

Bias: Sometimes encourages personal bias on the part of the inspector.

Sampling Inspection

Handling: You only handle a few of the parts; therefore, handling damage is less likely.

Bias: Decision making helps the inspector to be more objective. Bias is not often a problem here.

TABLE 12–2
Probability Distributions for Attributes

Distribution	Type	n/N ratio	Sample with replacement	$P_{\text{defectives}}$	Equation	Elements
Binomial	Discrete	$n < 10\% \ N$	Yes	$p \geq 10\%$ constant	$P_{(X)} = \dfrac{n!}{x!(n-x)!} \cdot p^x \cdot 1 - p^{n-x}$	*Given:* n = sample size p = constant proportion defective *Find:* x = occurrences
Poisson (approximates) binomial in some cases)	Discrete Continuous time or space (e.g., 500′ roll of steel)	$n > 10\% \ N$	Yes	$p < 10\%$ constant	$P_{(x)} = \dfrac{(np)^x \cdot e^{-np}}{x!}$	*Given:* e = 2.71828 constant n = sample size p = proportion defective *Find:* x = occurrence
Hypergeometric	Discrete (Finite universe, small lots, noncontinuous process)	High (n is small, N is also small) $n > 5\% \ N$	No	Number of defects are known	$P_{(x)} = \dfrac{C_x^d \cdot C_{n-x}^{N-d}}{C_n^N}$ (see note)	*Given:* n = sample d = occurrences in population N = population *Find:* x = occurrences

Note: The combinations of n items taken x at a time is: $C_x^n = \dfrac{n!}{x!(n-x)!}$

The previous information given on sampling and 100% inspection could lead a person to believe that there is no place for 100% inspection in industry. This is not true. There are some positive points to be made about 100% inspection that should be considered.

1. 100% inspection is more effective than sampling and is not prohibitive with regard to cost when it is *automated.* Machines do not get tired, bored, or biased.
2. 100% inspection (when done manually) can be made more effective by rotating the people who are doing the 100% inspection. Rotating (or having someone else take over the job for a while) offers less chance that the person will get tired or bored.
3. 100% inspection (done properly) gives you more information about quality levels than sampling does. This is because you had the opportunity to evaluate the quality of the entire lot being inspected.

All things considered, sampling inspection still comes out ahead so far as economic choices are concerned. All of the various plans for sampling have certain situations where they should be used, and if you choose the right one for the job and stick to it, the quality levels can be maintained.

Caution should be observed, however, when you make the decision to sample or 100% inspect; certain end items (especially those performing major or critical functions) should always be inspected and tested 100 percent to assure the customer maximum product quality and reliability. Product characteristics should be classified (e.g., critical, major, or minor) for best results.

LOT-BY-LOT ACCEPTANCE SAMPLING VERSUS PROCESS CONTROL SAMPLING

Important differences between lot-by-lot acceptance sampling and process control sampling should be considered when making the choice to sample. First is the effectiveness of the sampling plans. The overall effectiveness of sampling has already been discussed. Now the question is primarily where to use sampling or at what stages of the process. Process control sampling (refer to Chapter 13) involves the use of samples taken during the process and data are graphed (such as control charts and frequency distributions) in a manner that will measure and help monitor the variation of the process. Lot-by-lot acceptance sampling plans are designed to sample semifinished or finished lots for acceptance purposes. This method of sampling is reactive, in that the quality has already been built into the product by the time that acceptance sampling is used.

This is one of the primary reasons why process control sampling is more effective. Samples are taken during the processing of the product such that the identification of defects is closer to real time, and the process is controlled to prevent defects from being made. Lot-by-lot sampling plans are after the fact; when defects are found, it is usually too late (at this stage) to identify the true causes. When one considers the effect of finding problems at late stages of the process as opposed to finding problems at the time they happen, it is not too difficult to identify the most effective sampling plan.

The best decision is to find a balance between process control sampling during the process for preventive action and lot-by-lot sampling, which will help track the results of process control. Process control will usually be the primary choice as a tool for preventing defects from reaching the customer. The balance of this chapter is dedicated to how lot-by-lot sampling plans work and is intended to give the reader a basis for decisions regarding using the plans properly, or deciding which plan to use. A fundamental concept in lot-by-lot sampling is ensuring that the quality of a sample is representative of the quality of the lot. One aspect of representative samples is that samples are taken at random from the lot.

RANDOM SAMPLING

Many companies use sampling plans to inspect lots because of the advantages in sampling over 100% inspection. The goal of sampling is to accept good lots and reject bad lots a high percentage of the time. In addition to reading sampling tables and understanding them, the inspector must have the ability (and the desire) to take samples at *random*. Randomness is important when you are sampling a lot of parts and making a decision on the entire lot based on a small sample. That sample must be representative of the lot (not just the top layer).

Definition

Random sampling is giving every part in the lot an equal chance of being selected for the sample, regardless of its quality.

Drawbacks

Too often, it has been noted that when a company sample-inspects, emphasis is put on how to read the sampling tables and very little on the importance of random samples and how to take them. This is only one problem. Another is that some know how to take random samples but do not because it is too much work, or sometimes they cannot because of the way lots are packaged. It must be understood that when a company chooses to save money by sampling, they must also provide the means by which random samples can be taken. This may be in the form of different packaging or simply giving the inspector time to do so. Either way, it must be done. Sampling without randomness ruins the effectiveness of any sampling plan.

How Lots Are Packaged

This book will explain four main ways in which lots are delivered to inspection:

1. Single layer on a pallet
2. Boxes stacked on top of each other
3. One box where parts are in layers
4. Individual parts on a conveyor

Single Layer on a Pallet.   The easiest of all to random sample are those lots presented on a single layer on a pallet, as shown in Figure 12.1. In this case, every single box on the pallet should be given an equal chance of being selected (not just those nearest you). Then every part in those boxes chosen is given an equal chance of being selected for the sample.

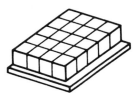

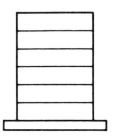

FIGURE 12.1
Single layer on a pallet.

FIGURE 12.2
Stacked layers in boxes.

Boxes Stacked on Top of Each Other. This situation is a little harder to sample randomly. To do so, you must unstack the boxes and draw your sample from the parts when all of them are exposed (giving them an equal chance) (see Figure 12.2). Often, this calls for a little muscle, but it is worth the effort.

One Box Where Parts Are in Layers. This one is the most difficult of all (especially if the parts are either very heavy or very small and numerous) (see Figure 12.3). This requires that the packaged layers are removable or that the inspector "dig" to obtain a random sample. It is very hard to dig down and get randomness because of the parts in the center and those at the very bottom. Sometimes "thief" sampling doors can be put into the box so that the inspector can look at the bottom layer.

Individual Parts on a Conveyor. This method of delivery makes it easy to get random samples. All of the parts are exposed, and your only decision is what sequence of parts to look at (every third one, or number 3, 9, 16, 24, etc.). There are many ways of obtaining random samples if this is the method of delivery.

1. *Numbers in a hat.* A hat with various numbers in it can be used. If your sample size for the entire lot is to be 13, draw 13 numbers from the hat and select the parts based on this sequence of numbers.
2. *Numbered balls in a jar.* This works the same way as the hat. For example, your sample size may be 5 out of 50 parts total. You draw 5 numbered balls from the jar. They are 3, 17, 21, 34, 48. Now, you simply inspect the 3rd, 17th, 21st, 34th, and 48th parts as they come down the conveyor.
3. *Table of random numbers.* The sample table in Table 12–3 is an example of a portion of a table of random numbers. These numbers have no predetermined sequence and are

FIGURE 12.3
Multilayers in a container.

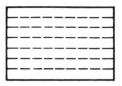

TABLE 12–3

Example Table of Random Numbers

Line \ Column	(1)	(2)	(3)	(4)	(5)
1	10480	15011	01536	02011	81647
2	23368	46573	25595	85393	30995
3	24130	48360	22527	97265	76393
4	42168	93093	06243	61680	07856
5	37570	39975	81837	16656	06121

completely random. The tables allow you to select at random to avoid bias in your sample. Random number tables are one of the best methods of assuring that random samples are taken.

Using the Table of Random Numbers

You can find random number tables in any book of math tables or a statistical text. If you get the chance to use one, the steps below basically explain how.

Step 1. Determine your sample size. (*Example:* an 80-piece lot and a five-piece sample.)

Step 2. Choose any line and column combination you want. (For this example, I chose line 1, column 1. The first number here is 10480.)

Step 3. Since 80 (the lot size) is two digits, you can use any combination from the five digits shown.

For example, in 10480 you can use 10, 04, 48, or 80 as your first number selected. I chose 10.

Step 4. Now go down the column using the first two numbers in each one. They are (line 1, column 1): 10, 23, 24, 42, and 37.

These are the numbered parts I will inspect. *On a conveyor,* I will inspect the 10th, 23rd, 24th, 37th, and 42nd pieces. *Parts in a box* must be set out and numbered prior to reading the table.

Note: If you reach a number that is repeated, skip that number and use the next one. If you reach a number that is too large (over 80), skip that number also, and select the next one.

Summary. There are numerous other ways to get random numbers (such as a deck of cards or rolling dice), but the main thing to remember is that *without randomness the risks of sampling error are greater.* Refer to Table 12–4 for the risks associated with sampling plans.

TABLE 12–4
Sampling Risks

Decision ⇓	Quality ⇒	Good Lot	Bad Lot
Accept		Correct decision	Beta risk (consumer's risk)
Reject		Alpha risk (producer's risk)	Correct decision

USING ANSI/ASQC Z1.4 FOR SAMPLING INSPECTION

Generally, Z1.4 is used as a basis for sampling to effectively sample-inspect products. Other sampling tables (such as Dodge–Romig tables) can also be used; these tables work well when the process average is known.

Before reading further, refer to the Glossary for sampling terms and definitions. See the Appendix for the actual tables taken from Z1.4.

With any sampling plans, the inspector must have certain pieces of information ahead of time:

1. *Lot size.* This is usually known on lots submitted for inspection.
2. *Inspection level.* This is usually given in inspection plans or procedures.
3. *AQL.* This is also usually given in inspection plans or procedures.
4. *Type of sampling to use.* This is single, double, or multiple.

When these things are known, it is simply a matter of reading the appropriate tables and understanding what to do with them.

These plans are primarily intended to be used to sample-inspect a continuing series of lots (or batches), as opposed to isolated lots.

Single Sampling Tables*

Shown in Table 12–5 is an example of *single sampling tables.* This type of sampling is the main area of discussion here. The actual code letter table appears in Table 12–6.

The thing you are looking for is called the *sampling plan.* This is simply the sample size (*n*), the accept number (Ac), and the reject number (Re). The steps for finding the sampling plan for a lot presented for inspection are as follows.

■ Example 12–12

Lot size = 510 pieces, AQL = 4.0 percent, general inspection level II.

Step 1. Starting with Table 12–6, find the range of numbers under "lot size" in which the lot size you are sampling (510) falls. This range is 501 to 1200.

---

* Refer to Table 12–5.

TABLE 12–5

Z1.4 Single Sampling Plans for Normal Inspection (Master Table)

Acceptable quality levels (normal inspection)

Sample size code letter	Sample size	0.010		0.015		0.025		0.040		0.065		0.10		0.15		0.25		0.40		0.65		1.0		1.5		2.5		4.0		6.5		10		15		25		40		65		100		150		250		400		650		1,000	
		Ac	Re	Ac	Re	Ac	Re	Ac	Re	Ac	Re	Ac	Re	Ac	Re	Ac	Re	Ac	Re	Ac	Re	Ac	Re	Ac	Re	Ac	Re	Ac	Re	Ac	Re	Ac	Re	Ac	Re	Ac	Re	Ac	Re	Ac	Re	Ac	Re	Ac	Re	Ac	Re	Ac	Re	Ac	Re
A	2	↓		↓		↓		↓		↓		↓		↓		↓		↓		↓		↓		↓		↓		↓		↓		↓		0	1	1	2	2	3	3	4	5	6	7	8	10	11	14	15	21	22	30	31
B	3	↓		↓		↓		↓		↓		↓		↓		↓		↓		↓		↓		↓		↓		↓		↓		0	1	1	2	2	3	3	4	5	6	7	8	10	11	14	15	21	22	30	31	44	45
C	5	↓		↓		↓		↓		↓		↓		↓		↓		↓		↓		↓		↓		↓		↓		0	1	1	2	2	3	3	4	5	6	7	8	10	11	14	15	21	22	30	31	44	45	↑	
D	8	↓		↓		↓		↓		↓		↓		↓		↓		↓		↓		↓		↓		↓		0	1	1	2	2	3	3	4	5	6	7	8	10	11	14	15	21	22	30	31	44	45	↑		↑	
E	13	↓		↓		↓		↓		↓		↓		↓		↓		↓		↓		↓		↓		0	1	1	2	2	3	3	4	5	6	7	8	10	11	14	15	21	22	30	31	44	45	↑		↑		↑	
F	20	↓		↓		↓		↓		↓		↓		↓		↓		↓		↓		↓		0	1	1	2	2	3	3	4	5	6	7	8	10	11	14	15	21	22	30	31	44	45	↑		↑		↑		↑	
G	32	↓		↓		↓		↓		↓		↓		↓		↓		↓		↓		0	1	1	2	2	3	3	4	5	6	7	8	10	11	14	15	21	22	30	31	44	45	↑		↑		↑		↑		↑	
H	50	↓		↓		↓		↓		↓		↓		↓		↓		↓		0	1	1	2	2	3	3	4	5	6	7	8	10	11	14	15	21	22	30	31	44	45	↑		↑		↑		↑		↑		↑	
J	80	↓		↓		↓		↓		↓		↓		↓		↓		0	1	1	2	2	3	3	4	5	6	7	8	10	11	14	15	21	22	30	31	44	45	↑		↑		↑		↑		↑		↑		↑	
K	125	↓		↓		↓		↓		↓		↓		↓		0	1	1	2	2	3	3	4	5	6	7	8	10	11	14	15	21	22	30	31	44	45	↑		↑		↑		↑		↑		↑		↑		↑	
L	200	↓		↓		↓		↓		↓		↓		0	1	1	2	2	3	3	4	5	6	7	8	10	11	14	15	21	22	30	31	44	45	↑		↑		↑		↑		↑		↑		↑		↑		↑	
M	315	↓		↓		↓		↓		↓		0	1	1	2	2	3	3	4	5	6	7	8	10	11	14	15	21	22	30	31	44	45	↑		↑		↑		↑		↑		↑		↑		↑		↑		↑	
N	500	↓		↓		↓		↓		0	1	1	2	2	3	3	4	5	6	7	8	10	11	14	15	21	22	30	31	44	45	↑		↑		↑		↑		↑		↑		↑		↑		↑		↑		↑	
P	800	↓		↓		↓		0	1	1	2	2	3	3	4	5	6	7	8	10	11	14	15	21	22	30	31	44	45	↑		↑		↑		↑		↑		↑		↑		↑		↑		↑		↑		↑	
Q	1,250	↓		↓		0	1	1	2	2	3	3	4	5	6	7	8	10	11	14	15	21	22	30	31	44	45	↑		↑		↑		↑		↑		↑		↑		↑		↑		↑		↑		↑		↑	
R	2,000	↓		0	1	1	2	2	3	3	4	5	6	7	8	10	11	14	15	21	22	30	31	44	45	↑		↑		↑		↑		↑		↑		↑		↑		↑		↑		↑		↑		↑		↑	

↓ = use first sampling plan below arrow. If sample size equals, or exceeds, lot or batch size, do 100% inspection.

↑ = use first sampling plan above arrow.

Ac = acceptance number.

Re = rejection number.

TABLE 12-6
Z1.4 Sample Size Code Letters

Lot or batch size	Special inspection levels				General inspection levels		
	S-1	S-2	S-3	S-4	I	II	III
2 to 8	A	A	A	A	A	A	B
9 to 15	A	A	A	A	A	B	C
16 to 25	A	A	B	B	B	C	D
26 to 50	A	B	B	C	C	D	E
51 to 90	B	B	C	C	C	E	F
91 to 150	B	B	C	D	D	F	G
151 to 280	B	C	D	E	E	G	H
281 to 500	B	C	D	E	F	H	J
501 to 1200	C	C	E	F	G	J	K
1201 to 3200	C	D	E	G	H	K	L
3201 to 10000	C	D	F	G	J	L	M
10001 to 35000	C	D	F	H	K	M	N
35001 to 150000	D	E	G	J	L	N	P
150001 to 500000	D	E	G	J	M	P	Q
500001 and over	D	E	H	K	N	Q	R

Step 2. Look straight across from the lot size range you found earlier until you are directly under the inspection level you are working with (which in this example is general level II). There, you will find the code letter J.

Step 3. Turn to the sampling table, Table 12–5, and find code letter J again.

Step 4. Look straight across the page from your code letter J and stop when you are under the AQL you've been given (4.0 percent). This is where you find the accept number (Ac) of 7 and the reject number (Re) of 8.

Step 5. Look in the same row J under "sample size" to find what size your sample should be. In this example, it is 80 pieces.

n	Ac	Re
80	7	8

This tells you to take a *random* sample of 80 pieces from the lot and if you find seven defectives (or fewer) in your sample, you set the defectives aside and accept the rest of the lot. But if you find eight (or more) defectives in the sample, this is cause for rejection of the entire lot and returning it for screening (or sorting).

The Arrows

In the sampling table in Table 12–5, you will see that there are arrows, some pointing up and some down. These arrows are used as follows. First, you follow the previous steps to find your sampling plan. If you come to a sampling plan that shows an arrow, this means that you use the sampling plan just above or just below (depending on which way the arrow is pointing).

■ **Example 12–13**

Lot size = 200, general level II, AQL = 1.0 percent.

The code letter is G, but under 1.0 percent you find an arrow pointing down. This means that you go to code letter H, and the sampling plan is

N	Ac	Re
50	1	2

Note: It is important to use the entire plan, including the new sample size, that coincides with letter H.

Remember that with any sampling plan, taking a random sample is important for the plan to be effective in accepting good quality and rejecting poor quality.

Double Sampling per ANSI/ASQC Z1.4

Double sampling is used from time to time by inspection. Generally, double sampling plans are used in inspection when product quality levels are "real good" or "real bad." Depending on the actual quality levels, double sampling can reduce the amount of inspection you do and still give you confidence that you will stop poor quality and accept good quality (per the AQL). Double sampling works similarly to the single sampling plans discussed previously. Double sampling plans, however, usually result in the least amount of inspection over time.

You still need to know:

- AQL
- Reduced, normal, or tightened
- Lot size
- Level I, II, III, or others

You still use the table of code letters to obtain a code letter for your plan. However, *you must use the double sampling tables.*

The actual tables are given in Table 12–7. When you have selected your code letter and followed it over to just below the AQL you are using, you will see two sample sizes, two numbers, and two reject numbers.

The example in Table 12–7 uses code letter M and 1.0 percent AQL. This simply tells you that you sample 200 pieces on the first sample. If you find three (or fewer) defectives, accept the lot and set the defectives aside. If you find four, five, or six defectives, take the second sample of 200 pieces. If you find seven or more defectives in the first sample, reject the entire lot for screening inspection.

If you do take the second sample, the accept numbers and reject numbers include the number of defects found in the first sample.

■ Example 12–14

You found six defectives in the first sample. This tells you to take the second sample of 200 pieces. If you find two more defectives, this will total eight defectives found; you still accept the lot. If you find three more defectives, reject the entire lot because the reject number is nine.

Normal, Tightened, and Reduced Inspection

Specific rules must be followed for switching between normal, tightened, and reduced levels.

Normal to Tightened:

When normal inspection is in effect, tightened inspection shall be instituted when 2 out of 2, 3, 4, or 5 consecutive lots or batches have been rejected on original inspection (ignoring resubmitted lots or batches).

TABLE 12-7

Z1.4 Double Sampling Plans for Normal Inspection (Master Table)

Each data cell below is shown as "Ac Re" (acceptance number, rejection number). Arrows: ↓ = use first sampling plan below arrow; ↑ = use first sampling plan above arrow; † = use corresponding single sampling plan.

Sample size code letter	Sample	Sample size	Cumulative sample size	0.010	0.015	0.025	0.040	0.065	0.10	0.15	0.25	0.40	0.65	1.0	1.5	2.5	4.0	6.5	10	15	25	40	65	100	150	250	400	650	1,000
A				↓	↓	↓	↓	↓	↓	↓	↓	↓	↓	↓	↓	↓	↓	↓	↓	†	↑	↑	↑	↑	↑	↑	↑	↑	↑
B	First	2	2	↓	↓	↓	↓	↓	↓	↓	↓	↓	↓	↓	↓	↓	↓	↓	†	0 2	0 3	1 4	2 5	3 7	5 9	7 11	11 16	17 22	25 31
	Second	2	4																	1 2	3 4	4 5	6 7	8 9	12 13	18 19	26 27	37 38	56 57
C	First	3	3	↓	↓	↓	↓	↓	↓	↓	↓	↓	↓	↓	↓	↓	↓	†	0 2	0 3	1 4	2 5	3 7	5 9	7 11	11 16	17 22	25 31	↑
	Second	3	6																1 2	3 4	4 5	6 7	8 9	12 13	18 19	26 27	37 38	56 57	
D	First	5	5	↓	↓	↓	↓	↓	↓	↓	↓	↓	↓	↓	↓	↓	†	0 2	0 3	1 4	2 5	3 7	5 9	7 11	11 16	17 22	25 31	↑	↑
	Second	5	10															1 2	3 4	4 5	6 7	8 9	12 13	18 19	26 27	37 38	56 57		
E	First	8	8	↓	↓	↓	↓	↓	↓	↓	↓	↓	↓	↓	↓	†	0 2	0 3	1 4	2 5	3 7	5 9	7 11	11 16	17 22	25 31	↑	↑	↑
	Second	8	16														1 2	3 4	4 5	6 7	8 9	12 13	18 19	26 27	37 38	56 57			
F	First	13	13	↓	↓	↓	↓	↓	↓	↓	↓	↓	↓	↓	†	0 2	0 3	1 4	2 5	3 7	5 9	7 11	11 16	17 22	25 31	↑	↑	↑	↑
	Second	13	26													1 2	3 4	4 5	6 7	8 9	12 13	18 19	26 27	37 38	56 57				
G	First	20	20	↓	↓	↓	↓	↓	↓	↓	↓	↓	↓	†	0 2	0 3	1 4	2 5	3 7	5 9	7 11	11 16	17 22	25 31	↑	↑	↑	↑	↑
	Second	20	40												1 2	3 4	4 5	6 7	8 9	12 13	18 19	26 27	37 38	56 57					
H	First	32	32	↓	↓	↓	↓	↓	↓	↓	↓	↓	†	0 2	0 3	1 4	2 5	3 7	5 9	7 11	11 16	17 22	25 31	↑	↑	↑	↑	↑	↑
	Second	32	64											1 2	3 4	4 5	6 7	8 9	12 13	18 19	26 27	37 38	56 57						
J	First	50	50	↓	↓	↓	↓	↓	↓	↓	↓	†	0 2	0 3	1 4	2 5	3 7	5 9	7 11	11 16	17 22	25 31	↑	↑	↑	↑	↑	↑	↑
	Second	50	100										1 2	3 4	4 5	6 7	8 9	12 13	18 19	26 27	37 38	56 57							
K	First	80	80	↓	↓	↓	↓	↓	↓	↓	†	0 2	0 3	1 4	2 5	3 7	5 9	7 11	11 16	17 22	25 31	↑	↑	↑	↑	↑	↑	↑	↑
	Second	80	160									1 2	3 4	4 5	6 7	8 9	12 13	18 19	26 27	37 38	56 57								
L	First	125	125	↓	↓	↓	↓	↓	↓	†	0 2	0 3	1 4	2 5	3 7	5 9	7 11	11 16	17 22	25 31	↑	↑	↑	↑	↑	↑	↑	↑	↑
	Second	125	250								1 2	3 4	4 5	6 7	8 9	12 13	18 19	26 27	37 38	56 57									
M	First	200	200	↓	↓	↓	↓	↓	†	0 2	0 3	1 4	2 5	3 7	5 9	7 11	11 16	17 22	25 31	↑	↑	↑	↑	↑	↑	↑	↑	↑	↑
	Second	200	400							1 2	3 4	4 5	6 7	8 9	12 13	18 19	26 27	37 38	56 57										
N	First	315	315	↓	↓	↓	↓	†	0 2	0 3	1 4	2 5	3 7	5 9	7 11	11 16	17 22	25 31	↑	↑	↑	↑	↑	↑	↑	↑	↑	↑	↑
	Second	315	630						1 2	3 4	4 5	6 7	8 9	12 13	18 19	26 27	37 38	56 57											
P	First	500	500	↓	↓	↓	†	0 2	0 3	1 4	2 5	3 7	5 9	7 11	11 16	17 22	25 31	↑	↑	↑	↑	↑	↑	↑	↑	↑	↑	↑	↑
	Second	500	1,000					1 2	3 4	4 5	6 7	8 9	12 13	18 19	26 27	37 38	56 57												
Q	First	800	800	↓	↓	†	0 2	0 3	1 4	2 5	3 7	5 9	7 11	11 16	17 22	25 31	↑	↑	↑	↑	↑	↑	↑	↑	↑	↑	↑	↑	↑
	Second	800	1,600				1 2	3 4	4 5	6 7	8 9	12 13	18 19	26 27	37 38	56 57													
R	First	1,250	1,250	↓	†	0 2	0 3	1 4	2 5	3 7	5 9	7 11	11 16	17 22	25 31	↑	↑	↑	↑	↑	↑	↑	↑	↑	↑	↑	↑	↑	↑
	Second	1,250	2,500			1 2	3 4	4 5	6 7	8 9	12 13	18 19	26 27	37 38	56 57														

Acceptable quality levels (normal inspection)

↓ = use first sampling plan below arrow. If sample size equals or exceeds lot or batch size, do 100 % inspection.
↑ = use first sampling plan above arrow.
Ac = acceptance number.
Re = rejection number.
† Use corresponding single sampling plan (or alternatively, use double sampling plan below, where available).

Tightened to Normal:

When tightened inspection is in effect, normal inspection shall be instituted when 5 consecutive lots or batches have been considered acceptable on original inspection.

Normal to Reduced:

When normal inspection is in effect, reduced inspection shall be instituted provided that all of the following conditions are satisfied:

a. The preceding 10 lots or batches have been on normal inspection and all have been accepted on original inspection; and

b. The total number of defectives (or defects) in the samples from the preceding 10 lots or batches is equal to or less than the applicable number given in Table VIII. If double or multiple sampling is in use, all samples inspected should be included, not "first" sample only; [*Note:* Table VIII is Limit Numbers for Reduced Inspection (see the standard).] and

c. Production is at a steady rate; and

d. Reduced inspection is considered desirable.

Reduced to Normal:

When reduced inspection is in effect, normal inspection shall be instituted if any of the following occur on original inspection:

a. A lot or batch is rejected; or

b. A lot or batch is considered acceptable using other procedures (see Mil-Std-105E); or

c. Production becomes irregular or delayed; or

d. Other conditions warrant that normal inspection shall be instituted.

INTRODUCTION TO DODGE–ROMIG SAMPLING TABLES

Dodge–Romig sampling tables are another method of sampling by attributes. These tables are most effective when you know the process average (or fraction defective) of the lot. This process average is called ($\bar{p}$). If Dodge–Romig tables are to be used with an unknown process average ($\bar{p}$), you should use the largest $\bar{p}$ in the tables until you gather enough data to establish a process average.

Dodge–Romig sampling tables are used with the understanding that rejected lots will be screened (or 100 percent inspected). When used properly, Dodge–Romig sampling plans reduce the amount of inspection without increasing the risk of poor outgoing quality. This reduced amount of inspection saves time and money and allows you to inspect lots quicker and still be effective in stopping poor quality.

Dodge–Romig tables come in four types.

1. Single sampling (LTPD) lot tolerance percent defective
2. Double sampling (LTPD)
3. Single sampling (AOQL)
4. Double sampling (AOQL)

AOQL plans are applicable to producer's outgoing inspection; LTPD plans are applicable to consumer's receiving inspection.

Percentages at the top of each table reflect the percent defective lot that will be rejected using any of the plans within that table. Refer to Table 12–8.

When double sampling (with Dodge–Roming tables) instead of single sampling, fewer units need to be inspected, provided that the decision to accept or reject the lot can be made on the first sample. See Table 12–9.

Table 12–9 is another example of a Dodge–Romig inspection table. It shows that with a lot size ranging from 801 pieces to 1000 pieces, your first sample will be 55 pieces and you can accept the lot if there are no defectives in the first sample. If there is one defective in the first sample, you must take another sample. If there is more than one defective, you reject the lot. If you take a second sample, your cumulative sample size ($n_1 + n_2$) is 85 pieces, and you can accept the lot if there is only one defective, but you must reject the lot if there is more than one defective part.

ANSI/ASQC Z1.9 VARIABLES SAMPLING TABLES

Variables sampling tables (per ANSI Z1.9) are based on the normal distribution.

Sampling by variables is different from sampling by attributes because here you are measuring actual sizes and variation on a continuous scale (such as inches or pounds) and are not simply concerned with defective or not defective.

ANSI/ASQC Z1.9 has four sections: A, B, C, and D.

Section A. Introduction.

Section B. Consists of sampling plans that are used when the variability is unknown, and the *standard deviation method* is used.

Section C. Consists of sampling plans that are used when the variability is unknown, and the *range method* is used.

Section D. Consists of sampling plans that are used when the variability is known.

There are five inspection levels: I, II, III, IV, V. If there is no level specified, use level IV.

The use of variables sampling requires that you

1. Choose the level.
2. Choose the method.
3. Know the AQL.
4. Know the lot size.

Table 12–10 shows the ranges of AQLs, and you must use this table to establish the AQL needed in the plans.

■ **Example 12–15**

Specified AQL Range	**Use This AQL**
.110 to .164	.15

Single Sampling Table for
Average Outgoing Quality Limit (AOQL) = 2.0%

Lot Size	Process Average 0 to 0.04%			Process Average 0.05 to 0.40%			Process Average 0.41 to 0.80%			Process Average 0.81 to 1.20%			Process Average 1.21 to 1.60%			Process Average 1.61 to 2.00%		
	n	c	p_t %	n	c	p_t %	n	c	p_t %	n	c	p_t %	n	c	p_t %	n	c	p_t %
1–15	All	0	–	All	0	–	All	0	–	All	0	–	All	0	–	All	0	–
16–50	14	0	13.6	14	0	13.6	14	0	13.6	14	0	13.6	14	0	13.6	14	0	13.6
51–100	16	0	12.4	16	0	12.4	16	0	12.4	16	0	12.4	16	0	12.4	16	0	12.4
101–200	17	0	12.2	17	0	12.2	17	0	12.2	17	0	12.2	35	1	10.5	35	1	10.5
201–300	17	0	12.3	17	0	12.3	17	0	12.3	37	1	10.2	37	1	10.2	37	1	10.2
301–400	18	0	11.8	18	0	11.8	38	1	10.0	38	1	10.0	38	1	10.0	60	2	8.5
401–500	18	0	11.9	18	0	11.9	39	1	9.8	39	1	9.8	60	2	8.6	60	2	8.6
501–600	18	0	11.9	18	0	11.9	39	1	9.8	39	1	9.8	60	2	8.6	60	2	8.6
601–800	18	0	11.9	40	1	9.6	40	1	9.6	65	2	8.0	65	2	8.0	85	3	7.5
801–1000	18	0	12.0	40	1	9.6	40	1	9.6	65	2	8.1	65	2	8.1	90	3	7.4
1001–2000	18	0	12.0	41	1	9.4	65	2	8.2	65	2	8.2	95	3	7.0	120	4	6.5
2001–3000	18	0	12.0	41	1	9.4	65	2	8.2	95	3	7.0	120	4	6.5	180	6	5.8
3001–4000	18	0	12.0	42	1	9.3	65	2	8.2	95	3	7.0	155	5	6.0	210	7	5.5
4001–5000	18	0	12.0	42	1	9.3	70	2	7.5	125	4	6.4	155	5	6.0	245	8	5.3
5001–7000	18	0	12.0	42	1	9.3	95	3	7.0	125	4	6.4	185	6	5.6	280	9	5.1
7001–10,000	42	1	9.3	70	2	7.5	95	3	7.0	155	5	6.0	220	7	5.4	350	11	4.8
10,001–20,000	42	1	9.3	70	2	7.6	95	3	7.0	190	6	5.6	290	9	4.9	460	14	4.4
20,001–50,000	42	1	9.3	70	2	7.6	125	4	6.4	220	7	5.4	395	12	4.5	720	21	3.9
50,001–100,000	42	1	9.3	95	3	7.0	160	5	5.9	290	9	4.9	505	15	4.2	955	27	3.7

Single Sampling Table for
Average Outgoing Quality Limit (AOQL) = 2.5%

2.5%

Lot Size	Process Average 0 to 0.05%			Process Average 0.06 to 0.50%			Process Average 0.51 to 1.00%			Process Average 1.01 to 1.50%			Process Average 1.51 to 2.00%			Process Average 2.01 to 2.50%		
	n	c	p_t %	n	c	p_t %	n	c	p_t %	n	c	p_t %	n	c	p_t %	n	c	p_t %
1–10	All	0	–	All	0	–	All	0	–	All	0	–	All	0	–	All	0	–
11–50	11	0	17.6	11	0	17.6	11	0	17.6	11	0	17.6	11	0	17.6	11	0	17.6
51–100	13	0	15.3	13	0	15.3	13	0	15.3	13	0	15.3	13	0	15.3	13	0	15.3
101–200	14	0	14.7	14	0	14.7	14	0	14.7	29	1	12.9	29	1	12.9	29	1	12.9
201–300	14	0	14.9	14	0	14.9	30	1	12.7	30	1	12.7	30	1	12.7	30	1	12.7
301–400	14	0	15.0	14	0	15.0	31	1	12.3	31	1	12.3	31	1	12.3	48	2	10.7
401–500	14	0	15.0	14	0	15.0	32	1	12.0	32	1	12.0	49	2	10.6	49	2	10.6
501–600	14	0	15.1	14	0	15.1	32	1	12.0	50	2	10.4	50	2	10.4	70	3	9.3
601–800	14	0	15.1	32	1	12.0	32	1	12.0	50	2	10.5	50	2	10.5	70	3	9.4
801–1000	15	0	14.2	33	1	11.7	33	1	11.7	50	2	10.6	70	3	9.4	90	4	8.5
1001–2000	15	0	14.2	33	1	11.7	55	2	9.3	75	3	8.8	95	4	8.0	120	5	7.6
2001–3000	15	0	14.2	33	1	11.8	55	2	9.4	75	3	8.8	120	5	7.6	145	6	7.2
3001–4000	15	0	14.3	33	1	11.8	55	2	9.5	100	4	7.9	125	5	7.4	195	8	6.6
4001–5000	15	0	14.3	33	1	11.8	75	3	8.9	100	4	7.9	150	6	7.0	225	9	6.3
5001–7000	33	1	11.8	55	2	9.7	75	3	8.9	125	5	7.4	175	7	6.7	250	10	6.1
7001–10,000	34	1	11.4	55	2	9.7	75	3	8.9	125	5	7.4	200	8	6.4	310	12	5.8
10,001–20,000	34	1	11.4	55	2	9.7	100	4	8.0	150	6	7.0	260	10	6.0	425	16	5.3
20,001–50,000	34	1	11.4	55	2	9.7	100	4	8.0	180	7	6.7	345	13	5.5	640	23	4.8
50,001–100,000	34	1	11.4	80	3	8.4	125	5	7.4	235	9	6.1	435	16	5.2	800	28	4.5

n = sample size; c = acceptance number
"All" indicates that each piece in the lot is to be inspected
p_t = lot tolerance per cent defective with a Consumer's Risk (P_C) of 0.10

TABLE 12–8
Section of a Dodge–Romig Sampling Table. Single Sampling Plans (AOQL = 2.0% and 2.5%)

Double Sampling Table for
Average Outgoing Quality Limit (AOQL) = 1.0%

Lot Size	Process Average 0 to 0.02%						Process Average 0.03 to 0.20%						Process Average 0.21 to 0.40%					
	Trial 1		Trial 2			p_t %	Trial 1		Trial 2			p_t %	Trial 1		Trial 2			p_t %
	n_1	c_1	n_2	n_1+n_2	c_2		n_1	c_1	n_2	n_1+n_2	c_2		n_1	c_1	n_2	n_1+n_2	c_2	
1–25	All	0	–	–	–	–	All	0	–	–	–	–	All	0	–	–	–	–
26–50	22	0	–	–	–	7.7	22	0	–	–	–	7.7	22	0	–	–	–	7.7
51–100	33	0	17	50	1	6.9	33	0	17	50	1	6.9	33	0	17	50	1	6.9
101–200	43	0	22	65	1	5.8	43	0	22	65	1	5.8	43	0	22	65	1	5.8
201–300	47	0	28	75	1	5.5	47	0	28	75	1	5.5	47	0	28	75	1	5.5
301–400	49	0	31	80	1	5.4	49	0	31	80	1	5.4	55	0	60	115	2	4.8
401–500	50	0	30	80	1	5.4	50	0	30	80	1	5.4	55	0	65	120	2	4.7
501–600	50	0	30	80	1	5.4	50	0	30	80	1	5.4	60	0	65	125	2	4.6
601–800	50	0	35	85	1	5.3	60	0	70	130	2	4.5	60	0	70	130	2	4.5
801–1000	55	0	30	85	1	5.2	60	0	75	135	2	4.4	60	0	75	135	2	4.4
1001–2000	55	0	35	90	1	5.1	65	0	75	140	2	4.3	75	0	120	195	3	3.8
2001–3000	65	0	80	145	2	4.2	65	0	80	145	2	4.2	75	0	125	200	3	3.7
3001–4000	70	0	80	150	2	4.1	70	0	80	150	2	4.1	80	0	175	255	4	3.5
4001–5000	70	0	80	150	2	4.1	70	0	80	150	2	4.1	80	0	180	260	4	3.4
5001–7000	70	0	80	150	2	4.1	75	0	125	200	3	3.7	80	0	180	260	4	3.4
7001–10,000	70	0	80	150	2	4.1	80	0	125	205	3	3.6	85	0	180	265	4	3.3
10,001–20,000	70	0	80	150	2	4.1	80	0	130	210	3	3.6	90	0	230	320	5	3.2
20,001–50,000	75	0	80	155	2	4.0	80	0	135	215	3	3.6	95	0	300	395	6	2.9
50,001–100,000	75	0	80	155	2	4.0	85	0	180	265	4	3.3	170	1	380	550	8	2.6

Lot Size	Process Average 0.41 to 0.60%						Process Average 0.61 to 0.80%						Process Average 0.81 to 1.00%					
	Trial 1		Trial 2			p_t %	Trial 1		Trial 2			p_t %	Trial 1		Trial 2			p_t %
	n_1	c_1	n_2	n_1+n_2	c_2		n_1	c_1	n_2	n_1+n_2	c_2		n_1	c_1	n_2	n_1+n_2	c_2	
1–25	All	0	–	–	–	–	All	0	–	–	–	–	All	0	–	–	–	-
26–50	22	0	–	–	–	7.7	22	0	–	–	–	7.7	22	0	–	–	–	7.7
51–100	33	0	17	50	1	6.9	33	0	17	50	1	6.9	33	0	17	50	1	6.9
101–200	43	0	22	65	1	5.8	43	0	22	65	1	5.8	47	0	43	90	2	5.4
201–300	55	0	50	105	2	4.9	55	0	50	105	2	4.9	55	0	50	105	2	4.9
301–400	55	0	60	115	2	4.8	55	0	60	115	2	4.8	60	0	80	140	3	4.5
401–500	55	0	65	120	2	4.7	60	0	95	155	3	4.3	60	0	95	155	3	4.3
501–600	60	0	65	125	2	4.6	65	0	100	165	3	4.2	65	0	100	165	3	4.2
601–800	65	0	105	170	3	4.1	65	0	105	170	3	4.1	70	0	140	210	4	3.9
801–1000	65	0	110	175	3	4.0	70	0	150	220	4	3.8	125	1	180	305	6	3.5
1001–2000	80	0	165	245	4	3.7	135	1	200	335	6	3.3	140	1	245	385	7	3.2
2001–3000	80	0	170	250	4	3.6	150	1	265	415	7	3.0	215	2	355	570	10	2.8
3001–4000	85	0	220	305	5	3.3	160	1	330	490	8	2.8	225	2	455	680	12	2.7
4001–5000	145	1	225	370	6	3.1	225	2	375	600	10	2.7	240	2	595	835	14	2.5
5001–7000	155	1	285	440	7	2.9	235	2	440	675	11	2.6	310	3	665	975	16	2.4
7001–10,000	165	1	355	520	8	2.7	250	2	585	835	13	2.4	385	4	785	1170	19	2.3
10,001–20,000	175	1	415	590	9	2.6	325	3	655	980	15	2.3	520	6	980	1500	24	2.2
20,001–50,000	250	2	490	740	11	2.4	340	3	910	1250	19	2.2	610	7	1410	2020	32	2.1
50,001–100,000	275	2	700	975	14	2.2	420	4	1050	1470	22	2.1	770	9	1850	2620	41	2.0

Trial 1: n_1 = first sample size; c_1 = acceptance number for first sample
"All" indicates that each piece in the lot is to be inspected
Trial 2: n_2 = second sample size. c_2 = acceptance number for first and second samples combined
p_t = lot tolerance per cent defective with a Consumer's Risk (P_C) of 0.10

TABLE 12–9
Section of Dodge–Romig Double Sampling Tables for Average Outgoing Quantity Limit (AOQL)

Sample Size Code Letters[1]

Lot Size		Inspection Levels				
		I	II	III	IV	V
3 to	8	B	B	B	B	C
9 to	15	B	B	B	B	D
16 to	25	B	B	B	C	E
26 to	40	B	B	B	D	F
41 to	65	B	B	C	E	G
66 to	110	B	B	D	F	H
111 to	180	B	C	E	G	I
181 to	300	B	D	F	H	J
301 to	500	C	E	G	I	K
501 to	800	D	F	H	J	L
801 to	1,300	E	G	I	K	L
1,301 to	3,200	F	H	J	L	M
3,201 to	8,000	G	I	L	M	N
8,001 to	22,000	H	J	M	N	O
22,001 to	110,000	I	K	N	O	P
110,001 to	550,000	I	K	O	P	Q
550,001 and over		I	K	P	Q	Q

[1] Sample size code letters given in body of table are applicable when the indicated inspection levels are to be used.

AQL Conversion Table

For specified AQL values falling within these ranges	Use this AQL value
—— to 0.049	0.04
0.050 to 0.069	0.065
0.070 to 0.109	0.10
0.110 to 0.164	0.15
0.165 to 0.279	0.25
0.280 to 0.439	0.40
0.440 to 0.699	0.65
0.700 to 1.09	1.0
1.10 to 1.64	1.5
1.65 to 2.79	2.5
2.80 to 4.39	4.0
4.40 to 6.99	6.5
7.00 to 10.9	10.0
11.00 to 16.4	15.0

TABLE 12–10
Z1.9 Tables for AQL Conversion and Sample Size Code Letters

If you are working with an AQL of 0.13 percent, you must use 0.15 percent in the sampling tables.

There are two methods to use in finding the variability of a lot when sampling per ANSI/ASQC Z1.9. These are the standard deviation method and the range method.

Standard Deviation Method

Q_u is the number of standard deviations that the mean $(\overline{X})$ is away from the upper tolerance limit.

Q_L is the number of standard deviations that the mean $(\overline{X})$ is away from the lower tolerance limit. s is the estimated standard deviation.

$$s = \sqrt{\frac{\Sigma(x - \overline{x})^2}{n - 1}}$$

$$Q = \frac{\text{tolerance limit} - \text{mean}}{\text{standard deviation}}$$

Range Method

You must find $\overline{R}$, which is the average range of the subgroups. (Sometimes there is only one subgroup; if so, R is equal to $\overline{R}$.)

$$Q = \frac{\text{tolerance limit} - \text{mean}}{\text{average range}}$$

or

$$Q = \frac{\text{tolerance limit} - \overline{X}}{\overline{R}}$$

Note: The mean $(\overline{X})$ is the average of the data.

These methods will be expanded on in the following pages.

There are three different severities for inspection: *normal, tightened,* and *reduced.* Each of these severities has rules about when to switch from one to the other. The severity must also be known for the sampling plan to be found.

As with most plans, when the sample size is equal to or larger than the lot size, every part must be inspected.

ANSI/ASQC Z1.9 Standard Deviation Method: Single Specification Limit

■ Example 12–16

A manufacturer produces 28 parts where datum A must have a maximum surface finish of AA32. The severity is normal inspection, level IV, and the AQL is 1.5 percent.

Step 1.   Find the AQL to be used in the conversation table. It is, in this case, 1.5 percent, since 1.5 percent falls between 1.10 and 1.64 range (see Table 12–10).

Step 2.   Find the sample size code letter according to the lot size of 28 pieces and level IV. This is letter D (see Table 12–10).

Step 3.   Go to the table for standard deviation method: single specification limit (Table 12–11). Read across from code letter D and find that the sample size is 5. Read also under the 1.5 percent AQL and find the K constant to be 1.40. This K value must be remembered for future steps.

TABLE 12–11

Table for Standard Deviation Method Single Specification Limit—Form 1 (Master Table for *Normal and Tightened* Inspection for Plans Based on Variability Unknown)

Sample size code letter	Sample size	.04 k	.065 k	.10 k	.15 k	.25 k	.40 k	.65 k	1.00 k	1.50 k	2.50 k	4.00 k	6.50 k	10.00 k	15.00 k
B	3	↓	↓	↓	↓	↓	↓	↓	▲	▲	1.12	.958	.765	.566	.341
C	4	↓	↓	↓	↓	↓	↓	↓	1.45	1.34	1.17	1.01	.814	.617	.393
D	5						↓	1.65	1.53	1.40	1.24	1.07	.874	.675	.455
E	7					↓	1.88	1.75	1.62	1.50	1.33	1.15	.955	.755	.536
F	10				↓	2.11	1.98	1.84	1.72	1.58	1.41	1.23	1.03	.828	.611
G	15	2.64	2.53	2.42	2.32	2.20	2.06	1.91	1.79	1.65	1.47	1.30	1.09	.886	.664
H	20	2.69	2.58	2.47	2.36	2.24	2.11	1.96	1.82	1.69	1.51	1.33	1.12	.917	.695
I	25	2.72	2.61	2.50	2.40	2.26	2.14	1.98	1.85	1.72	1.53	1.35	1.14	.936	.712
J	30	2.73	2.61	2.51	2.41	2.28	2.15	2.00	1.86	1.73	1.55	1.36	1.15	.946	.723
K	35	2.77	2.65	2.54	2.45	2.31	2.18	2.03	1.89	1.76	1.57	1.39	1.18	.969	.745
L	40	2.77	2.66	2.55	2.44	2.31	2.18	2.03	1.89	1.76	1.58	1.39	1.18	.971	.746
M	50	2.83	2.71	2.60	2.50	2.35	2.22	2.08	1.93	1.80	1.61	1.42	1.21	1.00	.774
N	75	2.90	2.77	2.66	2.55	2.41	2.27	2.12	1.98	1.84	1.65	1.46	1.24	1.03	.804
O	100	2.92	2.80	2.69	2.58	2.43	2.29	2.14	2.00	1.86	1.67	1.48	1.26	1.05	.819
P	150	2.96	2.84	2.73	2.61	2.47	2.33	2.18	2.03	1.89	1.70	1.51	1.29	1.07	.841
Q	200	2.97	2.85	2.73	2.62	2.47	2.33	2.18	2.04	1.89	1.70	1.51	1.29	1.07	.845
Acceptable Quality Levels (tightened inspection)		.065	.10	.15	.25	.40	.65	1.00	1.50	2.50	4.00	6.50	10.00	15.00	

Acceptable Quality Levels (normal inspection)

All AQL values are in percent defective.
↓ Use first sampling plan below arrow, that is, both sample size as well as k value. When sample size equals or exceeds lot size, every item in the lot must be inspected.

Step 4.   Randomly select five units from the lot and measure the actual surface finish. Record each measurement. Let's say that they were 5, 9, 10, 7, and 8.

Step 5.   Compute $\overline{X}$ and the standard deviation (s): $s = 1.924$ $\overline{X} = 7.8$.

Step 6.   Find Q_u (since it is only the upper limit you are concerned with). This upper limit is AA32.

$$Q_u = \frac{\text{upper tol.} - \overline{X}}{s} = \frac{32 - 7.8}{1.924} = 12.578$$

Step 7.   Compare Q_u (which is 12.578) to the K constant you found in step 3. If Q_u is equal to or larger than K, the lot is acceptable. If it is smaller, the lot must be rejected.

This lot is acceptable because Q_u (12.578) is larger than K (1.40).

Standard Deviation Method: Double Specification Limit

■ Example 12–17

A lot size of 35 manufactured parts must be within a surface finish tolerance of AA20 to AA10 on datum C. The severity is normal inspection, level III, and an AQL of 2 percent.

Step 1.   Find where the 2 percent AQL falls in the conversion table of AQLs. Here the AQL of 2 percent is within the range of 1.65 to 2.79, so you would use 2.5 percent (see Table 12–10).

Step 2.   Find the sample size code letter. Here the lot size is 35 and level III applies, so the code letter is B (see Table 12–10).

Step 3.   Go to the table for standard deviation method: double specification limit (see Table 12–12). Read across from code letter B and find the sample size to be 3 and the M constant number to be 7.59 (reading under the AQL of 2.5 percent).

Step 4.   Draw a random sample of three units and measure the surface finish of datum C. Record the actual measurements. Let's say that they were 15, 16, and 18.

Step 5.   Find $\overline{X}$ and the s (standard deviation). $\overline{X} = 16.33$; $s = 1.53$ using previously given formulas.

Step 6.   Find Q_u and Q_L (since it is a double specification limit). This upper limit is AA20.

$$Q_u = \frac{U - \overline{X}}{s} = \frac{20 - 16.33}{1.53} = 2.40$$

TABLE 12–12

Table for Standard Deviation Method Double Specification Limit—Form 2 (Master Table for *Normal and Tightened* Inspection for Plans Based on Variability Unknown)

Sample size code letter	Sample size	\.04 M	\.065 M	\.10 M	\.15 M	\.25 M	\.40 M	\.65 M	1.00 M	1.50 M	2.50 M	4.00 M	6.50 M	10.00 M	15.00 M
		\.04	\.065	\.10	\.15	\.25	\.40	\.65	1.00	1.50	2.50	4.00	6.50	10.00	15.00
B	3	→	→	→	→	→	→	↓	▶	▶	7.59	18.86	26.94	33.69	40.47
C	4	→	→	→	→	→	→	→	1.53	5.50	10.92	16.45	22.86	29.45	36.90
D	5	→	→	→	→	→	→	1.33	3.32	5.83	9.80	14.39	20.19	26.56	33.99
E	7	→	→	→	→	0.422	1.06	2.14	3.55	5.35	8.40	12.20	17.35	23.29	30.50
F	10	→	→	→	0.349	0.716	1.30	2.17	3.26	4.77	7.29	10.54	15.17	20.74	27.57
G	15	0.099	0.186	0.312	0.503	0.818	1.31	2.11	3.05	4.31	6.56	9.46	13.71	18.94	25.61
H	20	0.135	0.228	0.365	0.544	0.846	1.29	2.05	2.95	4.09	6.17	8.92	12.99	18.03	24.53
I	25	0.155	0.250	0.380	0.551	0.877	1.29	2.00	2.86	3.97	5.97	8.63	12.57	17.51	23.97
J	30	0.179	0.280	0.413	0.581	0.879	1.29	1.98	2.83	3.91	5.86	8.47	12.36	17.24	23.58
K	35	0.170	0.264	0.388	0.535	0.847	1.23	1.87	2.68	3.70	5.57	8.10	11.87	16.65	22.91
L	40	0.179	0.275	0.401	0.566	0.873	1.26	1.88	2.71	3.72	5.58	8.09	11.85	16.61	22.86
M	50	0.163	0.250	0.363	0.503	0.789	1.17	1.71	2.49	3.45	5.20	7.61	11.23	15.87	22.00
N	75	0.147	0.228	0.330	0.467	0.720	1.07	1.60	2.29	3.20	4.87	7.15	10.63	15.13	21.11
O	100	0.145	0.220	0.317	0.447	0.689	1.02	1.53	2.20	3.07	4.69	6.91	10.32	14.75	20.66
P	150	0.134	0.203	0.293	0.413	0.638	0.949	1.43	2.05	2.89	4.43	6.57	9.88	14.20	20.02
Q	200	0.135	0.204	0.294	0.414	0.637	0.945	1.42	2.04	2.87	4.40	6.53	9.81	14.12	19.92
		\.065	\.10	\.15	\.25	\.40	\.65	1.00	1.50	2.50	4.00	6.50	10.00	15.00	

Acceptable Quality Levels (normal inspection)

Acceptability Quality Levels (tightened inspection)

All AQL and table values are in percent defective.

↓ Use first sampling plan below arrow, that is, both sample size as well as M value. When sample size equals or exceeds lot size, every item in the lot must be inspected.

$$Q_L = \frac{L - \overline{X}}{s} = \frac{10 - 16.33}{1.53} = 4.14$$

(Do not be concerned here with minus numbers.)

Step 7.   Now you must find the *estimated lot percent defective* from the appropriate tables in Mil-Std-414. In these tables, a Q_u of 2.40 is estimated as 0.000 percent defective, and a Q_L of 4.14 is estimated also as 0.000 percent defective according to the sample size of 3. Samples of these tables are given in Table 12–13. Therefore, the total estimated percent defective here is

$$P = P_u + P_L = 0$$

Step 8.   The last step is to compare P with M. The lot is acceptable if P is equal to or less than M. M was 7.59 (in step 3 above) and P is 0.

Accept the lot.

ANSI/ASQCZ1.9 Range Method: Single Specification Limit

■ **Example 12–18**

A manufacturer produces 30 parts where datum B must have a surface finish of AA32 maximum. The inspection level is not specified (so use IV), normal inspection, and AQL of 1.5 percent.

Step 1.   Select the AQL to use from the AQL conversion table. Here the specified 1.5 percent falls between 1.10 and 1.64, and 1.5 percent is used (see Table 12–10).

Step 2.   Find the sample size code letter in the table according to lot size of 30 and level IV. It is code letter D (see Table 12–10).

Step 3.   Go to the table for range method: single specification limit and read across from code letter D and under 1.5 percent AQL to find that the sample size is 5, and the K constant number is 0.565. Remember this K number for later use (see Table 12–14).

Step 4.   Select (at random) five units from the lot, measure the actual surface finish of each, and record the results. Let's say that they are 5, 9, 10, 7, and 8.

Step 5.   Find $\overline{X}$ and $\overline{R}$.

$$\overline{X} = \frac{5 + 9 + 10 + 7 + 8}{5} = 7.8$$

$\overline{R}$ = average of the ranges

Q_U or Q_L	Sample Size															
	3	4	5	7	10	15	20	25	30	35	40	50	75	100	150	200
1.90	0.00	0.00	0.00	0.93	1.75	2.21	2.40	2.51	2.57	2.62	2.65	2.70	2.76	2.79	2.82	2.83
1.91	0.00	0.00	0.00	0.87	1.68	2.14	2.34	2.44	2.51	2.56	2.59	2.63	2.69	2.72	2.75	2.77
1.92	0.00	0.00	0.00	0.81	1.62	2.08	2.27	2.38	2.45	2.49	2.52	2.57	2.63	2.66	2.69	2.70
1.93	0.00	0.00	0.00	0.76	1.56	2.02	2.21	2.32	2.38	2.43	2.46	2.51	2.57	2.60	2.62	2.64
1.94	0.00	0.00	0.00	0.70	1.50	1.96	2.15	2.25	2.32	2.37	2.40	2.45	2.51	2.54	2.56	2.58
1.95	0.00	0.00	0.00	0.65	1.44	1.90	2.09	2.19	2.26	2.31	2.34	2.39	2.45	2.48	2.50	2.52
1.96	0.00	0.00	0.00	0.60	1.38	1.84	2.03	2.14	2.20	2.25	2.28	2.33	2.39	2.42	2.44	2.46
1.97	0.00	0.00	0.00	0.56	1.33	1.78	1.97	2.08	2.14	2.19	2.22	2.27	2.33	2.36	2.39	2.40
1.98	0.00	0.00	0.00	0.51	1.27	1.73	1.92	2.02	2.09	2.13	2.17	2.21	2.27	2.30	2.33	2.34
1.99	0.00	0.00	0.00	0.47	1.22	1.67	1.86	1.97	2.03	2.08	2.11	2.16	2.22	2.25	2.27	2.29
2.00	0.00	0.00	0.00	0.43	1.17	1.62	1.81	1.91	1.98	2.03	2.06	2.10	2.16	2.19	2.22	2.23
2.01	0.00	0.00	0.00	0.39	1.12	1.57	1.76	1.86	1.93	1.97	2.01	2.05	2.11	2.14	2.17	2.18
2.02	0.00	0.00	0.00	0.36	1.07	1.52	1.71	1.81	1.87	1.92	1.95	2.00	2.06	2.09	2.11	2.13
2.03	0.00	0.00	0.00	0.32	1.03	1.47	1.66	1.76	1.82	1.87	1.90	1.95	2.01	2.04	2.06	2.08
2.04	0.00	0.00	0.00	0.29	0.98	1.42	1.61	1.71	1.77	1.82	1.85	1.90	1.96	1.99	2.01	2.03
2.05	0.00	0.00	0.00	0.26	0.94	1.37	1.56	1.66	1.73	1.77	1.80	1.85	1.91	1.94	1.96	1.98
2.06	0.00	0.00	0.00	0.23	0.90	1.33	1.51	1.61	1.68	1.72	1.76	1.80	1.86	1.89	1.92	1.93
2.07	0.00	0.00	0.00	0.21	0.86	1.28	1.47	1.57	1.63	1.68	1.71	1.76	1.81	1.84	1.87	1.88
2.08	0.00	0.00	0.00	0.18	0.82	1.24	1.42	1.52	1.59	1.63	1.66	1.71	1.77	1.79	1.82	1.84
2.09	0.00	0.00	0.00	0.16	0.78	1.20	1.38	1.48	1.54	1.59	1.62	1.66	1.72	1.75	1.78	1.79
2.10	0.00	0.00	0.00	0.14	0.74	1.16	1.34	1.44	1.50	1.54	1.58	1.62	1.68	1.71	1.73	1.75
2.11	0.00	0.00	0.00	0.12	0.71	1.12	1.30	1.39	1.46	1.50	1.53	1.58	1.63	1.66	1.69	1.70
2.12	0.00	0.00	0.00	0.10	0.67	1.08	1.26	1.35	1.42	1.46	1.49	1.54	1.59	1.62	1.65	1.66
2.13	0.00	0.00	0.00	0.08	0.64	1.04	1.22	1.31	1.38	1.42	1.45	1.50	1.55	1.58	1.61	1.62
2.14	0.00	0.00	0.00	0.07	0.61	1.00	1.18	1.28	1.34	1.38	1.41	1.46	1.51	1.54	1.57	1.58
2.15	0.00	0.00	0.00	0.06	0.58	0.97	1.14	1.24	1.30	1.34	1.37	1.42	1.47	1.50	1.53	1.54
2.16	0.00	0.00	0.00	0.05	0.55	0.93	1.10	1.20	1.26	1.30	1.34	1.38	1.43	1.46	1.49	1.50
2.17	0.00	0.00	0.00	0.04	0.52	0.90	1.07	1.16	1.22	1.27	1.30	1.34	1.40	1.42	1.45	1.46
2.18	0.00	0.00	0.00	0.03	0.49	0.87	1.03	1.13	1.19	1.23	1.26	1.30	1.36	1.39	1.41	1.42
2.19	0.00	0.00	0.00	0.02	0.46	0.83	1.00	1.09	1.15	1.20	1.23	1.27	1.32	1.35	1.38	1.39
2.20	0.000	0.000	0.000	0.015	0.437	0.803	0.968	1.061	1.120	1.161	1.192	1.233	1.287	1.314	1.340	1.352
2.21	0.000	0.000	0.000	0.010	0.413	0.772	0.936	1.028	1.087	1.128	1.158	1.199	1.253	1.279	1.305	1.318
2.22	0.000	0.000	0.000	0.006	0.389	0.743	0.905	0.996	1.054	1.095	1.125	1.166	1.219	1.245	1.271	1.283
2.23	0.000	0.000	0.000	0.003	0.366	0.715	0.875	0.965	1.023	1.063	1.093	1.134	1.186	1.212	1.238	1.250
2.24	0.000	0.000	0.000	0.002	0.345	0.687	0.845	0.935	0.992	1.032	1.061	1.102	1.154	1.180	1.205	1.218
2.25	0.000	0.000	0.000	0.001	0.324	0.660	0.816	0.905	0.962	1.002	1.031	1.071	1.123	1.148	1.173	1.186
2.26	0.000	0.000	0.000	0.000	0.304	0.634	0.789	0.876	0.933	0.972	1.001	1.041	1.092	1.117	1.142	1.155
2.27	0.000	0.000	0.000	0.000	0.285	0.609	0.762	0.848	0.904	0.943	0.972	1.011	1.062	1.087	1.112	1.124
2.28	0.000	0.000	0.000	0.000	0.267	0.585	0.735	0.821	0.876	0.915	0.943	0.982	1.033	1.058	1.082	1.094
2.29	0.000	0.000	0.000	0.000	0.250	0.561	0.710	0.794	0.849	0.887	0.915	0.954	1.004	1.029	1.053	1.065

TABLE 12–13

Table for Estimating the Lot Percent Defective Using the Standard Deviation Method

Q_U or Q_L	Sample Size															
	3	4	5	7	10	15	20	25	30	35	40	50	75	100	150	200
2.30	0.000	0.000	0.000	0.000	0.233	0.538	0.685	0.769	0.823	0.861	0.888	0.927	0.977	1.001	1.025	1.037
2.31	0.000	0.000	0.000	0.000	0.218	0.516	0.661	0.743	0.797	0.834	0.862	0.900	0.949	0.974	0.997	1.009
2.32	0.000	0.000	0.000	0.000	0.203	0.495	0.637	0.719	0.772	0.809	0.836	0.874	0.923	0.947	0.971	0.982
2.33	0.000	0.000	0.000	0.000	0.189	0.474	0.614	0.695	0.748	0.784	0.811	0.848	0.897	0.921	0.944	0.956
2.34	0.000	0.000	0.000	0.000	0.175	0.454	0.592	0.672	0.724	0.760	0.787	0.824	0.872	0.895	0.915	0.930
2.35	0.000	0.000	0.000	0.000	0.163	0.435	0.571	0.650	0.701	0.736	0.763	0.799	0.847	0.870	0.893	0.905
2.36	0.000	0.000	0.000	0.000	0.151	0.416	0.550	0.628	0.678	0.714	0.740	0.776	0.823	0.846	0.869	0.880
2.37	0.000	0.000	0.000	0.000	0.139	0.398	0.530	0.606	0.656	0.691	0.717	0.753	0.799	0.822	0.845	0.856
2.38	0.000	0.000	0.000	0.000	0.128	0.381	0.510	0.586	0.635	0.670	0.695	0.730	0.777	0.799	0.822	0.833
2.39	0.000	0.000	0.000	0.000	0.118	0.364	0.491	0.566	0.614	0.648	0.674	0.709	0.754	0.777	0.799	0.810
2.40	0.000	0.000	0.000	0.000	0.109	0.348	0.473	0.546	0.594	0.628	0.653	0.687	0.732	0.755	0.777	0.787
2.41	0.000	0.000	0.000	0.000	0.100	0.332	0.455	0.527	0.575	0.608	0.633	0.667	0.711	0.733	0.755	0.766
2.42	0.000	0.000	0.000	0.000	0.091	0.317	0.437	0.509	0.555	0.588	0.613	0.646	0.691	0.712	0.734	0.744
2.43	0.000	0.000	0.000	0.000	0.083	0.302	0.421	0.491	0.537	0.569	0.593	0.627	0.670	0.692	0.713	0.724
2.44	0.000	0.000	0.000	0.000	0.076	0.288	0.404	0.474	0.519	0.551	0.575	0.608	0.651	0.672	0.693	0.703
2.45	0.000	0.000	0.000	0.000	0.069	0.275	0.389	0.457	0.501	0.533	0.556	0.589	0.632	0.653	0.673	0.684
2.46	0.000	0.000	0.000	0.000	0.063	0.262	0.373	0.440	0.484	0.516	0.539	0.571	0.613	0.634	0.654	0.664
2.47	0.000	0.000	0.000	0.000	0.057	0.249	0.359	0.425	0.468	0.499	0.521	0.553	0.595	0.615	0.635	0.646
2.48	0.000	0.000	0.000	0.000	0.051	0.237	0.344	0.409	0.452	0.482	0.505	0.536	0.577	0.597	0.617	0.627
2.49	0.000	0.000	0.000	0.000	0.046	0.226	0.331	0.394	0.436	0.466	0.488	0.519	0.560	0.580	0.600	0.609
2.50	0.000	0.000	0.000	0.000	0.041	0.214	0.317	0.380	0.421	0.451	0.473	0.503	0.543	0.563	0.582	0.592
2.51	0.000	0.000	0.000	0.000	0.037	0.204	0.304	0.366	0.407	0.436	0.457	0.487	0.527	0.546	0.565	0.575
2.52	0.000	0.000	0.000	0.000	0.033	0.193	0.292	0.352	0.392	0.421	0.442	0.472	0.511	0.530	0.549	0.558
2.53	0.000	0.000	0.000	0.000	0.029	0.184	0.280	0.339	0.379	0.407	0.428	0.457	0.495	0.514	0.533	0.542
2.54	0.000	0.000	0.000	0.000	0.026	0.174	0.268	0.326	0.365	0.393	0.413	0.442	0.480	0.499	0.517	0.527
2.55	0.000	0.000	0.000	0.000	0.023	0.165	0.257	0.314	0.352	0.379	0.400	0.428	0.465	0.484	0.502	0.511
2.56	0.000	0.000	0.000	0.000	0.020	0.156	0.246	0.302	0.340	0.366	0.386	0.414	0.451	0.469	0.487	0.496
2.57	0.000	0.000	0.000	0.000	0.017	0.148	0.236	0.291	0.327	0.354	0.373	0.401	0.437	0.455	0.473	0.482
2.58	0.000	0.000	0.000	0.000	0.015	0.140	0.226	0.279	0.316	0.341	0.361	0.388	0.424	0.441	0.459	0.468
2.59	0.000	0.000	0.000	0.000	0.013	0.133	0.216	0.269	0.304	0.330	0.349	0.375	0.410	0.428	0.445	0.454
2.60	0.000	0.000	0.000	0.000	0.011	0.125	0.207	0.258	0.293	0.318	0.337	0.363	0.398	0.415	0.432	0.441
2.61	0.000	0.000	0.000	0.000	0.009	0.118	0.198	0.248	0.282	0.307	0.325	0.351	0.385	0.402	0.419	0.428
2.62	0.000	0.000	0.000	0.000	0.008	0.112	0.189	0.238	0.272	0.296	0.314	0.339	0.373	0.390	0.406	0.415
2.63	0.000	0.000	0.000	0.000	0.007	0.105	0.181	0.229	0.262	0.285	0.303	0.328	0.361	0.378	0.394	0.402
2.64	0.000	0.000	0.000	0.000	0.005	0.099	0.172	0.220	0.252	0.275	0.293	0.317	0.350	0.366	0.382	0.390
2.65	0.000	0.008	0.000	0.000	0.005	0.094	0.165	0.211	0.243	0.265	0.282	0.307	0.339	0.355	0.371	0.379
2.66	0.000	0.000	0.000	0.000	0.004	0.088	0.157	0.202	0.233	0.256	0.273	0.296	0.328	0.344	0.359	0.367
2.67	0.000	0.000	0.000	0.000	0.003	0.083	0.150	0.194	0.224	0.246	0.263	0.286	0.317	0.333	0.348	0.356
2.68	0.000	0.000	0.000	0.000	0.002	0.078	0.143	0.186	0.216	0.237	0.254	0.277	0.307	0.322	0.338	0.345
2.69	0.000	0.000	0.000	0.000	0.002	0.073	0.136	0.179	0.208	0.229	0.245	0.267	0.297	0.312	0.327	0.335

TABLE 12–13, *continued*

Q_U or Q_L	\multicolumn{16}{c}{Sample Size}

Q_U or Q_L	3	4	5	7	10	15	20	25	30	35	40	50	75	100	150	200
3.50	0.000	0.000	0.000	0.000	0.000	0.000	0.000	0.002	0.003	0.005	0.007	0.009	0.013	0.015	0.018	0.019
3.51	0.000	0.000	0.000	0.000	0.000	0.000	0.000	0.002	0.003	0.005	0.006	0.009	0.013	0.015	0.017	0.018
3.52	0.000	0.000	0.000	0.000	0.000	0.000	0.000	0.002	0.003	0.005	0.006	0.008	0.012	0.014	0.017	0.018
3.53	0.000	0.000	0.000	0.000	0.000	0.000	0.000	0.001	0.003	0.004	0.006	0.008	0.012	0.014	0.016	0.017
3.54	0.000	0.000	0.000	0.000	0.000	0.000	0.000	0.001	0.003	0.004	0.005	0.008	0.011	0.013	0.015	0.016
3.55	0.000	0.000	0.000	0.000	0.000	0.000	0.000	0.001	0.003	0.004	0.005	0.007	0.011	0.012	0.015	0.016
3.56	0.000	0.000	0.000	0.000	0.000	0.000	0.000	0.001	0.002	0.004	0.005	0.007	0.010	0.012	0.014	0.015
3.57	0.000	0.000	0.000	0.000	0.000	0.000	0.000	0.001	0.002	0.003	0.005	0.006	0.010	0.011	0.013	0.014
3.58	0.000	0.000	0.000	0.000	0.000	0.000	0.000	0.001	0.002	0.003	0.004	0.006	0.009	0.011	0.013	0.014
3.59	0.000	0.000	0.000	0.000	0.000	0.000	0.000	0.001	0.002	0.003	0.004	0.006	0.009	0.010	0.012	0.013
3.60	0.000	0.000	0.000	0.000	0.000	0.000	0.000	0.001	0.002	0.003	0.004	0.006	0.008	0.010	0.012	0.013
3.61	0.000	0.000	0.000	0.000	0.000	0.000	0.000	0.001	0.002	0.003	0.004	0.005	0.008	0.010	0.011	0.012
3.62	0.000	0.000	0.000	0.000	0.000	0.000	0.000	0.001	0.002	0.003	0.003	0.005	0.008	0.009	0.011	0.012
3.63	0.000	0.000	0.000	0.000	0.000	0.000	0.000	0.001	0.001	0.002	0.003	0.005	0.007	0.009	0.010	0.011
3.64	0.000	0.000	0.000	0.000	0.000	0.000	0.000	0.001	0.001	0.002	0.003	0.004	0.007	0.008	0.010	0.011
3.65	0.000	0.000	0.000	0.000	0.000	0.000	0.000	0.001	0.001	0.002	0.003	0.004	0.007	0.008	0.010	0.010
3.66	0.000	0.000	0.000	0.000	0.000	0.000	0.000	0.000	0.001	0.002	0.003	0.004	0.006	0.008	0.009	0.010
3.67	0.000	0.000	0.000	0.000	0.000	0.000	0.000	0.000	0.001	0.002	0.003	0.004	0.006	0.007	0.009	0.010
3.68	0.000	0.000	0.000	0.000	0.000	0.000	0.000	0.000	0.001	0.002	0.002	0.004	0.006	0.007	0.008	0.009
3.69	0.000	0.000	0.000	0.000	0.000	0.000	0.000	0.000	0.001	0.002	0.002	0.003	0.005	0.007	0.008	0.009
3.70	0.000	0.000	0.000	0.000	0.000	0.000	0.000	0.000	0.001	0.002	0.002	0.003	0.005	0.006	0.008	0.008
3.71	0.000	0.000	0.000	0.000	0.000	0.000	0.000	0.000	0.001	0.001	0.002	0.003	0.005	0.006	0.007	0.008
3.72	0.000	0.000	0.000	0.000	0.000	0.000	0.000	0.000	0.001	0.001	0.002	0.003	0.005	0.006	0.007	0.008
3.73	0.000	0.000	0.000	0.000	0.000	0.000	0.000	0.000	0.001	0.001	0.002	0.003	0.005	0.006	0.007	0.007
3.74	0.000	0.000	0.000	0.000	0.000	0.000	0.000	0.000	0.001	0.001	0.002	0.003	0.004	0.005	0.007	0.007
3.75	0.000	0.000	0.000	0.000	0.000	0.000	0.000	0.000	0.001	0.001	0.002	0.002	0.004	0.005	0.006	0.007
3.76	0.000	0.000	0.000	0.000	0.000	0.000	0.000	0.000	0.001	0.001	0.001	0.002	0.004	0.005	0.006	0.007
3.77	0.000	0.000	0.000	0.000	0.000	0.000	0.000	0.000	0.001	0.001	0.001	0.002	0.004	0.005	0.006	0.006
3.78	0.000	0.000	0.000	0.000	0.000	0.000	0.000	0.000	0.000	0.001	0.001	0.002	0.004	0.004	0.005	0.006
3.79	0.000	0.000	0.000	0.000	0.000	0.000	0.000	0.000	0.000	0.001	0.001	0.002	0.003	0.004	0.005	0.006
3.80	0.000	0.000	0.000	0.000	0.000	0.000	0.000	0.000	0.000	0.001	0.001	0.002	0.003	0.004	0.005	0.006
3.81	0.000	0.000	0.000	0.000	0.000	0.000	0.000	0.000	0.000	0.001	0.001	0.002	0.003	0.004	0.005	0.005
3.82	0.000	0.000	0.000	0.000	0.000	0.000	0.000	0.000	0.000	0.001	0.001	0.002	0.003	0.004	0.005	0.005
3.83	0.000	0.000	0.000	0.000	0.000	0.000	0.000	0.000	0.000	0.001	0.001	0.002	0.003	0.004	0.004	0.005
3.84	0.000	0.000	0.000	0.000	0.000	0.000	0.000	0.000	0.000	0.001	0.001	0.001	0.003	0.003	0.004	0.005
3.85	0.000	0.000	0.000	0.000	0.000	0.000	0.000	0.000	0.000	0.001	0.001	0.001	0.002	0.003	0.004	0.004
3.86	0.000	0.000	0.000	0.000	0.000	0.000	0.000	0.000	0.000	0.000	0.001	0.001	0.002	0.003	0.004	0.004
3.87	0.000	0.000	0.000	0.000	0.000	0.000	0.000	0.000	0.000	0.000	0.001	0.001	0.002	0.003	0.004	0.004
3.88	0.000	0.000	0.000	0.000	0.000	0.000	0.000	0.000	0.000	0.000	0.001	0.001	0.002	0.003	0.004	0.004
3.89	0.000	0.000	0.000	0.000	0.000	0.000	0.000	0.000	0.000	0.000	0.001	0.001	0.002	0.003	0.003	0.004
3.90	0.000	0.000	0.000	0.000	0.000	0.000	0.000	0.000	0.000	0.000	0.001	0.001	0.002	0.003	0.003	0.004

TABLE 12–13, *continued*

TABLE 12–14

Table for Range Method Single Specification Limit—Form 1 (Master Table for *Normal and Tightened* Inspection for Plans Based on Variability Unknown)

Sample size code letter	Sample size	*Acceptable Quality Levels (normal inspection)*													
		.04	.065	.10	.15	.25	.40	.65	1.00	1.50	2.50	4.00	6.50	10.00	15.00
		k	k	k	k	k	k	k	k	k	k	k	k	k	k
B	3	↓	↓	↓	↓	↓	↓	↓	▼	▼	.587	.502	.401	.296	.178
C	4	↓	↓	↓	↓	↓	↓	↓	.651	.598	.525	.450	.364	.276	.176
D	5	↓	↓	↓	↓	↓	↓	.663	.614	.565	.498	.431	.352	.272	.184
E	7	↓	↓	↓	↓	.702	.659	.613	.569	.525	.465	.405	.336	.266	.189
F	10	↓	↓	↓	.916	.863	.811	.755	.703	.650	.579	.507	.424	.341	.252
G	15	1.09	1.04	.999	.958	.903	.850	.792	.738	.684	.610	.536	.452	.368	.276
H	25	1.14	1.10	1.05	1.01	.951	.896	.835	.779	.723	.647	.571	.484	.398	.305
I	30	1.15	1.10	1.06	1.02	.959	.904	.843	.787	.730	.654	.577	.490	.403	.310
J	35	1.16	1.11	1.07	1.02	.964	.908	.848	.791	.734	.658	.581	.494	.406	.313
K	40	1.18	1.13	1.08	1.04	.978	.921	.860	.803	.746	.668	.591	.503	.415	.321
L	50	1.19	1.14	1.09	1.05	.988	.931	.893	.812	.754	.676	.598	.510	.421	.327
M	60	1.21	1.16	1.11	1.06	1.00	.948	.885	.826	.768	.689	.610	.521	.432	.336
N	85	1.23	1.17	1.13	1.08	1.02	.962	.899	.839	.780	.701	.621	.530	.441	.345
O	115	1.24	1.19	1.14	1.09	1.03	.975	.911	.851	.791	.711	.631	.539	.449	.353
P	175	1.26	1.21	1.16	1.11	1.05	.994	.929	.868	.807	.726	.644	.552	.460	.363
Q	230	1.27	1.21	1.16	1.12	1.06	.996	.931	.870	.809	.728	.645	.553	.462	.364
		.065	.10	.15	.25	.40	.65	1.00	1.50	2.50	4.00	6.50	10.00	15.00	

Acceptable Quality Levels (tightened inspection)

All AQL values are in percent defective.

↓ Use first sampling plan below arrow, that is, both sample size as well as k value. When sample size equals or exceeds lot size, every item in the lot must be inspected.

386

But here there is only one range. Highest − lowest = range, so $10 - 5 = 5$.

Step 6.   Find Q_u.

$$Q_a = \frac{U - \overline{X}}{\overline{R}} = \frac{32 - 7.8}{5} = 4.84$$

Step 7.   Compare Q_u to K. If Q_u is larger than K, the lot is acceptable. Q_u is 4.84, K is 0.565.

Accept the lot.

ANSI/ASQCZ1.9 Range Method: Double Specification Limits

■ **Example 12–19**

A lot of 28 parts is to be inspected. The tolerance for surface finish on a surface is AA12 to AA8. We have for this example normal inspection, level IV, AQL = 1.68 percent.

Step 1.   Select the AQL to use from the table. The specified AQL of 1.68 percent falls between 1.65 and 2.79; therefore, use 2.5 percent (see Table 12–10).

Step 2.   Find the sample size code letter in the table according to the lot size (28) and level IV. It is D (see Table 12–10).

Step 3.   Refer to the table for range method: double specification limit. Read across from code letter D and under 2.5 percent AQL to find the sample size, which is 5, and the M number, which is 9.90. Remember this M number for later use (see Table 12–15).

Step 4.   Randomly select five units from the lot, measure the actual surface finish, and record the measurements. Let's say they are 9, 11, 10, 12, and 12.

Step 5.   Find $\overline{X}$ and R of the measurements.

$$\overline{X} = 10.8 \qquad R = 3$$

Step 6.   Find the c factor in the table. This is 2.474; the M number, remember, is 9.90 (see Table 12–15).

Step 7.   Find Q_u and Q_L.

$$Q_u = \frac{(U - \overline{X})c}{R} = \frac{(12 - 10.8)2.474}{3} = 0.99$$

$$Q_L = \frac{(L - \overline{X})c}{R} = \frac{(8 - 10.8)2.474}{3} = 2.3$$

TABLE 12-15

Table for Range Method Double Specification Limit—Form 2 (Master Table for *Normal and Tightened* Inspection for Plans Based on Variability Unknown)

Sample size code letter	Sample size	c factor	\.04 M	\.065 M	\.10 M	\.15 M	\.25 M	\.40 M	\.65 M	1.00 M	1.50 M	2.50 M	4.00 M	6.50 M	10.00 M	15.00 M
B	3	1.910	→	→	→	→	→	→	→	▼	▼	7.59	18.86	26.94	33.69	40.47
C	4	2.234	→	→	→	→	→	→	↓	1.53	5.50	10.92	16.45	22.86	29.45	36.90
D	5	2.474	→	→	→	→	→	→	1.42	3.44	5.93	9.90	14.47	20.27	26.59	33.95
E	7	2.830	→	→	→	→	.28	.89	1.99	3.46	5.32	8.47	12.35	17.54	23.50	30.66
F	10	2.405	→	→	→	.23	.58	1.14	2.05	3.23	4.77	7.42	10.79	15.49	21.06	27.90
G	15	2.379	.061	.136	.253	.430	.786	1.30	2.10	3.11	4.44	6.76	9.76	14.09	19.30	25.92
H	25	2.358	.125	.214	.336	.506	.827	1.27	1.95	2.82	3.96	5.98	8.65	12.59	17.48	23.79
I	30	2.353	.147	.240	.366	.537	.856	1.29	1.96	2.81	3.92	5.88	8.50	12.36	17.19	23.42
J	35	2.349	.165	.261	.391	.564	.883	1.33	1.98	2.82	3.90	5.85	8.42	12.24	17.03	23.21
K	40	2.346	.160	.252	.375	.539	.842	1.25	1.88	2.69	3.73	5.61	8.11	11.84	16.55	22.38
L	50	2.342	.169	.261	.381	.542	.838	1.25	1.60	2.63	3.64	5.47	7.91	11.57	16.20	22.26
M	60	2.339	.158	.244	.356	.504	.781	1.16	1.74	2.47	3.44	5.17	7.54	11.10	15.64	21.63
N	85	2.335	.156	.242	.350	.493	.755	1.12	1.67	2.37	3.30	4.97	7.27	10.73	15.17	21.05
O	115	2.333	.153	.230	.333	.468	.718	1.06	1.58	2.25	3.14	4.76	6.99	10.37	14.74	20.57
P	175	2.331	.139	.210	.303	.427	.655	.972	1.46	2.08	2.93	4.47	6.60	9.89	14.15	19.88
Q	230	2.330	.142	.215	.308	.432	.661	.976	1.47	2.08	2.92	4.46	6.57	9.84	14.10	19.82
Acceptable Quality Levels (tightened inspection)			.065	.10	.15	.25	.40	.65	1.00	1.50	2.50	4.00	6.50	10.00	15.00	

Acceptable Quality Levels (normal inspection)

All AQL and table values are in percent defective.

↓ Use first sampling plan below arrow, that is, both sample size as well as M value. When sample size equals or exceeds lot size, every item in the lot must be inspected.

388

Step 8. Find the estimated lot percent defective from the tables in Mil-Std-414 called "Estimated Percent Defective Using the Range Method."

Where Q_u is 0.99 (and sample size is 5), the estimated percent defective P_u is 16.66 (see Table 12–16).

Where Q_L is 2.31 (and sample size is 5), the estimated percent defective P_L is 0 (see Table 12–16).

$$P_u + P_L = P$$

So

$$16.66 + 0 = 16.66$$

Step 9. Compare P to M. If P is equal to or larger than M, reject the lot.

$$P = 16.66 \qquad M = 9.90$$

Reject the lot.

Note: All of the measured values of this lot were within specification limits, but the lot was rejected. It was rejected due to its inherent variation. Notice that the measured values were 9, 10, 11, 12, and 12, and that 12 is the upper tolerance limit. The spread, and the fact that the measurements were too near the high limit, caused the lot to be rejected.

OPERATING CHARACTERISTIC (OC) CURVES

All sampling plans (sample size, accept number, and reject number) have their own operating characteristic curves. These are called OC curves for short. The inspector is usually not concerned with using OC curves, but should know that they exist and understand what they are for. Figure 12.4 shows an example of what an OC curve looks like.

FIGURE 12.4
Sample operating characteristic (OC) curve.

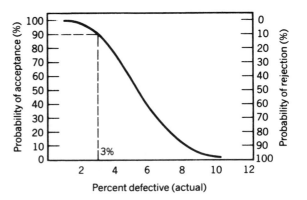

Qu or QL	Sample Size															
	3	4	5	7	10	15	25	30	35	40	50	60	85	115	175	230
0	50.00	50.00	50.00	50.00	50.00	50.00	50.00	50.00	50.00	50.00	50.00	50.00	50.00	50.00	50.00	50.00
.1	47.24	46.67	46.44	46.29	46.20	46.13	46.08	46.07	46.06	46.05	46.05	46.04	46.03	46.03	46.02	46.02
.2	44.46	43.33	42.90	42.60	42.42	42.29	42.19	42.17	42.16	42.15	42.13	42.12	42.10	42.10	42.08	42.08
.3	41.63	40.00	39.37	38.95	38.70	38.51	38.38	38.34	38.32	38.31	38.28	38.27	38.26	38.24	38.23	38.22
.31	41.35	39.67	39.02	38.59	38.33	38.14	38.00	37.96	37.94	37.93	37.90	37.89	37.88	37.86	37.85	37.84
.32	41.06	39.33	38.67	38.23	37.96	37.77	37.63	37.59	37.57	37.55	37.53	37.51	37.50	37.48	37.47	37.46
.33	40.77	39.00	38.32	37.87	37.60	37.39	37.25	37.21	37.19	37.18	37.15	37.14	37.12	37.11	37.09	37.09
.34	40.49	38.67	37.97	37.51	37.23	37.02	36.88	36.84	36.82	36.80	36.77	36.76	36.74	36.73	36.71	36.71
.35	40.20	38.33	37.62	37.15	36.87	36.65	36.50	36.46	36.44	36.43	36.40	36.39	36.37	36.36	36.34	36.33
.36	39.91	38.00	37.28	36.79	36.50	36.29	36.13	36.09	36.07	36.05	36.03	36.01	35.99	35.97	35.96	35.96
.37	39.62	37.67	36.93	36.43	36.14	35.92	35.76	35.72	35.70	35.68	35.65	35.64	35.62	35.61	35.59	35.59
.38	39.33	37.33	36.58	36.07	35.78	35.55	35.39	35.35	35.33	35.31	35.28	35.27	35.25	35.24	35.22	35.22
.39	39.03	37.00	36.23	35.72	35.41	35.19	35.02	34.98	34.96	34.94	34.92	34.90	34.88	34.87	34.85	34.85
.40	38.74	36.67	35.88	35.36	35.05	34.82	34.66	34.62	34.59	34.58	34.55	34.53	34.51	34.49	34.48	34.48
.41	38.45	36.33	35.54	35.01	34.69	34.46	34.29	34.25	34.23	34.21	34.18	34.17	34.14	34.12	34.11	34.11
.42	38.15	36.00	35.19	34.65	34.33	34.10	33.93	33.89	33.86	33.85	33.82	33.80	33.78	33.77	33.75	33.74
.43	37.85	35.67	34.85	34.30	33.98	33.74	33.57	33.53	33.50	33.48	33.45	33.44	33.41	33.39	33.38	33.38
.44	37.56	35.33	34.50	33.95	33.62	33.38	33.21	33.17	33.14	33.12	33.09	33.08	33.05	33.03	33.02	33.02
.45	37.26	35.00	34.16	33.60	33.27	33.02	32.85	32.81	32.78	32.76	32.73	32.72	32.69	32.67	32.66	32.66
.46	36.96	34.67	33.81	33.24	32.91	32.66	32.49	32.45	32.42	32.40	32.37	32.36	32.33	32.31	32.30	32.30
.47	36.66	34.33	33.47	32.89	32.56	32.31	32.13	32.09	32.06	32.04	32.01	32.00	31.97	31.95	31.94	31.94
.48	36.35	34.00	33.12	32.55	32.21	31.96	31.78	31.74	31.71	31.69	31.66	31.64	31.62	31.61	31.59	31.58
.49	36.05	33.67	32.78	32.20	31.86	31.60	31.42	31.38	31.35	31.33	31.30	31.29	31.26	31.24	31.23	31.23
.50	35.75	33.33	32.44	31.85	31.51	31.25	31.07	31.03	31.00	30.98	30.95	30.94	30.91	30.89	30.88	30.87
.51	35.44	33.00	32.10	31.51	31.16	30.90	30.72	30.68	30.65	30.63	30.60	30.59	30.55	30.55	30.53	30.52
.52	35.13	32.67	31.76	31.16	30.81	30.55	30.37	30.33	30.30	30.28	30.25	30.24	30.21	30.19	30.18	30.17
.53	34.82	32.33	31.42	30.82	30.46	30.21	30.02	29.98	29.95	29.93	29.90	29.89	29.86	29.84	29.83	29.83
.54	34.51	32.00	31.08	30.47	30.12	29.86	29.68	29.64	29.61	29.59	29.56	29.54	29.52	29.50	29.48	29.48
.55	34.20	31.67	30.74	30.13	29.78	29.52	29.33	29.29	29.26	29.24	29.21	29.20	29.17	29.15	29.14	29.14
.56	33.88	31.33	30.40	29.79	29.44	29.18	28.99	28.95	28.92	28.90	28.87	28.86	28.83	28.81	28.80	28.79
.57	33.57	31.00	30.06	29.45	29.09	28.83	28.65	28.61	28.58	28.56	28.53	28.52	28.49	28.47	28.46	28.45
.58	33.25	30.67	29.73	29.11	28.76	28.50	28.31	28.27	28.24	28.22	28.19	28.18	28.15	28.13	28.12	28.12
.59	32.93	30.33	29.39	28.77	28.42	28.16	27.97	27.93	27.91	27.89	27.86	27.84	27.82	27.80	27.78	27.78
.60	32.61	30.00	29.05	28.44	28.08	27.82	27.64	27.60	27.57	27.55	27.52	27.51	27.48	27.46	27.45	27.45
.61	32.28	29.67	28.72	28.10	27.75	27.49	27.31	27.27	27.24	27.22	27.19	27.17	27.15	27.14	27.12	27.11
.62	31.96	29.33	28.39	27.77	27.41	27.16	26.97	26.93	26.91	26.89	26.86	26.84	26.82	26.81	26.79	26.78
.63	31.63	29.00	28.05	27.44	27.08	26.82	26.64	26.60	26.58	26.56	26.53	26.51	26.49	26.48	26.46	26.45
.64	31.30	28.67	27.72	27.11	26.75	26.50	26.32	26.28	26.25	26.23	26.20	26.19	26.16	26.14	26.13	26.13
.65	30.97	28.33	27.39	26.78	26.42	26.17	25.99	25.95	25.92	25.90	25.87	25.86	25.84	25.83	25.81	25.80
.66	30.63	28.00	27.06	26.45	26.10	25.84	25.67	25.63	25.60	25.58	25.55	25.54	25.52	25.50	25.48	25.48
.67	30.30	27.67	26.73	26.12	25.77	25.52	25.34	25.30	25.28	25.26	25.23	25.22	25.20	25.18	25.16	25.16
.68	29.96	27.33	26.40	25.79	25.45	25.20	25.02	24.98	24.96	24.94	24.91	24.90	24.88	24.87	24.85	24.84
.69	29.61	27.00	26.07	25.47	25.12	24.88	24.71	24.67	24.64	24.62	24.59	24.58	24.56	24.55	24.53	24.53

[1] Values tabulated are read in percent.

TABLE 12–16

Mil-Std-414 Table for Estimating the Lot Percent Defective Using the Range Method

Q_U or Q_L	3	4	5	7	10	15	25	30	35	40	50	60	85	115	175	230
								Sample Size								
.70	29.27	26.67	25.74	25.14	24.80	24.56	24.39	24.35	24.32	24.31	24.28	24.27	24.25	24.24	24.22	24.21
.71	28.92	26.33	25.41	24.82	24.48	24.24	24.07	24.03	24.01	23.99	23.97	23.95	23.93	23.91	23.90	23.90
.72	28.57	26.00	25.09	24.50	24.17	23.93	23.76	23.72	23.70	23.68	23.66	23.64	23.62	23.60	23.59	23.59
.73	28.22	25.67	24.76	24.18	23.85	23.61	23.45	23.41	23.39	23.37	23.35	23.33	23.32	23.30	23.29	23.29
.74	27.86	25.33	24.44	23.86	23.54	23.30	23.14	23.10	23.08	23.07	23.04	23.03	23.01	23.00	22.98	22.98
.75	27.50	25.00	24.11	23.55	23.22	22.99	22.84	22.80	22.78	22.76	22.74	22.72	22.71	22.69	22.68	22.68
.76	27.13	24.67	23.79	23.23	22.91	22.69	22.53	22.49	22.47	22.46	22.43	22.42	22.41	22.39	22.38	22.38
.77	26.77	24.33	23.47	22.92	22.60	22.38	22.23	22.19	22.17	22.16	22.13	22.12	22.11	22.09	22.08	22.08
.78	26.39	24.00	23.15	22.60	22.30	22.08	21.93	21.90	21.88	21.86	21.85	21.83	21.81	21.80	21.78	21.78
.79	26.02	23.67	22.83	22.29	21.99	21.78	21.64	21.60	21.58	21.57	21.54	21.53	21.52	21.50	21.49	21.49
.80	25.64	23.33	22.51	21.98	21.69	21.48	21.34	21.30	21.28	21.27	21.26	21.24	21.22	21.22	21.20	21.20
.81	25.25	23.00	22.19	21.68	21.39	21.18	21.04	21.01	20.99	20.98	20.97	20.95	20.93	20.93	20.91	20.91
.82	24.86	22.67	21.87	21.37	21.09	20.89	20.75	20.72	20.70	20.69	20.68	20.66	20.64	20.64	20.62	20.42
.83	24.47	22.33	21.56	21.06	20.79	20.59	20.46	20.43	20.41	20.40	20.38	20.37	20.36	20.35	20.34	20.34
.84	24.07	22.00	21.24	20.76	20.49	20.30	20.17	20.15	20.13	20.12	20.10	20.09	20.08	20.06	20.06	20.06
.85	23.67	21.67	20.93	20.46	20.20	20.01	19.89	19.87	19.85	19.84	19.82	19.81	19.79	19.79	19.78	19.78
.86	23.26	21.33	20.62	20.16	19.90	19.73	19.60	19.58	19.57	19.56	19.54	19.54	19.52	19.51	19.50	19.50
.87	22.84	21.00	20.31	19.86	19.61	19.44	19.32	19.31	19.29	19.28	19.26	19.25	19.24	19.24	19.22	19.22
.88	22.42	20.67	20.00	19.57	19.33	19.16	19.04	19.03	19.01	19.00	18.98	18.98	18.97	18.96	18.95	18.95
.89	21.99	20.33	19.69	19.27	19.04	18.88	18.77	18.75	18.74	18.73	18.71	18.70	18.69	18.69	18.68	18.68
.90	21.55	20.00	19.38	18.98	18.75	18.60	18.50	18.48	18.47	18.46	18.44	18.43	18.42	18.42	18.41	18.41
.91	21.11	19.67	19.07	18.69	18.47	18.32	18.22	18.21	18.20	18.19	18.17	18.17	18.17	18.16	18.15	18.15
.92	20.66	19.33	18.77	18.40	18.19	18.05	17.96	17.95	17.93	17.92	17.92	17.90	17.89	17.89	17.88	17.88
.93	20.20	19.00	18.46	18.11	17.91	17.78	17.69	17.68	17.67	17.66	17.65	17.65	17.63	17.63	17.62	17.62
.94	19.74	18.67	18.16	17.82	17.64	17.51	17.43	17.42	17.41	17.40	17.39	17.39	17.37	17.37	17.36	17.36
.95	19.25	18.33	17.86	17.54	17.36	17.24	17.17	17.16	17.15	17.14	17.13	17.13	17.12	17.12	17.11	17.11
.96	18.76	18.00	17.56	17.26	17.09	16.98	16.91	16.90	16.89	16.88	16.88	16.87	16.86	16.86	16.86	16.86
.97	18.25	17.67	17.25	16.97	16.82	16.71	16.65	16.64	16.63	16.63	16.62	16.62	16.61	16.61	16.60	16.60
.98	17.74	17.33	16.96	16.70	16.55	16.45	16.39	16.38	16.38	16.37	16.37	16.37	16.36	16.36	16.36	16.36
.99	17.21	17.00	16.66	16.42	16.28	16.19	16.14	16.13	16.13	16.12	16.12	16.12	16.11	16.11	16.11	16.11
1.00	16.67	16.67	16.36	16.14	16.02	15.94	15.89	15.88	15.88	15.88	15.87	15.87	15.87	15.87	15.87	15.87
1.01	16.11	16.33	16.07	15.87	15.76	15.68	15.64	15.63	15.63	15.63	15.63	15.63	15.62	15.62	15.62	15.62
1.02	15.53	16.00	15.78	15.60	15.50	15.43	15.40	15.39	15.39	15.39	15.39	15.39	15.38	15.38	15.38	15.38
1.03	14.93	15.67	15.48	15.33	15.24	15.18	15.15	15.15	15.15	15.15	15.15	15.15	15.15	15.15	15.15	15.15
1.04	14.31	15.33	15.19	15.06	14.98	14.94	14.91	14.91	14.91	14.91	14.91	14.91	14.91	14.91	14.91	14.91
1.05	13.66	15.00	14.91	14.79	14.73	14.69	14.67	14.67	14.67	14.67	14.67	14.68	14.68	14.68	14.68	14.68
1.06	12.98	14.67	14.62	14.53	14.48	14.45	14.44	14.44	14.44	14.44	14.44	14.44	14.45	14.45	14.45	14.45
1.07	12.27	14.33	14.33	14.27	14.23	14.21	14.20	14.21	14.21	14.21	14.21	14.21	14.22	14.22	14.22	14.22
1.08	11.51	14.00	14.05	14.01	13.98	13.97	13.97	13.98	13.98	13.98	13.98	13.99	13.99	13.99	14.00	14.00
1.09	10.71	13.67	13.76	13.75	13.74	13.73	13.74	13.75	13.75	13.75	13.76	13.76	13.77	13.77	13.78	13.78

TABLE 12–16, *continued*

q/qc	Sample Size															
	3	4	5	7	10	15	25	30	35	40	50	60	85	115	175	230
1.90	0.00	0.00	0.00	0.67	1.45	1.99	2.38	2.47	2.53	2.57	2.64	2.68	2.74	2.77	2.81	2.83
1.91	0.00	0.00	0.00	0.62	1.38	1.93	2.32	2.41	2.47	2.51	2.58	2.61	2.67	2.70	2.74	2.76
1.92	0.00	0.00	0.00	0.56	1.32	1.86	2.25	2.34	2.41	2.45	2.51	2.55	2.61	2.64	2.68	2.70
1.93	0.00	0.00	0.00	0.51	1.26	1.80	2.19	2.28	2.34	2.38	2.45	2.49	2.55	2.58	2.61	2.63
1.94	0.00	0.00	0.00	0.46	1.20	1.74	2.13	2.22	2.28	2.32	2.39	2.43	2.49	2.52	2.55	2.57
1.95	0.00	0.00	0.00	0.42	1.15	1.68	2.07	2.16	2.22	2.26	2.33	2.37	2.43	2.46	2.49	2.51
1.96	0.00	0.00	0.00	0.37	1.09	1.62	2.01	2.10	2.16	2.20	2.27	2.31	2.37	2.40	2.43	2.45
1.97	0.00	0.00	0.00	0.33	1.04	1.57	1.95	2.04	2.10	2.14	2.21	2.25	2.31	2.34	2.38	2.40
1.98	0.00	0.00	0.00	0.30	0.99	1.51	1.90	1.99	2.05	2.09	2.15	2.19	2.25	2.28	2.32	2.34
1.99	0.00	0.00	0.00	0.26	0.94	1.46	1.84	1.93	1.99	2.03	2.10	2.14	2.20	2.23	2.26	2.28
2.00	0.00	0.00	0.00	0.23	0.89	1.41	1.79	1.88	1.94	1.98	2.05	2.08	2.14	2.17	2.21	2.23
2.01	0.00	0.00	0.00	0.20	0.84	1.36	1.74	1.83	1.89	1.93	1.99	2.03	2.09	2.12	2.16	2.18
2.02	0.00	0.00	0.00	0.17	0.80	1.31	1.69	1.78	1.83	1.87	1.94	1.98	2.04	2.07	2.10	2.12
2.03	0.00	0.00	0.00	0.14	0.75	1.26	1.64	1.73	1.78	1.82	1.89	1.93	1.99	2.02	2.05	2.07
2.04	0.00	0.00	0.00	0.12	0.71	1.21	1.59	1.68	1.73	1.77	1.84	1.88	1.94	1.97	2.00	2.02
2.05	0.00	0.00	0.00	0.10	0.67	1.17	1.54	1.63	1.69	1.73	1.79	1.83	1.89	1.92	1.95	1.97
2.06	0.00	0.00	0.00	0.08	0.63	1.12	1.49	1.58	1.64	1.68	1.74		1.84	1.87	1.91	1.93
2.07	0.00	0.00	0.00	0.06	0.60	1.08	1.45	1.54	1.59	1.63	1.70	1.74	1.79	1.82	1.86	1.88
2.08	0.00	0.00	0.00	0.05	0.56	1.04	1.40	1.49	1.55	1.59	1.65	1.69	1.75	1.78	1.81	1.83
2.09	0.00	0.00	0.00	0.03	0.53	1.00	1.36	1.45	1.50	1.54	1.61	1.64	1.70	1.73	1.77	1.79
2.10	0.00	0.00	0.00	0.02	0.49	0.96	1.32	1.41	1.46	1.50	1.56	1.60	1.66	1.69	1.72	1.74
2.11	0.00	0.00	0.00	0.01	0.46	0.92	1.28	1.36	1.42	1.46	1.52	1.56	1.61	1.64	1.68	1.70
2.12	0.00	0.00	0.00	0.00	0.43	0.88	1.24	1.32	1.38	1.42	1.48	1.52	1.57	1.60	1.64	1.66
2.13	0.00	0.00	0.00	0.00	0.40	0.85	1.20	1.28	1.34	1.38	1.44	1.48	1.53	1.56	1.60	1.62
2.14	0.00	0.00	0.00	0.00	0.38	0.81	1.16	1.25	1.30	1.34	1.40	1.44	1.49	1.52	1.56	1.58
2.15	0.00	0.00	0.00	0.00	0.35	0.78	1.13	1.21	1.26	1.30	1.36	1.40	1.45	1.48	1.52	1.54
2.16	0.00	0.00	0.00	0.00	0.32	0.75	1.09	1.17	1.22	1.26	1.32	1.36	1.41	1.44	1.48	1.50
2.17	0.00	0.00	0.00	0.00	0.30	0.71	1.06	1.13	1.18	1.22	1.29	1.32	1.38	1.41	1.44	1.46
2.18	0.00	0.00	0.00	0.00	0.28	0.68	1.02	1.10	1.15	1.19	1.25	1.28	1.34	1.37	1.40	1.41
2.19	0.00	0.00	0.00	0.00	0.26	0.65	0.99	1.06	1.11	1.15	1.22	1.25	1.30	1.33	1.37	1.39
2.20	0.000	0.000	0.000	0.000	0.236	0.625	0.954	1.030	1.083	1.122	1.178	1.214	1.267	1.299	1.330	1.346
2.21	0.000	0.000	0.000	0.000	0.217	0.597	0.922	0.997	1.058	1.089	1.144	1.180	1.233	1.265	1.295	1.311
2.22	0.000	0.000	0.000	0.000	0.199	0.570	0.891	0.966	1.018	1.056	1.111	1.147	1.199	1.231	1.261	1.277
2.23	0.000	0.000	0.000	0.000	0.182	0.544	0.861	0.935	0.986	1.025	1.079	1.115	1.167	1.197	1.228	1.244
2.24	0.000	0.000	0.000	0.000	0.166	0.519	0.831	0.905	0.956	0.994	1.048	1.083	1.135	1.165	1.195	1.211
2.25	0.000	0.000	0.000	0.000	0.150	0.495	0.802	0.875	0.926	0.964	1.018	1.052	1.104	1.134	1.163	1.179
2.26	0.000	0.000	0.000	0.000	0.136	0.471	0.775	0.847	0.897	0.935	0.987	1.022	1.073	1.103	1.132	1.148
2.27	0.000	0.000	0.000	0.000	0.123	0.449	0.748	0.819	0.869	0.906	0.958	0.993	1.043	1.073	1.103	1.118
2.28	0.000	0.000	0.000	0.000	0.111	0.427	0.722	0.792	0.841	0.878	0.930	0.964	1.014	1.044	1.073	1.088
2.29	0.000	0.000	0.000	0.000	0.099	0.406	0.697	0.766	0.814	0.851	0.902	0.936	0.986	1.015	1.044	1.059

TABLE 12–16, *continued*

Q_U or Q_L	Sample Size															
	3	4	5	7	10	15	25	30	35	40	50	60	85	115	175	230
2.30	0.000	0.000	0.000	0.000	0.089	0.386	0.672	0.741	0.789	0.825	0.875	0.909	0.959	0.988	1.016	1.031
2.31	0.000	0.000	0.000	0.000	0.079	0.367	0.648	0.716	0.763	0.799	0.849	0.882	0.931	0.960	0.988	1.003
2.32	0.000	0.000	0.000	0.000	0.070	0.348	0.624	0.691	0.739	0.774	0.823	0.856	0.905	0.934	0.962	0.976
2.33	0.000	0.000	0.000	0.000	0.061	0.330	0.601	0.668	0.715	0.750	0.798	0.831	0.879	0.908	0.935	0.950
2.34	0.000	0.000	0.000	0.000	0.054	0.313	0.579	0.645	0.691	0.720	0.774	0.807	0.854	0.882	0.909	0.924
2.35	0.000	0.000	0.000	0.000	0.047	0.296	0.558	0.623	0.669	0.703	0.750	0.782	0.829	0.857	0.884	0.899
2.36	0.000	0.000	0.000	0.000	0.040	0.280	0.538	0.602	0.646	0.680	0.728	0.759	0.806	0.833	0.860	0.874
2.37	0.000	0.000	0.000	0.000	0.035	0.265	0.518	0.580	0.624	0.658	0.705	0.736	0.782	0.809	0.836	0.850
2.38	0.000	0.000	0.000	0.000	0.029	0.250	0.498	0.560	0.604	0.637	0.683	0.714	0.759	0.787	0.813	0.827
2.39	0.000	0.000	0.000	0.000	0.025	0.236	0.479	0.541	0.584	0.616	0.662	0.693	0.737	0.764	0.791	0.804
2.40	0.000	0.000	0.000	0.000	0.021	0.223	0.461	0.521	0.564	0.596	0.641	0.671	0.715	0.742	0.769	0.782
2.41	0.000	0.000	0.000	0.000	0.017	0.210	0.443	0.503	0.545	0.577	0.621	0.651	0.695	0.721	0.747	0.760
2.42	0.000	0.000	0.000	0.000	0.014	0.198	0.426	0.485	0.526	0.557	0.601	0.631	0.674	0.701	0.726	0.739
2.43	0.000	0.000	0.000	0.000	0.011	0.186	0.410	0.467	0.508	0.539	0.582	0.611	0.654	0.679	0.705	0.718
2.44	0.000	0.000	0.000	0.000	0.009	0.175	0.393	0.450	0.491	0.521	0.564	0.593	0.635	0.660	0.685	0.698
2.45	0.000	0.000	0.000	0.000	0.007	0.165	0.378	0.434	0.473	0.503	0.545	0.573	0.616	0.641	0.665	0.678
2.46	0.000	0.000	0.000	0.000	0.005	0.154	0.362	0.417	0.456	0.486	0.528	0.556	0.597	0.622	0.646	0.659
2.47	0.000	0.000	0.000	0.000	0.004	0.145	0.348	0.403	0.441	0.470	0.511	0.538	0.579	0.604	0.627	0.640
2.48	0.000	0.000	0.000	0.000	0.003	0.136	0.333	0.387	0.425	0.454	0.494	0.522	0.562	0.586	0.609	0.622
2.49	0.000	0.000	0.000	0.000	0.002	0.127	0.321	0.372	0.409	0.438	0.478	0.504	0.545	0.569	0.593	0.605
2.50	0.000	0.000	0.000	0.000	0.001	0.118	0.307	0.358	0.395	0.423	0.463	0.489	0.528	0.552	0.575	0.587
2.51	0.000	0.000	0.000	0.000	0.001	0.111	0.294	0.345	0.381	0.409	0.447	0.473	0.512	0.536	0.558	0.570
2.52	0.000	0.000	0.000	0.000	0.000	0.103	0.282	0.331	0.367	0.394	0.432	0.458	0.497	0.519	0.542	0.553
2.53	0.000	0.000	0.000	0.000	0.000	0.096	0.270	0.319	0.354	0.381	0.418	0.444	0.481	0.503	0.526	0.537
2.54	0.000	0.000	0.000	0.000	0.000	0.089	0.258	0.306	0.340	0.367	0.404	0.428	0.466	0.488	0.510	0.522
2.55	0.000	0.000	0.000	0.000	0.000	0.083	0.247	0.294	0.328	0.354	0.390	0.415	0.451	0.473	0.495	0.506
2.56	0.000	0.000	0.000	0.000	0.000	0.077	0.237	0.283	0.316	0.341	0.377	0.401	0.437	0.459	0.480	0.491
2.57	0.000	0.000	0.000	0.000	0.000	0.071	0.227	0.272	0.304	0.328	0.364	0.388	0.424	0.445	0.466	0.477
2.58	0.000	0.000	0.000	0.000	0.000	0.066	0.217	0.261	0.292	0.317	0.352	0.376	0.411	0.432	0.452	0.463
2.59	0.000	0.000	0.000	0.000	0.000	0.061	0.207	0.251	0.282	0.305	0.340	0.363	0.397	0.418	0.439	0.449
2.60	0.000	0.000	0.000	0.000	0.000	0.056	0.198	0.240	0.271	0.294	0.328	0.351	0.385	0.406	0.426	0.436
2.61	0.000	0.000	0.000	0.000	0.000	0.052	0.189	0.231	0.260	0.283	0.317	0.339	0.372	0.393	0.413	0.423
2.62	0.000	0.000	0.000	0.000	0.000	0.048	0.181	0.221	0.250	0.273	0.306	0.327	0.360	0.381	0.400	0.410
2.63	0.000	0.000	0.000	0.000	0.000	0.044	0.173	0.212	0.241	0.263	0.295	0.316	0.349	0.368	0.388	0.398
2.64	0.000	0.000	0.000	0.000	0.000	0.040	0.164	0.203	0.232	0.253	0.285	0.306	0.338	0.357	0.376	0.386
2.65	0.000	0.000	0.000	0.000	0.000	0.037	0.157	0.195	0.223	0.244	0.274	0.295	0.327	0.346	0.365	0.375
2.66	0.000	0.000	0.000	0.000	0.000	0.034	0.149	0.186	0.213	0.234	0.265	0.285	0.316	0.335	0.353	0.363
2.67	0.000	0.000	0.000	0.000	0.000	0.031	0.143	0.179	0.205	0.225	0.255	0.275	0.305	0.324	0.342	0.352
2.68	0.000	0.000	0.000	0.000	0.000	0.028	0.136	0.171	0.197	0.217	0.246	0.266	0.296	0.314	0.332	0.342
2.69	0.000	0.000	0.000	0.000	0.000	0.025	0.129	0.164	0.190	0.209	0.238	0.257	0.286	0.304	0.321	0.331

TABLE 12–16, *continued*

Definition of an OC Curve

An OC curve indicates the percentage of lots or batches that may be expected to be accepted under the various sampling plans for a given process quality.

An OC curve gives you an advantage for choosing the sampling plan that you can expect good results from. For example, let's say that using the sampling plan connected with the OC curve in Figure 12.4, you inspect a lot that is 3 percent defective. The curve shows that when a lot that is 3 percent defective is inspected with this plan, there is a 90 percent probability that it will be accepted and a 10 percent probability that it will be rejected.

It is simply a matter of reading the OC curves and choosing the sampling plan that will accept (or reject according to your requirements).

■ **Example 12–20**

You want a plan that will reject any lot worse than 2 percent defective. The OC curve above is not the one that will do this because it will accept lots that are 2 percent defective for 100 percent of the time. Therefore, you must look for another plan. OC curves are constructed primarily according to the sampling plan (sample size, accept number, reject number). The lot size is least important when working with OC curves; however, the lot size should be considered, since once the OC curve and plan are chosen, the lot size will be used to determine the sample size.

As stated earlier, OC curves are not often used by the inspector, but it is good for the inspector to know that they exist and understand how they help in the selection of sampling plans that are effective in maintaining a maximum allowable percent defective.

Another thing about OC curves is the fact that when the curve is steeper (nearly vertical), its ability to discriminate between good lots and bad lots is better. Sample sizes, however, are usually larger.

OC curves are widely used to help the quality engineer or manager select sampling plans that are effective in reducing the risks of sampling (accepting poor quality or rejecting good quality) and that help to keep the high cost of inspection down.

REVIEW QUESTIONS

Probability

1. If only one part is drawn, what is the probability of drawing a defective part from a lot that has 52 parts and 4 of them are defective?
2. If the probability of defectives in a lot is 0.06, what is the probability of good product?
3. A box contains 3 defective parts and 7 acceptable parts. If samples are taken one at a time (and they are not replaced), what is the probability of drawing 2 defective parts in a row?
4. What is the probability, on one draw from a deck of 52 cards, that an ace *or* king will be drawn?
5. A system has 3 resistors. The probability of each one working is R1 = 0.90, R2 = 0.85, and R3 = 0.95. If the system depends on each individual resistor working, what is the probability that the system will work?

Probability Distributions

6. A process is in control and the constant process average of defectives (p) is 0.15. A sample of 30 parts will be taken from a lot. What is the probability of finding exactly 3 defectives in the sample?
7. Refer to question 6. What is the probability of finding exactly 1 defective unit in the sample?
8. A process has a probability of defectives of $np = 0.05$. What is the probability of finding exactly 2 defectives in a sample?
9. A process has a probability of defectives of $np = 8$. What is the probability of finding exactly 3 defectives in a sample?
10. Small samples are being taken from small lots, and the number of defects (d) occurring in the population is known to be $d = 3$. If a sample of 13 parts is taken from a lot of 50 parts, what is the probability of finding exactly 1 defect in the sample?
11. Given the following information, what is the probability of finding exactly 2 defects in the sample:

$$N = 40 \qquad n = 8 \qquad d = 12$$

Sampling

12. For sample quality to be representative of lot quality, it is important to draw samples from the lot
    a. from the bottom
    b. at random
    c. from the top
    d. from the sides
13. Using ANSI/ASQC Z1.4, double sampling is generally more economical than single sampling when the process average (percent defective) is
    a. very good
    b. very bad
    c. typical
    d. none of the above
    e. a or b are correct
14. Using ANSI/ASQC Z1.4, single sampling, normal inspection, level II, AQL = 1.0%, find the sampling plan for a lot of 2,000 parts.
    a. 80, 1, 2
    b. 32, 2, 3
    c. 125, 3, 4
    d. 125, 2, 3
15. Refer to question 14. Find the appropriate double sampling plan.
    a. Sample 1 = 80, 1, 2   Sample 2 = 80, 4, 5
    b. Sample 1 = 80, 1, 4   Sample 2 = 80, 4, 5
    c. Sample 1 = 80, 0, 3   Sample 2 = 80, 3, 4
    d. Sample 1 = 80, 2, 4   Sample 2 = 80, 3, 4

16. Find the range of the following data:

0.500   0.505   0.509   0.501   0.505

    a. .002
    b. .003
    c. .090
    d. .009

17. Find the sample standard deviation for the data in question 16.
    a. .0036
    b. .005
    c. .002
    d. .0038

18. When using Dodge–Romig sampling tables, what process average ($\bar{p}$) should be used when the actual process average is not known?
    a. .10
    b. .05
    c. .01
    d. The highest process average in the tables until history is known

19. Using Dodge–Romig sampling tables and the following information, what is the appropriate sampling plan?

AOQL = 2.0%, lot size = 700, single sampling, actual $\bar{p}$ = 0.9%

    a. n = 60, c = 3
    b. n = 65, c = 2
    c. n = 32, c = 1
    d. n = 32, c = 2

20. Using ANSI/ASQC Z1.9 and the following information, is the lot acceptable (yes or no)?
Range method—Single specification limit
Normal inspection—Level is not specified
Lot size = 50, AQL = 1.0
Tolerance = .500 max.
R = .002, $\bar{X}$ = .492

21. Attributes data fall under what type of probability distribution?
    a. continuous
    b. exponential
    c. discrete
    d. normal

22. When lot sizes and sample sizes are very small, and the products come from a non-continuous process, which probability distribution applies?
    a. Poisson
    b. binomial
    c. normal
    d. hypergeometric

23. Which of the following distributions is often used to approximate the binomial distribution?
    a. hypergeometric when lots are large
    b. Poisson
    c. normal when data are not discrete
    d. none; the binomial cannot be approximated
24. What is the most effective lot-by-lot sampling plan for variables?
    a. ANSI/ASQC Z1.4
    b. Dodge–Romig plans
    c. ANSI/ASQC Z1.9
    d. Mil-Std-120
25. Which of the following is one of the best methods for obtaining a random sample?
    a. All parts on top have an equal chance of being selected.
    b. Use a table of random numbers.
    c. Make sure the sample is biased toward the defective parts.
    d. None of the above.
26. A lot of 45 parts must be sampled per ANSI/ASQC Z1.4, single sampling, normal inspection. The applicable AQL is 1.0%. Find the sampling plan.
    a. 8, 0, 1
    b. 13, 1, 2
    c. 13, 0, 1
    d. 32, 1, 2
27. A lot of 85 parts must be sampled per ANSI/ASQC Z1.4, single sampling, normal inspection, level III, AQL 1.0%. Find the appropriate sampling plan.
    a. 20, 0, 1
    b. 13, 0, 1
    c. 13, 1, 2
    d. 20, 5, 6
28. A small lot of 12 parts must be sampled per ANSI/ASQC, double sampling plan, normal inspection, level II, AQL 4.0%. Find the appropriate sampling plan.
    a. Sample 1: 3, 2, 5   Sample 2: 3, 6, 7
    b. Sample 1: 2, 1, 4   Sample 2: 2, 4, 5
    c. 3, 0, 1
    d. 5, 0, 1
29. A lot of 1,000 parts must be sampled per ANSI/ASQC Z1.4, double sampling plan, normal inspection, level II, AQL 1.0%. Find the appropriate sampling plan.
    a. 32, 0, 2   32, 1, 2
    b. 50, 0, 3   50, 3, 4
    c. 20, 0, 2   20, 1, 2
    d. 13, 1, 2

30. The same lot in question 29 must be sampled per Dodge–Romig single sampling plans, AQL 1.0%, LTPD plans at receiving inspection. The process average $(\bar{p})$ is 0.09%. Find the appropriate sampling plan.
    a. 205, 0
    b. 205, 1
    c. 335, 1
    d. 335, 0
31. A lot of 500 parts must be inspected at receiving inspection where the AQL is 1.0%, using LTPD plans, and the process average is unknown at the time of inspection. Find the appropriate LTPD plan.
    a. 180, 1
    b. 190, 1
    c. 175, 0
    d. 180, 0
32. When lot quality levels are very good or very bad, ANSI/ASQC Z1.4 double sampling plans can be more economical than the corresponding single sampling plans because
    a. the lot gets two chances to be accepted
    b. the accept numbers are lower
    c. the decision is reached at the first sample
    d. none of the above
33. Nonrandom sampling was the root cause of a lot that was rejected when it was better than the required AQL. This error is referred to as
    a. bad luck
    b. beta risk
    c. type II error
    d. alpha risk
34. A lot of 510 parts is to be inspected to a Dodge–Romig, LTPD, double sampling plan with an AQL of 0.5%. Find the appropriate LTPD sampling plan.
    a. 340, 0   110, 1
    b. 350, 1   130, 1
    c. 350, 0   130, 1
    d. 360, 0   150, 1
35. A lot of 300 parts must be sampled by variables sampling plans in ANSI/ASQC Z1.9, range method, single specification limit, normal inspection, AQL 1.0%. The maximum specification limit is 63 surface finish. The data collected from the sample of 10 parts is as follows:

    25   27   30   35   24   29   32   33   28   35

    Is the lot acceptable (yes or no)?
36. Refer to question 35. If the maximum specification were 35 surface finish, would the lot be acceptable (yes or no)?

Statistical Process Control

Statistical process control (SPC) can be thought of as a funnel for quality and productivity (Figure 13.1).

Process. Any collection of things that gets you from one point to another. A process could be anything from using a washline or a machine to driving to work in the morning.

All processes vary. No two things are ever alike. Even though the difference between the sizes of two machined parts may be very small, they are different!
There are two kinds of variation with regard to process control.

Random variation. This is variation that has several common causes. *Common* causes include vibration, heat, and humidity that either individually or collectively cause variation. These common causes are very difficult to identify and correct.

Nonrandom variation. This is variation that is not built into the process. Causes of this type of variation are called *special* causes. Special causes of variation are those that can be directly identified and corrected (particularly when you know exactly *when* they are occurring).

Variation is the reason we must have tolerances on the engineering drawing.

Drawing tolerance. These are the tolerance limits on a blueprint (or other specification). These limits define whether or not the product is acceptable.

Natural (process) tolerance. This is the measure of the ability of the process itself according to its own inherent variation. Keep in mind that a process will produce parts according to its own variation regardless of what the specifications say. The important thing is to make the process do what it can do consistently (control), then make sure that what it can do matches the blueprint requirements (capability).

OBJECTIVES OF SPC

The first objective of SPC is to get the process in control, which means the identification and elimination of *special* causes of variation.

A process in control. A process that is in control will consistently produce parts within its own *natural tolerance limits*. This is done by eliminating all of the special causes of variation that may exist. Some examples of special causes are tool wear, loose workholding devices, and power surges. There are many more. Process control is the ability of a process to hold a constant level of variation.

FIGURE 13.1
Statistical process control—a funnel for quality and productivity.

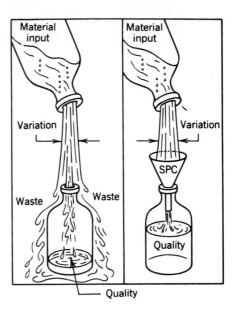

After a process is in control (stable) and producing consistently within its natural tolerance, it can then (and only then) be compared to the engineering tolerance limits to see if it is *capable* of meeting those limits.

A capable process. The capability of a process is directly related to the *ability* of the process to produce parts consistently within the drawing tolerance limits. Capability cannot be studied until the process is in *control,* because it is not consistent enough to trust the results of the study. We discuss capability in more detail later.

TOOLS FOR PROCESS CONTROL

The best tools for process control are *control charts.* These simple statistical charts are powerful tools for:

1. Detecting special causes of variation in the process *at the time* they exist.
2. Measuring the natural tolerance of the process that is due to *normal variation* (or common causes).
3. Assisting you in getting the process in control and reaching a capability of meeting drawing tolerance limits consistently.

Normal variation. When the variation in the process (due to common causes only) is exhibited by a bell-shaped curve (called the *normal distribution*).

Frequency distribution. A graphical technique for studying measured data to see the shape of the distribution is called a *frequency distribution.*

THE IMPORTANCE OF DISTRIBUTION

One example of distribution is when a group of measured values is put on a scale and an ✕ is placed for each time a value occurs (see Figure 13.2). If you draw a line around the outside of the plotted ✕'s you will see the shape of the distribution and if it is bell shaped, it is a normal distribution. Figure 13.2 is called a frequency distribution. A frequency distribution describes the variation of any process.

There are times when you can plot measured values on a distribution and the shape of the curve will not be normal. Some examples are shown in Figure 13.3.

One of the most important things to remember is that if you have a stable process (in control), you can *accurately predict the output* of the process. This is the reason why control charts are such powerful tools for process control. Even though the individual measurement may not be normally distributed, the *averages* plotted on a control chart will tend to take the shape of the normal distribution.

Note: Another tool that describes the variation of a process is a histogram (shown in Figure 13.4).

Definitions and Symbols

A few definitions and symbols are important to understand at this point.

Definitions

Variables. Variables are measurable characteristics of the part, such as inches of length, weight in pounds, or diameters in millimeters.

FIGURE 13.2
Frequency distribution and data.

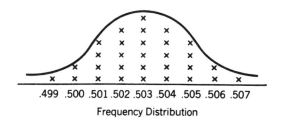

Frequency Distribution

Actual Diameters	Number Of Pieces
.499	1
.500	2
.501	4
.502	5
.503	6
.504	5
.505	4
.506	2
.507	1
	Total 30

FIGURE 13.3
Nonnormal distributions.

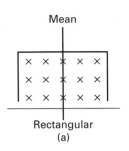

Mean

Rectangular
(a)

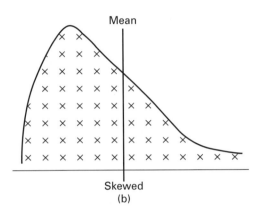

Mean

Skewed
(b)

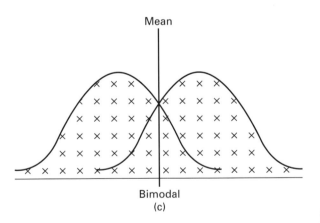

Mean

Bimodal
(c)

Attributes. Attributes are characteristics that must be (or are chosen to be) studied in terms of whether they are good or bad, on or off, Go or NoGo, and so forth.

Variable control charts. These charts require variable measurements. The X-bar (X) or range (R) charts are examples. The X-bar and R chart are two charts in one set.

FIGURE 13.4
Histogram of the data in
Figure 13.2.

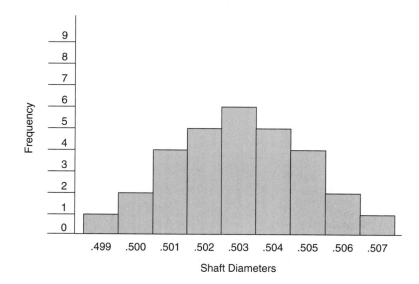

(This is the prime subject of this chapter.) Other variables charts include *X*-bar sigma, median charts, charts for individuals, charts for moving ranges, and regression charts.

Subgroup (sample) (*n*). A few parts are measured consecutively at certain times during the process. *X*-bar and *R* charts are usually five pieces.

Symbols

Σ means sum (or add).

$\hat{\sigma}$ is the estimated standard deviation using $\bar{R}$.

k means the total number of values added.

n means the sample size.

X_i means one "individual" measurement.

X-bar $(\bar{X})$ is the average.

X-double-bar $(\bar{\bar{X}})$ is the grand average of all of the $\bar{X}$ values.

s is the sample standard deviation.

R is the range.

$\bar{R}$ (*R*-bar) is the average range.

$A_2\ D_4\ D_3\ d_2$ are certain constant numbers depending on the sample size. See Table 13–1.

R_m is the moving range.

$\tilde{X}$ is the median or midpoint.

TABLE 13–1

Factors for Control Chart Limits

Sample Size "n"	A_2	A_3	A_5	$\bar{\bar{A}}_2$	d_2	D_3	D_4	D_5	D_6	c_4	B_3	B_4	E_2
2	1.880	2.659	2.224	1.88	1.128	0	3.267	0	3.865	.798	0	3.267	2.66
3	1.023	1.954	1.265	1.19	1.693	0	2.574	0	2.745	.886	0	2.568	1.77
4	.729	1.628	.829	.80	2.059	0	2.282	0	2.375	.921	0	2.266	1.46
5	.577	1.427	.712	.69	2.326	0	2.114	0	2.179	.940	0	2.089	1.29
6	.483	1.287	.562	.55	2.534	0	2.004	0	2.055	.952	.030	1.970	1.18
7	.419	1.182	.520	.51	2.704	.076	1.924	.078	1.967	.959	.118	1.882	1.11
8	.373	1.099	.441	.43	2.847	.136	1.864	.139	1.901	.965	.185	1.815	1.05
9	.337	1.032	.419	.41	2.970	.184	1.816	.187	1.850	.969	.239	1.761	1.01
10	.308	.975	.369	.36	3.078	.223	1.777	.227	1.809	.973	.284	1.716	.98

Normal Distribution Practice Problems and Solutions

1. The following surface finish values were obtained and are shown with the frequency at which each value occurred. Is the distribution of these values a normal distribution?

Surface Finish	Frequency
2	1
3	2
4	3
5	4
6	5
7	6
8	6
9	4
10	3
11	2
12	1

Solution

Arrange the values in a distribution (as shown in Figure 13.5).

2. The following shaft diameters were obtained. Is the distribution normal?

Shaft Dia.	Frequency	Solution
.500	2	.500 xx
.501	3	.501 xxx
.502	4	.502 xxxx

FIGURE 13.5
Frequency Distribution

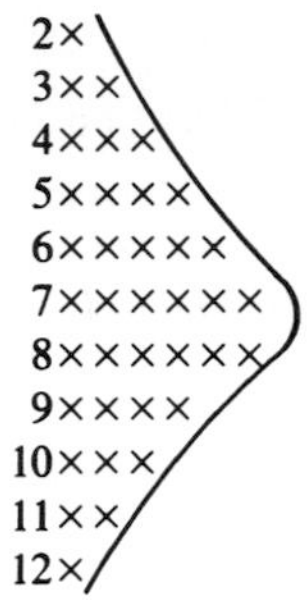

```
 2×
 3× ×
 4× × ×
 5× × × ×
 6× × × × ×
 7× × × × × ×
 8× × × × × ×
 9× × × ×
10× × ×
11× ×
12×
```

.503	5	.503 xxxxx
.504	7	.504 xxxxxxx
.505	8	.505 xxxxxxxx
.506	9	.506 xxxxxxxxx
.507	10	.507 xxxxxxxxxx
.508	11	.508 xxxxxxxxxxx
.509	5	.509 xxxxx
.510	3	.510 xxx

Answer: No, the distribution is not normal.

3. The following data have been collected. Is the distribution normal or not? (Answer yes or no.)

Shaft Dia.	No. of Pieces
.499	2
.506	12
.500	3
.507	10
.501	4
.505	16
.502	3
.504	14
.503	13

Solution

Arrange the data in order (either ascending or descending order). I chose descending (or from the largest number down). Then put an × beside the data point for each number of times it occurred.

Descending Order

.507	xxxxxxxxxx	(10 times)
.506	xxxxxxxxxxxx	(12 times)
.505	xxxxxxxxxxxxxxxx	(16 times)

            .504 xxxxxxxxxxxxxx   (14 times)
            .503 xxxxxxxxxxxxx   (13 times)
            .502 xxx   (3 times)
            .501 xxxx   (4 times)
            .500 xxx   (3 times)
            .499 xx   (2 times)

Answer: No, the distribution is not normal (or equally distributed about the mean.)

MEASURES OF CENTRAL TENDENCY

In statistical quality control, there are various measures of *central tendency.* Only three of these will be discussed here: mean, mode, and median.

Measures of central tendency are simply one number (or value) that represents the center of a group of data. These measures are vital to SPC in many ways. They are often used in conjunction with measures of *dispersion* (or variation), which will be discussed later.

The Mean, Called *X*-Bar ($\overline{X}$)

The mean is simply the arithmetic average of a group of numbers. For example, if there are five numbers in all, you simply add them up and divide the sum by 5.

$$\text{Formula: } \overline{X} = \frac{\Sigma X_i}{n}$$

where $\overline{X}$ is the mean (or average)
X_i represents each number
Σ means summation (or add)
n is the sample size

■ Example 13–1

Given 10 numbers:

$$4 \quad 3 \quad 7 \quad 9 \quad 8 \quad 3 \quad 2 \quad 4 \quad 6 \quad 8$$

Find $\overline{X}$.

Solution

$$\overline{X} = \frac{4 + 3 + 7 + 9 + 8 + 3 + 2 + 4 + 6 + 8}{10}$$

$$\overline{X} = 5.4$$

The Mode

The mode of a group of numbers is the easiest of all to find. It is simply the one single number that occurs most often in a group of numbers.

■ **Example 13–2**

Given 10 numbers:

$$8 \quad 2 \quad 3 \quad 6 \quad 7 \quad 9 \quad 4 \quad 2 \quad 3 \quad 2$$

Find the mode.

Solution

The mode here is 2 because the number 2 occurs three times above, more times than any of the other numbers.

Note: Groups of data can have more than one mode (or can be multimodal).

■ **Example 13–3**

Bimodal (2 modes). Given the following numbers, which are the modes?

$$\underline{2} \quad \underline{2} \quad \underline{3} \quad 5 \quad \underline{3} \quad 6 \quad 8$$

Solution

The modes are 2 and 3.

■ **Example 13–4**

Trimodal (3 modes). Given the following numbers, which are the modes?

$$\underline{3} \quad \underline{2} \quad \underline{5} \quad \underline{2} \quad \underline{5} \quad \underline{3} \quad 6$$

Solution

The modes are 2, 3, and 5.

The Median (or Midpoint)

The median is the value in the middle. One-half of the data are above it and the other half below it when the data have been *ordered*.

Ordering data. This means rearranging the data and listing them in ascending or descending order.

Example of Ordered Data.   Given these data:

$$7, \ 3, \ 2, \ 4, \ 5$$

Ascending order is 2, 3, 4, 5, 7. Descending order is 7, 5, 4, 3, 2. There are two basic methods to find the median, depending on whether the sample size (n) is an even number or an odd number. Remember, the median is the midpoint (or middle value) of ordered data.

■ **Example 13–5: Odd Samples**

Given these data:

$$9, \ 2, \ 5, \ 13, \ 6, \ 1, \ 11$$

Identify the median.

Solution

The median is the mid-value when the data are ordered, so order the data:

$$1, \ 2, \ 5, \ 6, \ 9, \ 11, \ 13 \text{ (the median is 6)}$$

■ **Example 13–6: Even Samples**

Given these data:

$$9, \ 3, \ 7, \ 5, \ 2, \ 1, \ 5, \ 3, \ 5, \ 8$$

Determine the median.

Solution

For even samples the median is the average of the two values in the middle of ordered data, so order the data:

$$1, \ 2, \ 3, \ 3, \ 5, \ 5, \ 5, \ 7, \ 8, \ 9 \text{ (the two middle values are both 5)}$$

The average of the two middle values (5 and 5) is 5.

Note: There are more complex methods for computing medians of even samples of data, but for practical purposes, the method in Example 13–6 will suffice.

Central Tendency Practice Problems and Solutions

1. Find the mean of the following data: 8, 6, 5, 9, 4, 7, 8, 10, 2, 4.

Solution

$$Mean\ (\overline{X}) = \frac{8 + 6 + 5 + 9 + 4 + 7 + 8 + 10 + 2 + 4}{10} = \frac{63}{10} = 6.3$$

2. Arrange these numbers in descending order: 5, 4, 8, 1, 6, 4, 3, 5, 6.

Solution
Descending order is from the largest down to the smallest number. The numbers are 8, 6, 6, 5, 5, 4, 4, 3, 1.

3. Arrange these numbers in ascending order: 3, 9, 3, 5, 6, 1, 5, 6, 8, 8, 2.

Solution
Ascending order is from the smallest to the largest number. The numbers are 1, 2, 3, 3, 5, 5, 6, 6, 8, 8, 9.

4. Find the mode of these numbers: 1, 6, 3, 8, 3, 6, 8, 5, 3, 6, 5, 3, 4, 3.

Solution
The mode is the number that occurs more often than any other number. The number above that occurs most often is 3.

5. Find the median of the following numbers: 3, 8, 6, 5, 1, 7, 9.

Solution
The median of an *odd* set of numbers is found by arranging them in ascending (or descending) order, and it is the value where there are the same amount of numbers above it as there are below it. The numbers are 1, 3, 5, 6, 7, 8, 9. The median here is 6.

6. Find the median of the following even set of numbers: 5, 3, 7, 2, 5, 9, 4, 6, 8, 7.

Solution
There are 10 numbers ($n = 10$).

Median = the average of the middle two values
Ordered data: 2, 3, 4, 5, 5, 6, 7, 9
Median = 5

7. What is the mode of these numbers (or data)? 4, 6, 7, 2, 4, 2, 6, 8, 9, 10, 4, 2, 6.

Solution
The numbers 2, 4 and 6 are each listed three times; therefore, the mode of this group of data is trimodal (or has three modes).

MEASURES OF DISPERSION (VARIATION)

In statistics three measures of variation are commonly used: *range, variance,* and *standard deviation.* The most common are the range and standard deviation. For calculations (involving the variance and standard deviation especially) a statistical calculator or software is recommended.

Range (*R*)

The range of any set of data is the largest data point minus the smallest data point.

$$\text{Range } (R) = \text{Max}_{.Xi} - \text{Min}_{.Xi}$$

■ Example 13–7

Rockwell hardness readings are taken on five shafts. The data for these hardness readings are 58, 61, 60, 64, and 55. Find the range of the data.

$$\begin{aligned}
\text{Range } (R) &= \text{Max}_{.Xi} - \text{Min}_{.Xi} \\
&= 64 - 55 \\
&= 9
\end{aligned}$$

Variance (*V*)

The variance is primarily used in cases where variability of different samples must be combined (e.g., added, subtracted, etc.) to calculate overall variances. Variances can be combined where standard deviations cannot. The formula for variance is:

Population Variance

$$Variance = \frac{\Sigma(X_i - \mu)^2}{N}$$

Sample Variance

$$Variance = \frac{\Sigma(X_i - \overline{X})^2}{n - 1}$$

where μ = the population mean
$\overline{X}$ = the sample mean
X_i = each individual data point
n = the sample size
Σ = means sum
N = the population size

The variance, as we will soon see, is the square of the standard deviation. So if the standard deviation is known, the variance is more easily computed by squaring the standard deviation.

$$Variance = s^2 \text{ (Samples)}$$

or

$$Variance = \sigma^2 \text{ (Population)}$$

■ **Example 13–8**

Find the sample variance of the following data: 58, 61, 60, 64, and 55.

Solution

Using a statistical calculator, find the sample standard deviation, then square that standard deviation, or calculate it as shown below:

$$Variance = \frac{\Sigma(X_i - \overline{X})^2}{n - 1} = 11.3$$

■ **Example 13–9**

Find the population variance of the data in Example 13–8.

Solution

Using a statistical calculator, find the population standard deviation, then square that standard deviation, or calculate it as shown below:

$$Variance = \frac{\Sigma(X_i - \mu)^2}{n} = 9.04$$

Standard Deviation

The standard deviation of a set of data is the square root of the variance. Referring to the preceding formulas for variance, the formulas for standard deviation are as follows:

Population Standard Deviation (σ)	**Sample Standard Deviation (s)**
$$\sigma = \sqrt{\frac{\Sigma(X_i - \mu)^2}{N}}$$	$$s = \sqrt{\frac{\Sigma(X_i - \overline{X})^2}{n - 1}}$$

where μ = the population mean
$\overline{X}$ = the sample mean
X_i = each individual data point
n = the sample size
Σ = means sum
σ = the symbol for population standard deviation
s = the symbol for sample standard deviation
N = the population size

■ **Example 13–10**

Find the sample standard deviation of the following data: 12, 14, 11, 10, 9, 8.

Solution

Using a statistical calculator, find the same standard deviation.

$$s = \sqrt{\frac{\Sigma(X_i - \overline{X})^2}{n - 1}} = 2.16$$

■ **Example 13–11**

Find the population standard deviation of the data in Example 13–10.

Solution

Using a statistical calculator, find the population standard deviation.

$$\sigma = \sqrt{\frac{\Sigma(X_i - \mu)^2}{n}} = 1.97$$

In cases where the variance is known, the standard deviation is computed by finding the square root of the variance:

$$\sigma = \sqrt{\text{Population Variance}}$$

or

$$s = \sqrt{\text{Sample Variance}}$$

■ **Example 13–12**

What is the standard deviation if the variance is 9?

Solution

The standard deviation is the square root of the variance, so

$$s = \sqrt{V} = \sqrt{9} = 3$$

Notes Regarding Statistical Calculators

For reasons unknown to the author, various manufacturers of statistical calculators use *different* symbols to represent population and sample standard deviation. The following is a brief example of those symbols:

Population Standard Deviation (σ)	**Sample Standard Deviation (s)**
σ	s
σ_n	σ_{n-1}
σ'	s_X
s_{Xn}	s_{Xn-1}

Consult User's Manual for your calculator to make sure which symbols represent the population versus sample standard deviation.

COMPUTING THE AVERAGE, RANGE, AND STANDARD DEVIATION WITH NEGATIVE NUMBERS IN THE DATA

Process control charts (such as short production run charts) are often structured in such a way that negative numbers appear in the data. If this occurs, computations must be carefully made to avoid error. The following suggestions can be used to make those computations.

Computing the Average ($\overline{X}$)

The average is the sum of all of the values divided by the total number of values. Consider the following data.

$$5, \ 3, \ 4, \ -2, \ 0$$

The addition of these numbers can be confusing if the correct rules are not followed. A quick rule to use is: Add all of the positive numbers and subtract the negative number from the total. Therefore, the sequence of the problem above is

$$5 + 3 + 4 + 0 - 2 = 10$$

To compute the average, remember that the 0 is a value. Therefore, there are five numbers in all. The average is $\frac{10}{5}$ or 2.

Computing the Range (R)

The range is the largest value minus the smallest value in the data. To compute the range with negative numbers in the data, simply follow this rule: To subtract a negative number from a positive number, change its sign and *add* it. For an example we will use the same values previously mentioned (5, 3, 4, -2, 0).

$$\begin{aligned} R &= 5 - (-2) \\ &= 5 + 2 \\ &= 7 \end{aligned}$$

For further explanation of this rule, refer to the discussion of the real number line in Chapter 14.

Computing the Standard Deviation (σ)

The steps to compute the standard deviation (σ) were covered earlier in this chapter (using positive values). Using negative values, the problem surfaces in the step where $X - \overline{X}$ is required. In this step, positive numbers must often be subtracted from negative numbers. In this case, the following rule applies: To subtract a positive value from a negative value, add the two values and the answer will still be negative. For example,

$$(-2) - (+3) = -5$$

After this obstacle is overcome, the calculation data can continue as was shown earlier in this chapter.

Dispersion Practice Problems and Solutions

1. Find the range of the following numbers: 3, 6, 1, 6, 8, 2, 5, 8, 9.

Solution
Range is the largest number minus the smallest. The largest above is 9, and the smallest is 1.

$$\text{Range } (R) = 9 - 1 = 8 \text{ answer}$$

2. Find the population variance of the following numbers: 4, 6, 8, 2, 4, 5.

Solution
The variance is the sum of X_i minus $\overline{X}$ squared, then divided by the sample size.

$$\overline{X} = \frac{4 + 6 + 8 + 2 + 4 + 5}{6} = \frac{29}{6} = 4.8$$

X_i	$X_i - \overline{X}$	$(X_i - \overline{X})^2$
4	−0.8	0.64
6	1.2	1.44
8	3.2	10.24
2	−2.8	7.84
4	−0.8	0.64
5	0.2	0.04

Sum is 20.84; then 20.84 ÷ 6 = 3.5, the variance.
3. Find the population standard deviation of the numbers above. The standard deviation is the square root of the variance. The variance above is 3.5 and the square root of 3.5 is 1.87.
4. Find the population standard deviation of these numbers: 5, 2, 8, 6, 4.

Solution

The population standard deviation is the square root of the variance; *or* the square root of the sum of $X_i - \overline{X}$ squared, then divided by the sample size.

$$\overline{X} = \frac{5 + 2 + 8 + 6 + 4}{5} = \frac{25}{5} = 5$$

X_i	$X_i - \overline{X}$	$(X_i - \overline{X})^2$
5	0	0
2	−3	9
8	3	9
6	1	1
4	−1	1

Sum is $20 \div 5 = 4$, the variance. Four is the variance, so the square root of 4 is the standard deviation. *Answer:* 2.

5. Process A has a 6 sigma spread of .0003 in. and process B has a 6 sigma spread of .0009 in. Process A costs $50 to produce a part, and process B costs $10 to produce a part. If the total tolerance of the dimension is .006 in., what process would you choose to produce the part?

Solution

Both processes will produce the part easily, simply by comparing their 6 sigma spread to the tolerance (total). Therefore, use process B and save $40 per piece produced.

DEFINITION OF A PROCESS IN CONTROL

A process is said to be in a state of control when special causes of variation have been eliminated to the extent that the points plotted on a control chart remain within the control limits and exhibit random (normal) variation between control limits. (A state of control does not indicate a capable process.)

Natural Patterns of Variation

When points are plotted on a control chart (Figure 13.6), the following guidelines can be used to check for control.

FIGURE 13.6

A process in control.

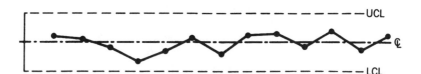

1. No points are outside the control limits.
2. Points are almost equally disposed on both sides of the central line.
3. About two-thirds of all points are near the central line.
4. Only a few points are near either control limit.
5. There are no continuous runs of about seven or more points on one side of the centerline.
6. There are no groups of continuous points heading in a trend toward either control limit.
7. There are no hugging patterns (e.g., all points lie within the plus or minus 1σ zones).
8. There are no straight-line patterns.

Some Advantages of a Process in Control

Some advantages of having a process in control are given in the list that follows:

1. There is more uniformity (less variation) between units, which decreases the chances of producing a defective product.
2. Fewer samples are necessary to judge product quality, since they are more uniform.
3. Inspection costs are reduced.
4. A more accurate definition of capability and sound business decisions can be made such as:
   a. Selection of specification limits
   b. Knowledge of the yield of the process
   c. Selection of the appropriate process
5. The percentage of product that will be produced within specification limits can more accurately be calculated (% yield).
6. There will be considerably less scrap, rework, and other wastes of productive time and money.

CONTROL LIMITS VERSUS SPECIFICATION LIMITS

Control Limits

1. Must be calculated from data gathered during the process.
Calculate grand average $(\overline{X})$ and average range (R)
Example: Subgroup size = 5
    Number of subgroups = 10
$$\overline{X} = 4.001$$
$$\overline{R} = .007$$
UCL (upper control limit)
LCL (lower control limit)
$\text{UCL} = \overline{\overline{X}} + A_2(\overline{R})$
(when the sample size is 5, $A_2 = .577$)
    $= 4.001 + (.577)(.007)$
    $= 4.001 + .004$
    $= \underline{4.005}$

Specification Limits

1. Are given on the print, operation sheet, or other specifications.
Example: $4.000 \pm .010$
Specification limits are
    4.010 and 3.990
UTL (upper tolerance limit) = 4.010
LTL (lower tolerance limit) = 3.990

$$LCL = \overline{\overline{X}} - A_2(\overline{R})$$
$$= 4.001 - (.577)(.007)$$
$$= \underline{4.001 - .004}$$
$$= \underline{3.997}$$

2. Are used to determine whether or not the *process* is in *statistical control.*

2. Are used to determine whether or not the parts inspected are per the specification.

3. For process control and capability, the control limits must be within the specification limits.

3. To produce parts at minimum costs, the specification limits should be outside the control limits. This indicates that the process has the capability to produce parts per specifications.

4. If a plotted point is outside the control limits, it indicates that the process is *not* in statistical control, and assignable causes exist at the time the sample was drawn for inspection, as shown in Figure 13.7.

4. If a plotted point is within the control limits, but outside the specification limits, this indicates that the process is in statistical control, but *does not have* the capability of producing the parts to specifications. So the process variation should be improved (see Figure 13.8).

FIGURE 13.7
Control chart limits.

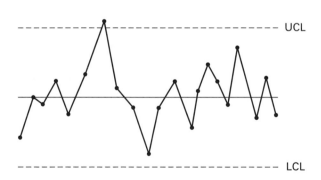

FIGURE 13.8
Control chart limits.

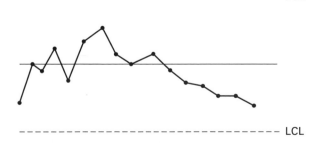

TABLE FOR THE INITIAL SELECTION OF SUBGROUP SIZES AND INTERVALS

There is no absolute rule for how often we should sample. The realities of the factory layout and the cost of sampling must be balanced with the value of the data obtained. In general, it is best to sample quite often at the beginning, and reduce the sampling frequency when the data permit. Table 13–2 (taken from Mil-Std-414) can be used for estimating the amount of initial sampling required.

For example, if the process is expected to produce 3000 pieces per shift, then from the table we should sample 50 pieces per shift. If we use the typical subgroup size of five, then $\frac{50}{5}$ (or 10 samples need to be taken during the shift).

On an 8-hour shift there are 480 minutes. Therefore, we need to take a sample every 48 minutes ($\frac{480}{10}$). Thus we take a sample of five pieces every 48 minutes in this case.

Note: Large sample sizes tend to make control limits more sensitive than smaller samples.

VARIOUS PLOTTING ERRORS TO AVOID

A control chart can be misplotted in several ways. A misplotted chart can and will cause a poor decision to be made on the process. The following are some specific errors to avoid.

1. Instrument errors
   a. Wrong discrimination (use the 10 percent rule).
   b. Wrong instrument for the measurement.
   c. Instrument is not calibrated.
   d. Instrument is not properly used.
   e. Many other possible measurement errors such as heat, dirt, and other problems.
2. Miscalculations
   a. $\overline{X}$ miscalculated on the $\overline{X}$ chart.
   b. Range miscalculated on the range chart.
   c. *p, np, c,* or *u* miscalculated in the case of attribute charts.*

---

* *p* chart, control chart for fraction defective; *np* chart, control chart for number defective; *c* chart, control chart for number of defects; *u* chart, control chart for defects per unit.

TABLE 13–2
Estimated Sample Sizes

Production Rate per Shift	Total Number of Pieces to Be Sampled (per Shift)
1 to 65	5
66 to 110	10
111 to 180	15
181 to 300	25
301 to 500	30
501 to 800	35
801 to 1,300	40
1,301 to 3,200	50
3,201 to 8,000	60
8,001 to 22,000	85

3. Plotting the point
   a. Plotting the point on the wrong line.
4. Sampling errors
   a. Using one chart for two processes/machines.
   b. Biased data (as a result of bias in measurement).
   c. Taking samples too often.
   d. Not taking samples often enough.

HOW MEASUREMENT ERRORS AFFECT CHARTS

Measurement Error	How It Affects the Chart
Wrong selection of the measuring tool	
a. Discrimination	**a.** Large distance between plotted points
b. Application	
c. Technique	**b.** Rounding error in plotting
	c. Appearance of an out-of-control condition when there is none
	d. Action based on meaningless data
	e. Misguiding chart results
	f. Incorrect control limits
Untrained personnel	**a.** Plotted points questionable
	b. Calculations questionable
	c. Meaningless data
Poor precision of the measuring instrument	**a.** Appearance of in control when it is actually out of control, and vice versa
Incorrect subgroup size and inspection frequency	**a.** No meaningful data with respect to time
	b. Control conditions witnessed too late to take action

IMPORTANCE OF RATIONAL SUBGROUPS

The purpose of statistical control charts is to detect assignable (special) causes of variation in the process so that we have the chance to eliminate those causes and get the process into control. There are three sources of variation:

Lot-to-Lot Variation. The variation between lots or batches.

Part-to-Part Variation. The variation between parts (or subgroups of parts).

Within-the-Part Variation. The variation within one part such as out of roundness or taper.

Rational subgrouping is necessary to provide assurance that the signals of special cause variation in a process are due to the process, not poor subgrouping methods.

Rational subgroups. Subgroup samples containing consecutively made products that are taken from the process in a manner that would cause us to expect little variation in the measurements within the subgroup, yet noticeable variation from subgroup to subgroup.

The most rational subgroups are taken in order of production (small continuous samples taken at regular time intervals). The following are examples of irrational subgrouping.

Within-the-Piece Variation

Within-the-piece variation such as taper, out of roundness, or out of parallelism can cause significant problems in data collection and control charting. An example of this is a boring machine that bores diameters in the part that are significantly out of round. If measurements were taken at different locations while control charting, the roundness error (within the piece) could cause false signals on the control chart.

Multiple Stations and Multiple Characteristics

A multiple drill press drills three different holes in the part that have the same specifications. But they are drilled with three different drills that may wear differently. If the three holes were combined on the chart as one subgroup, this could cause false signals on the control chart due to hole-to-hole variation.

Order of Production

Rational subgroups should also be representative of production over a given period of time; therefore, it is important that the time period between subgroups is identified and followed. For example, an operator should have taken 5 parts consecutively produced every hour during the shift, but instead the operator measured and plotted 40 parts made at the end of the shift. When the process went "out of control" it may have been due to irrational subgrouping, not special causes in the process.

Lot-to-Lot Variation

Various heat lots of materials are used on a process. Each lot is rolled sheets of metal. The variation within the roll and from roll to roll is fairly consistent, but the variation from heat lot to heat lot is not consistent. A rational subgroup would be one that signaled the lot-to-lot variation (if that were the purpose of the chart). If subgroup samples were only taken at random, it would not be a rational subgroup and might cause the process to appear to be in control when it is not in control.

HOW CALCULATION ERRORS AFFECT CHARTS

Error	How It Affects the Chart
$\overline{X}$ calculated too low or too high, and plotted	a. Machine adjustment when not necessary (overcontrol)
	b. Consistently in and out of control
	c. Appearance of two universes

R calculated too high or too low, and plotted

a. Action taken to correct dispersion when dispersion is not a problem

b. No action taken to correct dispersion when dispersion is a problem

GUIDELINES FOR STARTING AN AVERAGE AND RANGE CONTROL CHART

The following pages discuss how to start a control chart. The examples use an $\overline{X}$–R chart. Once a characteristic has been selected for a control chart application, the following guidelines will be helpful in starting the chart.

Note: Decide firsthand the exact purpose of the chart.

1. *Measurement.* Make sure the measuring instrument discriminates to at least 10 percent of the total tolerance to be measured.
2. *Characteristic selected.* Make sure the characteristic is clearly defined with respect to the technique used for measurement. If not, include any required special instructions.
3. *Related information.* Fill in the appropriate information at the top of the control chart.
4. *Sample size.* Select the subgroup size and time interval between subgroups. At first, this should be stringent. It can be relaxed later.
5. *Data collection.* A minimum of 25 subgroups or 125 pieces are required to start a control chart. With these subgroups, the "trial" control limits and centerline can be calculated.
6. *Summation of subgroups.* Add each subgroup and enter the sum in the "sum" column.
7. *Averages and ranges.* Find the average $\overline{X}$ (X-bar) and the range (R) and enter them in the appropriate space below each subgroup.
8. *Average range $\overline{R}$ (R-bar).* Find the average range, which is the sum of all the subgroup ranges divided by the total number of subgroups.
9. *Grand average (X-grand bar) $\overline{\overline{X}}$).* Calculate the grand average $\overline{\overline{X}}$ (X-double bar), which is the sum of all the averages $\overline{X}$ (X-bars) divided by the total number of X-bars.
10. *Plot trial centerlines.* $\overline{R}$ (R-bar) is the centerline of the range chart. X-grand bar ($\overline{\overline{X}}$) is the centerline of the X-bar chart. These should be centered in the chart and drawn in solid lines.
11. *Range chart "trial" control limits.* Calculate the trial control limits for the range chart first. This will help you to select the scale for the X-bar chart later. The range chart control limits are calculated as follows. The upper control limit (UCL) = R-bar multiplied by the D_4 factor (for the selected sample size). The lower control limit (LCL) = R-bar multiplied by the D_3 factor.

Notes:

- See Table 13–1 for these factors.
- Use the UCL and centerline to select the scale that fits the chart.
- For sample size (n) less than 7, the lower control limit for a range chart is 0.

12. Draw the trial control limits for the range chart on the chart in dashed lines.
13. *X-bar chart "trial" control limits.* Calculate the trial control limits for the X-bar chart. These are

$$\text{UCL}_{\bar{x}} = \overline{\overline{X}} + A_2(\overline{R})$$
$$\text{LCL}_{\bar{x}} = \overline{\overline{X}} - A_2(\overline{R})$$

UCL equals *X*-grand bar + (*A2* times *R*-bar)
LCL equals *X*-grand bar − (*A2* times *R*-bar)

Note: Multiply the *A2* factor by *R*-bar first, then add or subtract.

14. Draw the trial control limits on the *X*-bar chart. (Select a comfortable scale on the chart.)
15. *Plotting points (X-bar chart).* Plot the *X*-bar ($\overline{X}$) for each subgroup on the $\overline{X}$ chart just below the subgroup, and connect the points.
16. *Plotting (range chart).* Plot the range for each subgroup on the range chart (directly under each subgroup).
17. *Initial decisions on the chart.*
   a. Look for any out-of-control conditions. (See the discussion of pattern analysis later in this chapter.)
   b. *If there are out-of-control conditions,* attempt to find an assignable (special) cause. If assignable cause is found, recalculate the control limits and centerlines for each chart excluding those data points that were out of control.
   c. *If there are no out-of-control conditions,* use the previously calculated control limits and centerlines to start the chart.
18. *Action on out-of-control conditions.* Act on all out-of-control conditions. *Remember:* If you find assignable cause for these points, exclude these data when revising control limits.
19. *Revision of control limits frequently.* Control limits need to be revised frequently. As the process tends to improve or deteriorate, the control limits will subsequently narrow or widen, respectively.

See Figure 13.9 for an example.
Note: Always mark the chart to indicate

- Out-of-control conditions and the cause
- Reason the chart was stopped (for example, the machine is down)
- Reason the chart was started again (for example, a new setup)
- When control limits were recalculated
- Any adjustments made to the process
- Any other pertinent information about the process

Average and Range Chart Practice Problems and Solutions

1. The following *X*-bars are given: 3.2, 3.4, 3.8, 2.9, 3.7. Find the grand average.

Solution
The grand average (*X*-double-bar) is found simply by adding all of the *X*-bars and dividing the sum by the number of *X*-bars.

$$\overline{\overline{X}} \text{ (grand average)} = \frac{3.2 + 3.4 + 3.8 + 2.9 + 3.7}{5 \text{ of them}} = \frac{17}{5} = 3.4$$

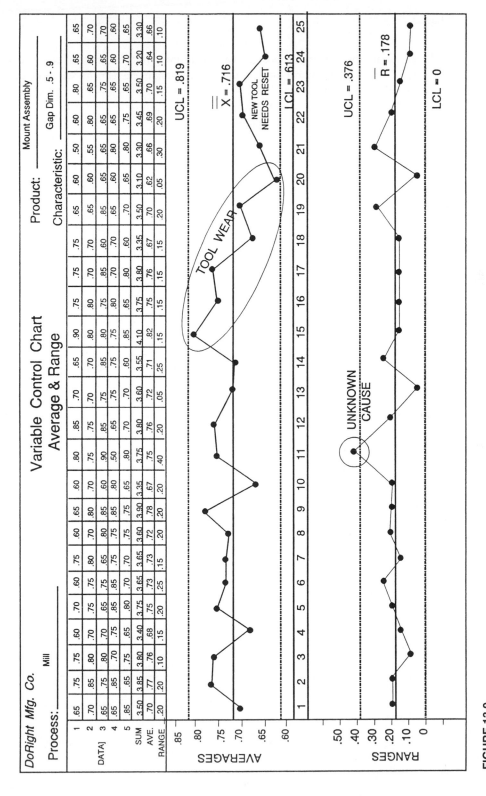

FIGURE 13.9

Average and range chart.

423

2. Three subgroups of five pieces each were measured after the grand average was found to be 6. Compute the control limits if the average range was 1.5 and see if the process is in control or not (*X*-bar chart).

Subgroups:
$$
\begin{array}{ccc}
3 & 4 & 4 \\
4 & 8 & 6 \\
7 & 7 & 3 \\
6 & 5 & 4 \\
4 & 4 & 5
\end{array}
$$

Solution
Control limits $= \overline{\overline{X}} \pm A_2(\overline{R}) = 6 \pm .577(1.5) =$ UCL 6.866
LCL 5.135

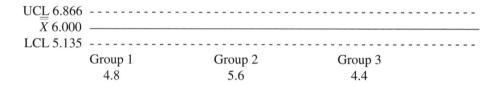

UCL 6.866
$\overline{\overline{X}}$ 6.000
LCL 5.135

Group 1	Group 2	Group 3
4.8	5.6	4.4

No, the process is not in control. It is out of control considerably on the low side.

3. For a range of 1.0, compute the range chart limits and see if the ranges are in control. The sample size (*n*) is 5.

Solution
R-chart limits are

$$
\text{UCL} = \overline{R}(D_4) = 1.0(2.114) = 2.114
$$
$$
\text{UCL} = \overline{R}(D_3) = 1.0(0) = 0
$$

Range is out of control at group 2 for some reason (see Figure 13.10).

4. On an *X*-bar chart, there is one single point that falls out of the control limits. What do you do?

FIGURE 13.10
One point out at subgroup 2. Note: Values plotted are for example only.

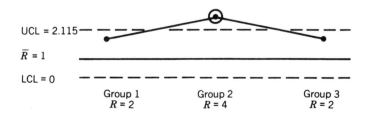

UCL = 2.115
$\overline{R} = 1$
LCL = 0

Group 1	Group 2	Group 3
R = 2	R = 4	R = 2

Solution
A single point falling outside the control limits is not something that should cause wide concern and panic. You should attempt to find the cause, or continue to run the process and monitor the next plotted point closely to see if there is a true problem or if there is just an outside source of the single point being out.

AVERAGE AND SIGMA CHARTS

Average ($\overline{X}$) and sigma (s) charts are often used for increased sensitivity to variation (especially when larger sample sizes are used). These charts are more difficult to work with than the $\overline{X}$–R charts due to the tedious calculation of the sample standard deviation (s).

Sample Standard Deviation

The formula is

$$s = \sqrt{\frac{\Sigma(X_i - \overline{X})^2}{n - 1}}$$

Note: For population standard deviation, use n, not $n - 1$.

where Σ means the sum
X_i is the individual measurements
$\overline{X}$ is the average
n is the sample size

The longhand calculation of (s) is performed in the following manner.

■ **Example 13–13**
Given these data—5, 6, 7, 6, 8—find (s).

Solution

	$n = 5$	$\overline{X} = 6.4$
List the X_i	**List $X_i - \overline{X}$**	**List $(X_i - \overline{X})^2$**
5	−1.4	1.96
6	−0.4	0.16
7	0.6	0.36
6	−0.4	0.16
8	1.6	2.56
		Sum (Σ) = 5.20

$$s = \sqrt{\frac{\Sigma(X_i - \overline{X})^2}{n - 1}}$$

$$= \sqrt{\frac{5.20}{4}}$$

$$= \sqrt{1.30}$$

$$= 1.140 \text{ rounded}$$

In most cases a calculator or computer should be used to avoid the time-consuming mathematics. It must be noted that at times the $\overline{X}-s$ chart is not as practical to use as the $\overline{X}-R$ chart, for many reasons.

The $\overline{X}$ chart is constructed in the same way as it was described earlier, except that since sigma (s) is used, the control limits are calculated using different factors.

The control limits for an $\overline{X}$ chart using sigma (s) are calculated using the following formula.

$$\mathrm{UCL}_{\bar{x}} = \overline{\overline{X}} + A_3(\bar{s}) \qquad \mathrm{LCL}_{\bar{x}} = \overline{\overline{X}} - A_3(\bar{s})$$

where $\overline{\overline{X}}$ is the grand average

A_3 is a factor depending upon the sample size (see Table 13–3)

$\bar{s}$ the average sample standard deviation

Note: Multiply first, then add or subtract.

The control limits for the sigma (s) chart are calculated using the following formula and Table 13–3.

$\mathrm{UCL}_S = B_4(\bar{s})$
$\mathrm{LCL}_S = B_3(\bar{s})$
$\bar{s}$ is the average sample standard deviation. It is the centerline of the ($\bar{s}$) chart.

$$\bar{s} = \frac{\text{sum of the individual sigmas}}{\text{total number of sigmas}}$$

TABLE 13–3

Table of control chart factors B_3, B_4, A_3.

n	2	3	4	5	6	7	8	9	10
B_4	3.267	2.568	2.266	2.089	1.970	1.882	1.815	1.761	1.716
B_3	0	0	0	0	0.030	0.118	0.185	0.239	0.284
A_3	2.659	1.954	1.628	1.427	1.287	1.182	1.099	1.032	0.975

Basic Steps to Construction (Refer to Figure 13.11 for an example)

Step 1.   Decide the purpose of the chart.

Step 2.   Select the characteristic to be controlled.

Step 3.   Collect the data and record the subgroups on the chart. (Select sample size and frequency.)

Step 4.   Calculate $\overline{X}$ for each subgroup and record it.

Step 5.   Calculate (s) for each subgroup and record it.

Step 6.   Find $\overline{\overline{X}}$ and record it.

Step 7.   Find $\overline{s}$ (the average of the sample standard deviations) and record it.

Step 8.   Find the control limits for both charts and draw them on the chart. Select the proper scale.

Step 9.   Plot the $\overline{X}$ values on the $\overline{X}$ chart and connect the points.

Step 10.   Plot the s values on the sigma chart and connect the points.

Step 11.   Interpret the chart for "control" the same way as the $\overline{X}$–R chart.

Step 12.   Revise the control limits if necessary.

Step 13.   Continue to use the chart for control.

Capability on an X–s Chart

The estimated standard deviation ($\hat{\sigma}$), called sigma hat, can be calculated by

$$\hat{\sigma} = \frac{\overline{s}}{c_4}$$

Refer to Table 13–4 for the c_4 factors.
Note: For a range chart:

$$\hat{\sigma} = \frac{\overline{R}}{d_2}$$

If both $\overline{X}$ and s charts are in control, and the individual measurements are normally distributed, process capability can be assessed (see process capability on page 434).

TABLE 13–4

Table of c_4 factors.

n	2	3	4	5	6	7	8	9	10
c_4	0.798	0.886	0.921	0.940	0.952	0.959	0.965	0.969	0.973

MEDIAN CONTROL CHARTS

Median control charts are alternative charts to $\overline{X}$–R charts for measured data.

Median. The middle value of grouped data.

The specific advantages of a median chart are that it

1. Is easy to use.
2. Does not require day-to-day calculations.
3. Shows both the median and the spread of the process.

Both individual values and medians are plotted.

Basic Steps to Construction (Refer to Figure 13.12 for an example)

The following steps can be followed to start a median chart.

Step 1.   Gather data (usually 10 or fewer in the sample). Odd sizes are more convenient. Only a single graph is plotted. Set the scale to indicate the larger of either
 a. The specification tolerance (plus some extra space) *or*
 b. $1\frac{1}{2}$ to 2 times the range
   The gage used should be at least accurate to 5 percent of the tolerance being measured.

Step 2.   Plot *all* of the individual measurements for each group on a vertical line above the group.

Step 3.   Circle the median of each group and connect them with lines.

Step 4.   Enter the median and range for each subgroup in the data column.

Step 5.   Find the average of the sample medians ($\tilde{\overline{X}}$) and draw it as the centerline of the chart. Find $\overline{R}$ and record it and calculate the UCL and LCL of range and median (use Table 13–5).

$$\text{UCL}_R = \overline{R}D_4 \qquad \text{LCL}_R = \overline{R}D_3 \qquad (D_3 \text{ is 0 if } n \text{ is less than 7})$$
$$\text{UCL}_{\tilde{X}} = \tilde{\overline{X}} + \tilde{A}_2\overline{R} \qquad \text{LCL}_{\tilde{X}} = \tilde{\overline{X}} - \tilde{A}_2\overline{R} \qquad (\tilde{\overline{X}} \text{ is the average median value.})$$

Step 6.   Plot the control limits for medians on the chart.

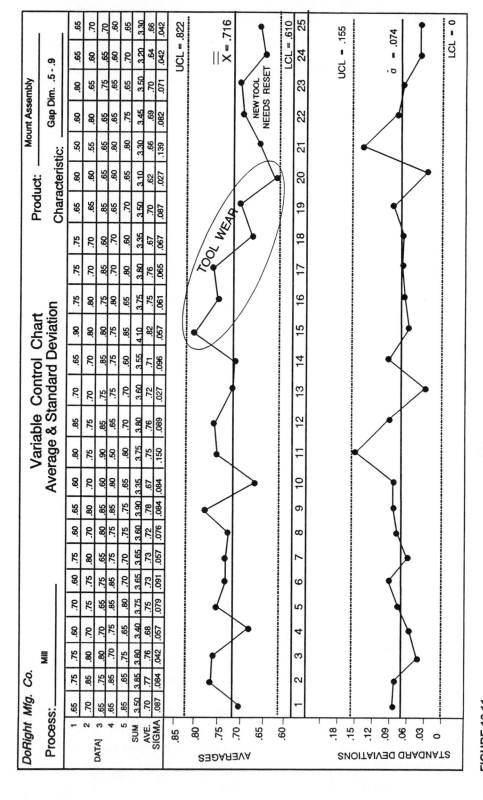

FIGURE 13.11
Average and standard deviation chart.

TABLE 13–5
Table of control chart factors D_3, D_4, and $\tilde{A}_2$.

n	2	3	4	5	6	7	8	9	10
D_4	3.267	2.574	2.282	2.114	2.004	1.924	1.864	1.816	1.777
D_3	0	0	0	0	0	0.076	0.136	0.184	0.223
$\tilde{A}_2$	1.88	1.19	0.80	0.69	0.55	0.51	0.43	0.41	0.36

Interpretation

Compare the UCL_R and LCL_R with each range. You can mark an index card or piece of plastic with the centerline of range and range control limits and compare this to the chart values plotted. Put a narrow vertical box around any subgroup that has excessive range.

Mark any median that is beyond the median control limits and note the spread of medians within the control limits. (Two-thirds of the points should be within the middle third area of the control limits.) Take action on out-of-control conditions.

CONTROL CHARTS FOR INDIVIDUALS AND MOVING RANGES

Individuals' Control Chart

Control charts for individuals use individual readings (X_i) instead of subgroups. Some examples of when an individual's control chart may be used are

1. In short product runs
2. In destructive testing

There are some drawbacks to using an individual's control chart. Some of these are:

1. You must take care in the interpretation if the distribution of the data is *not* normal.
2. Individuals' charts do not separate the piece-to-piece repeatability of the process.
3. Since you have only one value considered in the subgroup, even if the process is in control, the values of the average ($\overline{X}$) and the standard deviation ($\hat{\sigma}$) can have wide variability until you reach about 100 pieces or more.
4. Individuals' charts are not as sensitive to changes in the process as the $\overline{X}$–R chart.

In certain cases it may be better to use the $\overline{X}$–R chart with small sample sizes and increased time intervals as opposed to using an individual's control chart.

Steps in Starting the Individual's Chart (Refer to Figure 13.13 for example)

Step 1. Collect the data. Individual readings are recorded on the chart from left to right. This can be done on the same format as the X-bar chart except the title of the chart must be changed.

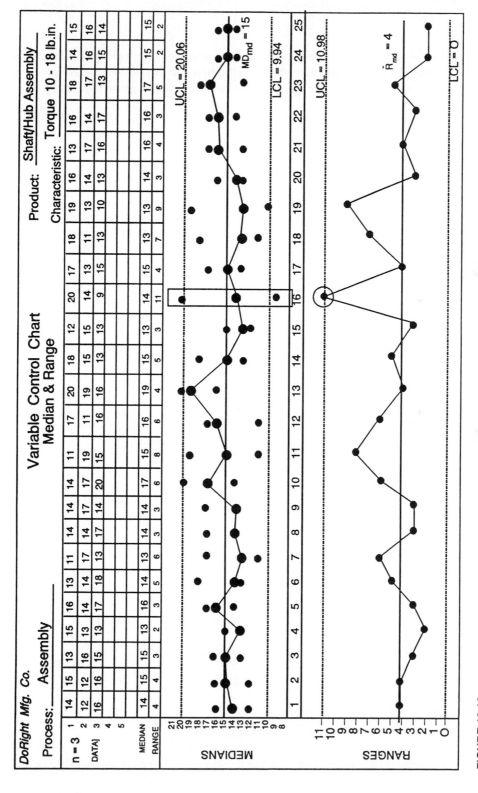

FIGURE 13.12
Median and range chart.

Step 2. Calculate the moving range (*Rm*) between the individual measurements. The moving range is the difference between the first and second values, the second and third values, the third and fourth values, and so on. For example, given five measurements: 7, 8, 10, 9, and 10

Rm between 7 and 8 is 1.
Rm between 8 and 10 is 2.
Rm between 10 and 9 is 1.
Rm between 9 and 10 is 1.

The four moving ranges are 1, 2, 1, and 1.

Note: There is no moving range for the first number. Since you calculated *Rm* on each of two values, *n* = 2.

Step 3. Select the scale for the individual's chart (*X*). This scale could be the largest of two values:
  a. The blueprint tolerance, *or*
  b. About two times the difference between the highest and lowest readings. Make sure that you number the scale on the chart.

Step 4. Select the scale for the moving range (*Rm*) chart. This scale should be the same scale as the individual's chart. Number the selected scale on the chart.

Step 5. Calculate and plot the process average ($\overline{\overline{X}}$), which is the sum of the individual measurements (X_i) divided by the total number of individual measurements (*k*).

$$\overline{\overline{X}} = \frac{\Sigma X_i}{k}$$

where Σ = sum or add
       X_i = individual measurements
       k = number of measurements

Step 6. Calculate and plot the average moving range ($\overline{Rm}$), which is the sum of the moving ranges (*Rm*) divided by the total number of moving ranges (*k* − 1).

$$\overline{Rm} = \frac{\Sigma Rm}{k - 1}$$

where Σ = sum or add
        Rm = each moving range
     $k - 1$ = total number of moving ranges

Remember: There is always one less moving range than there are individual measurements.

TABLE 13–6
Table of control chart factors D_4, D_3, and E_2.

n	2	3	4	5	6	7	8	9	10
D_4	3.267	2.574	2.282	2.114	2.004	1.924	1.864	1.816	1.777
D_3	0	0	0	0	0	.076	.136	.184	.223
E_2	2.66	1.77	1.46	1.29	1.18	1.11	1.05	1.01	0.98

Step 7. Calculate and plot the control limits for the individual's chart.

$$UCL_X = \overline{X} + E_2(\overline{Rm})$$
$$LCL_X = \overline{X} - E_2(\overline{Rm})$$

Note: Always multiply first, then add or subtract. (See Table 13–6 for E_2 factors.)

Step 8. Calculate and plot the moving range chart control limits:

$$UCL_{Rm} = D_4(\overline{Rm})$$
$$LCL_{Rm} = D_3(\overline{Rm})$$

Note: There is no lower control limit for an *Rm* chart if the sample size (*n*) is less than seven pieces.

Step 9. Plot the *X* values measured on the individual's chart and connect the points.

Step 10. Plot the moving ranges (*Rm*) on the moving range chart and connect the points.

Notes: Make sure the chart form:

1. Is filled out at the top with related information.
2. Is identified as an individual's (*X*) and *Rm* chart.
3. Has *X* values recorded in the data blocks.
4. Has *Rm* values recorded in the data blocks.

Interpretation of Individuals' and Moving Range Charts

Control. The moving range chart should be reviewed for points that are beyond control limits. These are signs that special causes exist. Be careful, however, in attempting to act on "trends" on a moving range chart, since this chart uses successive moving ranges.

The individual's chart should be interpreted for

1. Points beyond control limits.
2. The spread between points within the limits.
3. Trends.

TABLE 13–7
Table of d_2 factors for estimating the standard deviation.

n	2	3	4	5	6	7	8	9	10
d_2	1.128	1.693	2.059	2.326	2.534	2.704	2.847	2.970	3.078

Note: The distribution of the data must be normal or some false signals of special causes may occur.

Process Capability.   The estimated standard deviation ($\hat{\sigma}$), called "sigma hat," can be calculated in the same way as the range chart, except that you are using the average moving range ($\overline{Rm}$). Divide the $\overline{Rm}$ by d_2 to get sigma hat ($\hat{\sigma}$) (refer to Table 13–7).

If the process is in control and distributed normally, process capability can be assessed.

ATTRIBUTE CONTROL CHARTS

These are types of control charts other than the variable charts discussed previously. The following control charts control *attributes.*

Attributes. Attributes are involved in any situation where you have to (or choose to) say that a quality characteristic is good or bad, on or off, Go or NoGo, and the like. An example of this is a part that is gaged with a plug gage (Go–NoGo) or a light that is either off or on.

Various kinds of attribute control charts may be used. The chart used depends on what specific attribute you want to control. The attribute charts covered in this chapter are

- *p* **chart,** for controlling the *fraction defective.*
- **100***p* **chart,** for controlling the *percent defective.*
- *c* **chart,** for controlling the *number of defects.*
- *np* **chart,** for controlling the *number of defectives.*
- *u* **chart,** for controlling the *number of defects per unit.*

Defect. A single nonconforming characteristic on a part. Each part can have several defects.

Defective. A part that has one or more defects is called a defective part.

For example, a washer has many characteristics (we will use only three here). The washer has an inside diameter, outside diameter, and a thickness. If these three characteristics were all out of specification limits, then the washer has three defects. However, it is still only *one defective washer.*

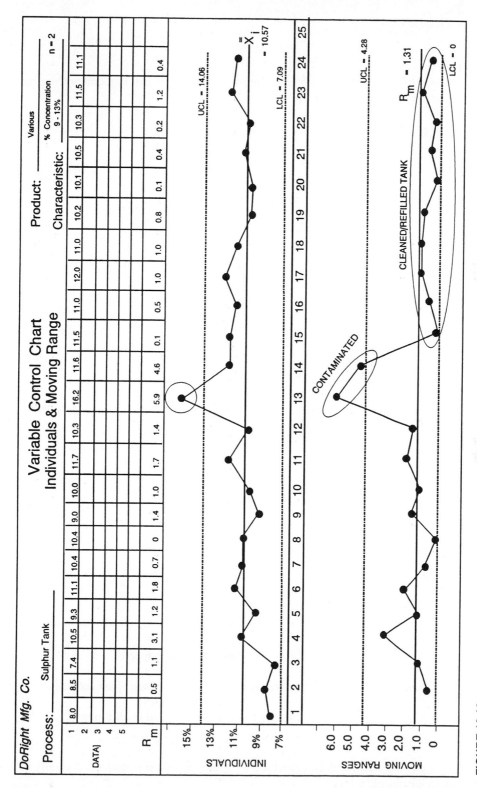

FIGURE 13.13
Individuals and moving range chart.

The *p* Chart

The most commonly used attribute control chart is the *p* chart. Remember that *p* means fraction defective, not percent.

Fraction Defective (*p*). The fraction defective (*p*) is the ratio of the number of defectives found divided by the number of parts inspected.

For example,

$$\text{Number of parts inspected} = 20$$

$$\text{Number of parts defective} = 3$$

$$p = \frac{3}{20} = .15$$

The *p* chart is used in the shop for several reasons, some of which are:

1. On the processes that use Go–NoGo gaging, such as plug gages, ring gages, or functional gages
2. At inspection areas to monitor or control the fraction defective
3. For any other attribute situation you want to control

p Chart Construction (Refer to Figures 13.14 and 13.15 for examples)

Step 1.   Have a clear understanding of the purpose of the chart.

Step 2.   Decide what area you want to control.

Step 3.   Decide what the sample size will be.

Note: The sample size should be the same if at all possible. The following formulas for control limits only work if it is the same.

Step 4.   Record the sample size (*n*) and the number of defectives found (*np*).

Step 5.   Calculate *p* for each sample.

$$p = \frac{np}{n} = \frac{\text{number of defectives found}}{\text{number of parts inspected}}$$

Step 6.   Find the average fraction defective ($\bar{p}$), which is the centerline of the chart.

$$\bar{p} = \frac{\text{total number of defectives found in all groups}}{\text{total number of pieces inspected}}$$

Note: $\bar{p}$ should not be calculated by adding all of the *p* values (especially if you have varying sample sizes).

Step 7. Calculate the upper and lower control limits (UCL–LCL).

Formula: If n is constant,

$$\bar{p} \pm 3\left(\sqrt{\frac{\bar{p}(1-\bar{p})}{n}}\right)$$

Note: If n is not constant,

$$\bar{p} \pm 3\left(\sqrt{\frac{\bar{p}(1-\bar{p})}{\bar{n}}}\right) \text{ average sample size}$$

Or, the first formula must be used to calculate the control limits for each point.

    Note: If any n value varies more than 25 percent from the average of n values $(\bar{n})$, then control limits should be calculated for each plot point. If not, $\bar{n}$ can be used.

■ **Example 13–14**

The sample size is constant $(n = 100)$ and $\bar{p}$ is calculated to be .26. Calculate the control limits.

$$\text{UCL, LCL} = \bar{p} \pm 3\left(\sqrt{\frac{\bar{p}(1-\bar{p})}{n}}\right) \text{ (since } n \text{ is a constant 100 pieces)}$$

$$= .26 \pm 3\left(\sqrt{\frac{.26(1-.26)}{100}}\right)$$

$$= .26 \pm 3\left(\sqrt{\frac{.1924}{100}}\right)$$

$$= .26 \pm 3\,(.044)$$
$$= .26 \pm .132$$
$$= \text{UCL} = .26 + .132 = .392$$
$$= \text{LCL} = .26 - .132 = .128$$

    Note: If the lower control limit is a minus number, always use 0 instead.

    Step 8. Plot the $\bar{p}$ (centerline) and the control limits (UCL = .392 and LCL = .128) on the control chart.

    Note: $\bar{p}$ goes in the center of the chart (solid line) and the scale is then selected to fit the chart.

    Step 9. Plot each p on the control chart and connect those points.

    Step 10. Look for out-of-control conditions. If you find them and identify the cause, recalculate the $\bar{p}$ and control limits using the remaining data.

Step 11. This shows (since $\bar{p}$ = .26) that the percent defective on the average is .26 × 100 = 26 percent. Therefore, the yield of good parts that can be expected is 74 percent.

Important: Remember that the formula for control limits using *n* only works when your sample sizes are all the same size. If your sample sizes are different, you need to

a. Divide by $(\bar{n})$ the average sample size in the formula, *or*
b. Calculate control limits for every plotted point.

Interpretation of the *p* Chart.   A point above the upper control limit (UCL) means:

1. A mistake *may* have been made in calculating the control limit or plotting the point,
2. The process has worsened either at that point in time or as part of a trend, or
3. Something in the measuring system may have changed (the gage or the inspection method, for instance).

A point below the lower control limit (LCL) means:

1. The control limit or the plotted point may be in error,
2. The process has improved (this should be studied for improvements that might be made permanent practice), or
3. The measuring system (gage, method, inspector) has changed.

p Chart Practice Problems and Solutions

1. Five production runs were completed. Each had some defectives. Compute the control limits for a *p* chart based on the information given in Table 13–8.

Solution
First calculate $\bar{p}$. *P*-bar is the sum of the quantity defective divided by the sum of the quantity produced (since the quantities produced are not the same).

TABLE 13–8
Data

Run No.	Quantity Made	Quantity Defective	Percent Defective
1	500	4	.8
2	100	1	1.0
3	300	2	.7
4	250	2	.8
5	400	3	.75
	Sum 1550	Sum 12	

$$\bar{p} = \frac{\text{sum defective}}{\text{sum produced}} = \frac{12}{1550} = .008$$

$$\text{Control limits} = \bar{p} \pm 3\left(\sqrt{\frac{\bar{p}(1 - \bar{p})}{n}} \right)$$

$$= .008 \pm 3\sqrt{\frac{.008(1 - .008)}{310}}$$

$$= .008 \pm 3\sqrt{.0000256}$$

$$= .008 \pm 3\,(.00506)$$

$$= .008 \pm .01518$$

$$\text{UCL} = .008 + .01518 = .023$$

$$\text{LCL} = .008 - .01518 = -.007 \text{ or } 0$$

2. Using the chart above, would six defective pieces out of a 400-piece run of parts be out of control?

Solution
No, because six defectives out of 400 is .015 fraction defective, and the upper control limit is .023.

The *np* Chart

The *np* chart is similar to the *p* chart except you are considering the number of defectives (*np*) rather than the fraction defective (*p*). As with the *p* chart, the sample size (*n*) should be constant. The *np* chart is used when:

1. The number of defectives is easier or more meaningful to report.
2. The sample size (*n*) remains constant from period to period.

Steps in using the *np* chart are:
Refer to Figure 13.16 for an example.

Step 1.   Record the constant sample sizes (*n*) on the chart.

Step 2.   Record the *number of defectives* found in each sample.

Step 3.   Plot the number of defectives found on the chart.

Step 4.   Calculate the average (*n$\bar{p}$*).

$$n\bar{p} = \frac{\text{sum of all } np}{\text{total number of subgroups}}$$

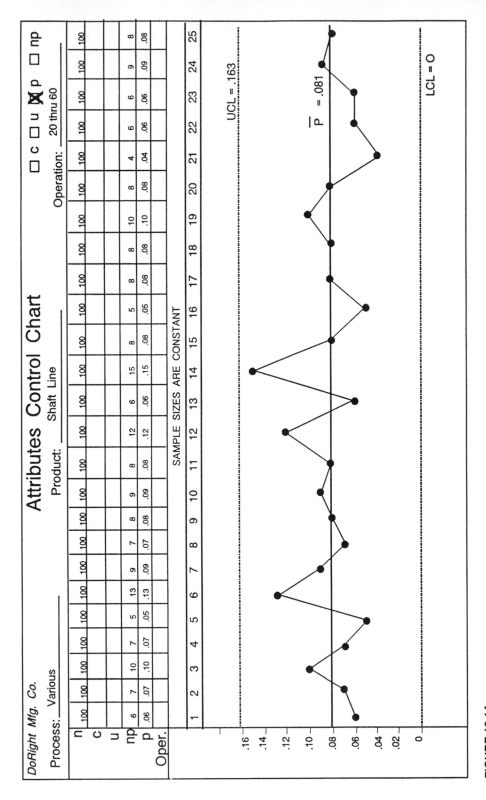

FIGURE 13.14

A *p* chart for constant sample sizes.

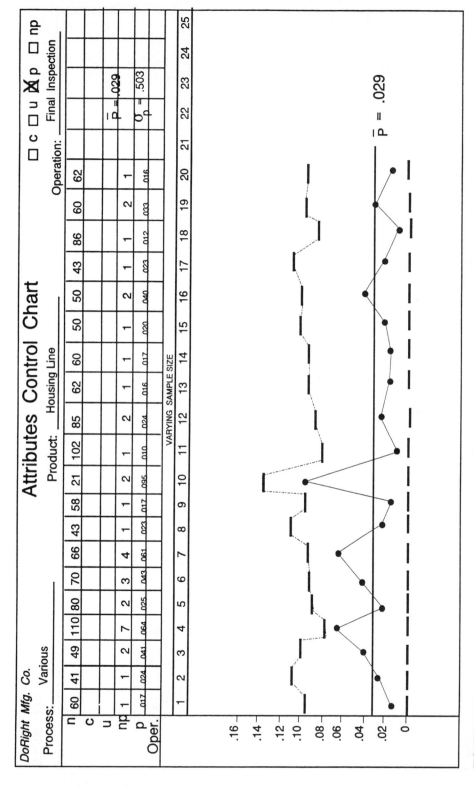

FIGURE 13.15

A *p* chart for varying sample sizes.

441

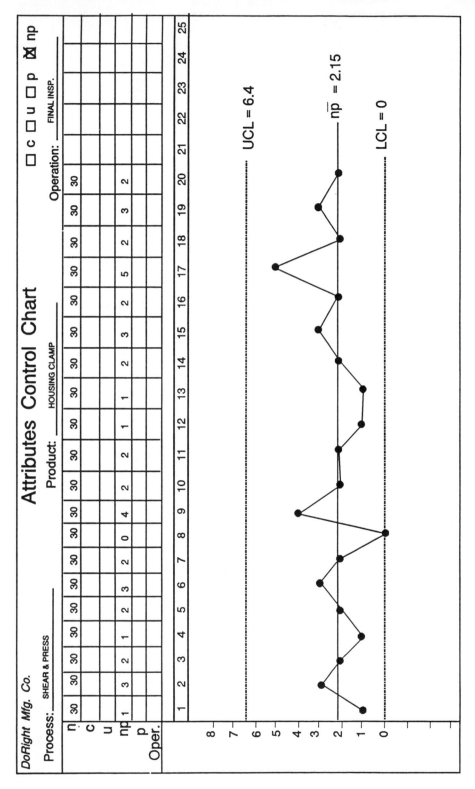

FIGURE 13.16
An *np* chart for defectives.

Step 5. Calculate the upper and lower control limits.

$$\begin{matrix} \text{UCL} \\ \text{LCL} \end{matrix} = n\bar{p} \pm 3\left(\sqrt{n\bar{p}\left(1 - \frac{n\bar{p}}{n}\right)} \right)$$

Interpretation of the *np* Chart

Control. The *np* chart has the same interpretation as the *p* chart except that you are talking in terms of the *number of defectives (np)* instead of the fraction defective *(p)*.

Capability. The process capability is the average number of defectives $(n\bar{p})$.

The 100*p* Chart

The 100*p* chart is the same as the *p* chart except that you convert the fraction defective *(p)* to percent defective (100*p*) by multiplying it times 100. For example,

$$p = \frac{\text{number of defectives found}}{\text{number of units inspected}}$$

$$100p = p \times 100 \text{ (answer is in percent, \%)}$$

$$\bar{p} = \frac{\text{total number of defectives found}}{\text{total number of parts inspected}}$$

$$100\bar{p} = \bar{p} \times 100 \text{ (answer is the average percent defective, \%)}$$

Interpretation of the 100*p* Chart.   The interpretation of the 100*p* chart is exactly the same as the *p* chart except that you are talking in terms of percent defective (100*p*) not fraction defective *(p)*.

Capability. The capability of the process is $100\bar{p}$, which is the average percent defective.

The *c* Chart

The *c* chart is an attribute chart that is used for controlling the number of defects. As mentioned earlier, a defect is a single nonconforming quality characteristic on the part. One part could have several defects, but it is only *one defective part.*

The centerline of the *c* chart is the average number of defects $(\bar{c})$, called *c*-bar. This is found by adding the total number of defects found and dividing by the number of units inspected.

■ **Example 13–15**

The following number of defects per sample was recorded.

Subgroup	Number of Defects
1	12
2	7
3	10
4	3
5	1

Therefore, c for each subgroup is 12, 7, 10, 3, and 1. The $\bar{c}$ (or the average number of defects) is found by adding each quantity of defects and dividing by 5.

$$\bar{c} = \frac{12 + 7 + 10 + 3 + 1}{5}$$

$$= \frac{33}{5}$$

$$= 6.6 \text{ (this is the centerline of the chart)}$$

The control limits for the c chart are calculated using the following formulas:

$$\text{UCL} = \bar{c} + 3\sqrt{\bar{c}} \qquad\qquad \text{LCL} = \bar{c} - 3\sqrt{\bar{c}}$$
$$= 6.6 + 3\sqrt{6.6} \qquad\qquad = 6.6 - 3\sqrt{6.6}$$
$$= 6.6 + 3(2.569) \qquad\qquad = 6.6 - 3(2.569)$$
$$= 6.6 + 7.707 \qquad\qquad = 6.6 - 7.707$$
$$= 14.307 \qquad\qquad = -1.107$$

Note: If this is a minus number, use 0 instead.

After the centerline and control limits are calculated and plotted, you can plot the individual c values on the chart and connect these points.

When out-of-control conditions are found, you should go back and investigate the particular defects that caused this condition. It is likely that you may find the cause upon investigation.

Interpretation of the c Chart (Refer to Figure 13.17 for an example of a c chart.)

Control. The c chart is interpreted the same way as the p chart except that you are discussing the number of *defects* (not defectives).

c Chart Practice Problems and Solutions

1. Find the centerline of a $\bar{c}$ chart for the following numbers:

Subgroup	Number of Defects Found
1	12
2	8
3	2
4	9
5	4

Solution

Centerline ($\bar{c}$) is equal to the number of defects added, then divided by the number of samples taken.

$$\bar{c} = \frac{12 + 8 + 2 + 9 + 4}{5} = 7$$

2. For the data above, compute the upper and lower control limits of a *c* chart.

Solution

Control limits (*c*-chart) = $\bar{c} + 3\sqrt{\bar{c}}$.

Upper limit = $7 + 3\sqrt{7}$ Lower limit = $7 - 3\sqrt{7}$

$= 7 + 3\,(2.6)$ $= 7 - 3\,(2.6)$

$= 7 + 7.78$ $= 7 - 7.8$

$= 14.8$ *Answer* $= 0$ *Answer*

3. After the control chart above was made, five more subgroups were plotted. Are the new subgroup plot points in control or not?

Subgroup	Number of Defects Found
6	12
7	10
8	9
9	15
10	16

Solution

No, the defects found at assembly numbers 9 and 10 are out of control.

The *u* Chart

The *u* chart is applicable when we want to view the data in terms of the number of *defects per unit* (or defects on a per unit basis). In cases where the subgroup sample sizes (*n*) are constant, the *u* chart has straight control limits. Be careful, however, when sample sizes per subgroup are not constant. As with the *p* chart and the *np* chart, if sample sizes vary significantly (any sample size varies more than 25 percent away from $\bar{n}$), control limits should be

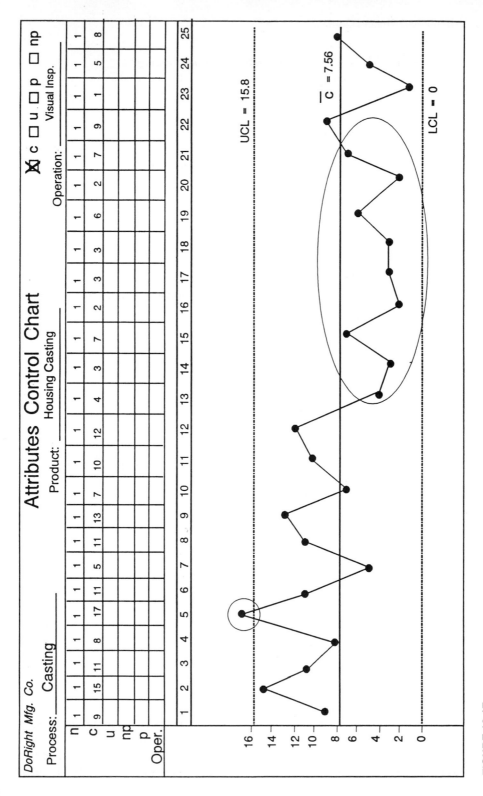

FIGURE 13.17

A *c* chart for defects.

calculated around every plot point. If sample sizes vary less than 25 percent from $\bar{n}$, then $\bar{n}$ can be used in the equation for the control limits.

Steps in Starting the u Chart (Refer to Figures 13.18 and 13.19 for examples).

Step 1. Collect the data.

Step 2. Record the number of units inspected (n) and the number of defects found (c).

Step 3. Calculate the u for each subgroup and plot the values.

$$u = \frac{c}{n}$$

where c = the number of defects found in the subgroup
n = the number of units inspected in the subgroup

Note: Units can mean parts, assemblies, subassemblies, etc.

Step 4. Calculate $\bar{u}$ (the centerline of the u chart).

$$\bar{u} = \frac{\Sigma c}{\Sigma n}$$

where Σc = the sum of all defects found in all subgroups
Σn = the sum of all units inspected in all subgroups

Step 5. Calculate the upper and lower control limits for the u chart.

$$UCL = \bar{u} + 3\frac{\sqrt{\bar{u}}}{\sqrt{n}}$$

$$LCL = \bar{u} - 3\frac{\sqrt{\bar{u}}}{\sqrt{n}}$$

Remember: If sample sizes vary less than 25 percent, use $\bar{n}$ in the formulas. If they vary more than 25 percent, calculate control limits around each plot point.

Interpretation of the u Chart.   The u chart is interpreted the same as the c chart except that you are talking in terms of defects found per unit, not just the total defects found. For example, 50 defects were found when five units were inspected.

$$\text{Defects per unit } (u) = \tfrac{50}{5} = 10 \text{ defects per unit}$$

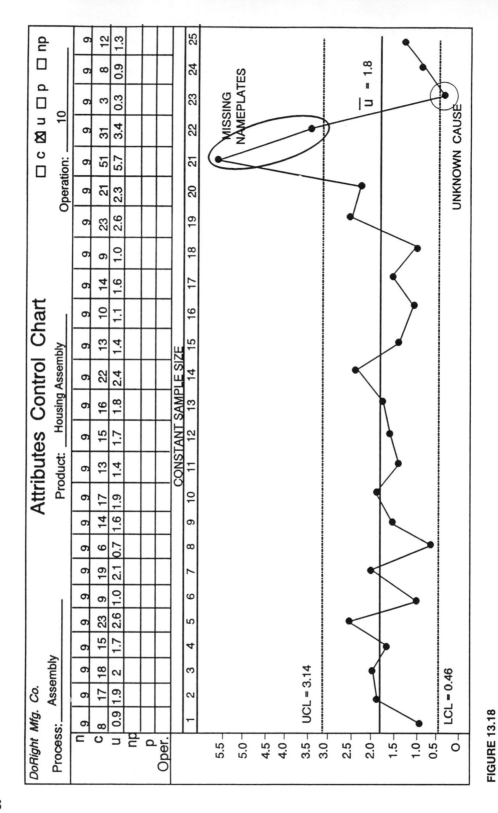

FIGURE 13.18

A *u* chart for defects per unit (constant sample sizes).

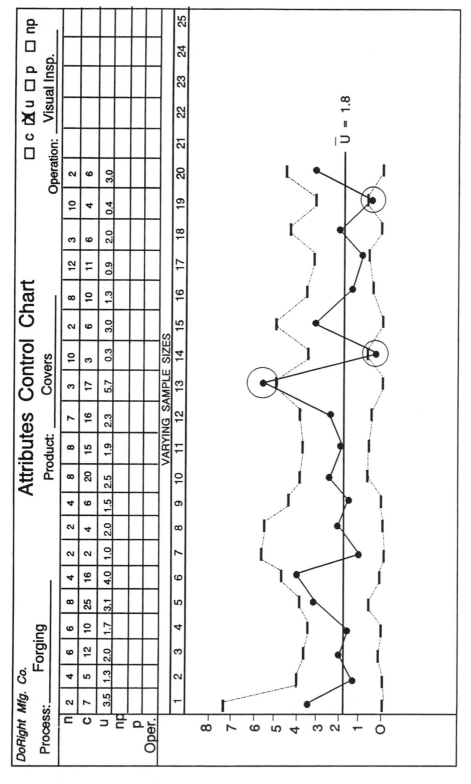

Attributes Control Chart

DoRight Mfg. Co.

Process: _____ Forging   Product: _____ Covers   Operation: _____ Visual Insp.

n	2	4	6	6	8	4	2	2	4	8	8	7	3	10	2	8	12	3	10	2					
c	7	5	12	10	25	16	2	4	6	20	15	16	17	3	6	10	11	6	4	6					
u	3.5	1.3	2.0	1.7	3.1	4.0	1.0	2.0	1.5	2.5	1.9	2.3	5.7	0.3	3.0	1.3	0.9	2.0	0.4	3.0					
np																									
p																									
Oper.																									

VARYING SAMPLE SIZES

$\bar{U} = 1.8$

FIGURE 13.19

A *u* chart for varying sample sizes.

449

INTERPRETING CONTROL CHARTS: PATTERN ANALYSIS

There are a wide variety of patterns that may be seen on a control chart, and an even wider variety of special causes that make these patterns appear on a given chart. Each process is different. The number of *special* causes that exist and the *natural tolerance* of the process depends largely on the condition of that process (wear, tear, methods used, training of operator, etc.). For example, a new process (new machine, equipment, tooling, etc.) would be expected to have fewer special and common causes (and a smaller natural tolerance) than a very old process. Therefore, the new process is expected to be better. Most of the time this is true. An exception to this rule is when the wrong process has been selected to produce the end result.

For another example, suppose a brand-new drill press is selected to produce hole sizes that really require the accuracy of a broaching machine. This process will generally have wider control limits (natural tolerance) than the broach. In addition, the drill press would likely be determined as an incapable process for the required holes.

> **Control limits.** Control limits are established by the inherent variation in a process. These limits are *not* related to blueprint (specification) limits.

In this case, that new drill press could be found to be incapable of producing the required accuracy of the holes; yet it is a brand-new machine.

The following examples show some control chart trends for the $\overline{X}$–R charts with some possible causes (not all of them) for these trends.

The Eight Basic Statements of Control

Note: Refer to Figure 13.20. If you can answer "true" to all eight of the statements that follow, the process is stable (in control). A false answer to any statement means that the process is out of control. Use the chart in Figure 13.20 for practice. Several false answers usually mean *trouble*.

1. There are no plotted points that are outside the control limits. True False
2. There is a relatively close balance between the total number of points that are above and below the centerline. True False
3. There are no consecutive runs of seven or more points on one side of the centerline. True False
4. The plotted points, in general, seem to be randomly falling over and under the centerline. True False
5. There are no steady trends of points heading directly toward either control limit. True False
6. Only a few of all points are near either control limit. True False
7. The plotted points do not appear to be "hugging" closely to the centerline with very little distance between them. True False
8. There are no straight-line patterns. True False

These are the basic features of control. Further control interpretation comes from pattern analysis.

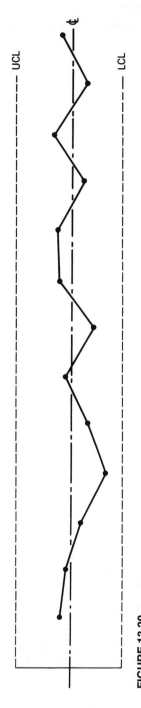

FIGURE 13.20
State of statistical control.

$\bar{X}$-Chart Conditions and Possible Causes

This section lists some examples of $\bar{X}$ chart conditions and some possible causes. Refer to Figures 13.21 through 13.25.

Condition/Description

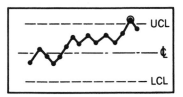

FIGURE 13.21
Jump shift (run) pattern (high or low).

Possible Causes

1. Change in machine setting
2. Different operator
3. Different material, method, process
4. Minor failure of a machine part
5. Measuring equipment setting/technique
6. Adjustment on individuals (overcontrol)
7. New gage
8. Fixture change
9. Parts changed

Note: Recurring run patterns can indicate shift-to-shift variation.

Condition/Description

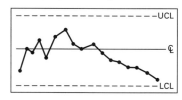

FIGURE 13.22
Trend pattern (either direction).

Possible Causes

1. Tool wear
2. Gradual equipment wear
3. Seasonal effects (temperature/humidity)
4. Dirt/chip buildup on work-holding devices
5. Operator fatigue
6. Change in coolant temperature

Condition/Description

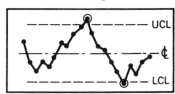

FIGURE 13.23
Recurring cycles pattern.

Possible Causes

1. Different incoming materials
2. Cold startup
3. Seasonal effects (temperature/humidity)
4. Voltage fluctuations
5. Merging of different processes
6. Chemical or mechanical properties
7. Periodic rotation of operators
8. Measuring equipment not precise
9. Calculation and plotting mistakes
10. Machine not holding setting

Condition/Description

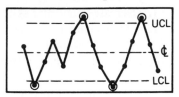

FIGURE 13.24
Two-universe pattern.

Possible Causes

1. Large differences in material quality
2. Two or more machines using the same chart
3. Within the piece variation not considered (such as taper/roundness)
4. Large differences in the method of measurement of the product

Condition/Description

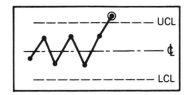

FIGURE 13.25
One point out (freak).

Possible Causes

1. Power surge
2. Hardness on a single part
3. Broken tool
4. Gage setting jumped

Range Chart Conditions and Possible Causes

This section lists some range chart conditions and some of the possible causes for those conditions. Refer to Figures 13.26 through 13.31.

Condition/Description

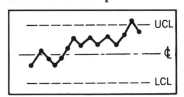

FIGURE 13.26
Jump shift (run) patterns (above the centerline).

Possible Causes

1. Sudden increase in gear play
2. Greater variation in incoming material
3. Inexperienced operator
4. Miscalculation of ranges
5. Excessive speeds and feeds
6. New operator
7. Change in methods
8. Long-term increase in process variability
9. Gage drift
10. New tools
11. Fixture change

Condition/Description

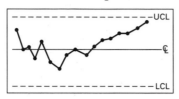

FIGURE 13.27
Trend pattern (increasing).

Possible Causes

1. Decrease in operator skill due to fatigue
2. A gradual decline in the homogeneity of incoming material
3. Some machine part or fixture loosening
4. Gage drift
5. Deterioration of maintenance
6. Tool wear (e.g., the outside diameter on a lathe trends to the high side)

Condition/Description

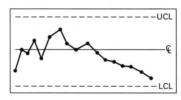

FIGURE 13.28
Trend pattern (decreasing).

Possible Causes

1. Improved operator skill
2. A gradual improvement in the homogeneity/ uniformity of incoming material
3. Better maintenance intervals and program
4. Previous operation more uniform in its output

Condition/Description

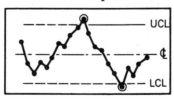

FIGURE 13.29
Recurring cycles pattern.

Possible Causes

1. Operator fatigue and rejuvenation due to periodic breaks
2. Lubrication cycles
3. Rotation of operators, fixtures, and gages
4. Differences between shifts
5. Worn tools
6. Differences between machine needs

Condition/Description

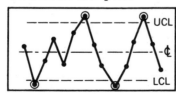

FIGURE 13.30
Two-universe pattern.

Possible Causes

1. Different machines or operators using the same chart
2. Materials used from different suppliers

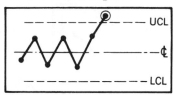

Condition/Description

FIGURE 13.31
One point out (no tread).

Possible Causes

1. Power surge
2. Hardness of a single part
3. Tool broke
4. Gage setting jumped

PROCESS CAPABILITY

Once a process has been brought into *statistical control* (all special causes of variation have been found and eliminated), you can bring out the blueprint (or other specifications) and study the *capability*.

Standard Deviation (σ)

The standard deviation is a measure of variability. Just as weight is measured in pounds and length is measured in inches, the variation of a process is measured in terms of standard deviation units. The Greek letter sigma (σ) is used to indicate *one standard deviation* of a population. Lower case s is used to indicate a sample standard deviation. The area under the normal curve is $\overline{X}$ (the average) plus and minus three standard deviations (see Figure 13.32).

Example of a Process That Is Capable to $+4\sigma$ Limits. The standard deviation can easily be estimated ($\hat{\sigma}$) on a control chart (Figure 13.33) by dividing the average range ($\overline{R}$) by

FIGURE 13.32
Areas under the normal curve.

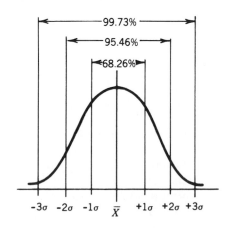

FIGURE 13.33
Example of ±4 (or 8) sigma
capability.

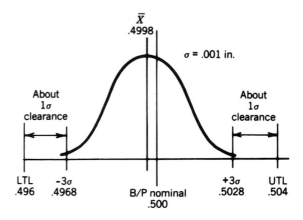

TABLE 13–9
Table of d_2 Factors

Sample Size (*n*)	d_2
2	1.128
3	1.693
4	2.059
5	2.326
6	2.534

the factor d_2 (found in Table 13–9). The d_2 factor depends on the sample size of the control chart (e.g., five pieces per hour). For example,

$$\overline{R} = .0023 \qquad n = 5 \text{ per hour}$$

$$\hat{\sigma}' = \frac{\overline{R}}{d_2} = \frac{.0023}{2.326} = .001 \text{ (rounded off)}$$

Definition of a Capable Process

Three primary conditions are needed for a capable process.

1. The process must be in a state of statistical control.
2. The control limits should be well inside the tolerance limits. The goal is usually a process capability index (C_p) of 1.33 min.

$$C_p = \frac{\text{total tolerance}}{6\sigma}$$

where $\hat{\sigma} = \dfrac{\bar{R}}{d_2}$ for control charts, or

σ = one standard deviation for samples
$\bar{R}$ = average range
d_2 = constant for a given sample size

For example,

$$\sigma = .0005$$

$$\text{Specified tolerance} = \pm.0025$$

$$\text{Total tolerance} = .005$$

$$C_p = \frac{.005}{6(.0005)} = 1.67$$

3. Prior to calculating the C_p, test the distribution of individuals for normality and central-ity (which means that $\bar{X}$ should be close to the nominal dimension).

Distribution Comparisons

Figures 13.34 through 13.36 show some comparisons between normal distributions and tol-erance limits. It is clear that a capable process requires centering and a narrow six-sigma spread to the extent that each three-sigma limit is well within tolerance limits.

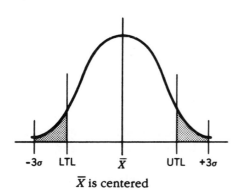

$\bar{X}$ is centered
R is the problem

FIGURE 13.34
Process spread is too wide.

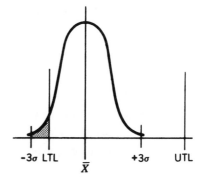

$\bar{X}$ is not centered
R is not the problem

FIGURE 13.35
Process center needs adjustment.

FIGURE 13.36

Process spread is centered and comfortably within tolerance limits.

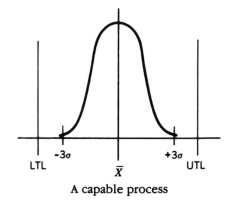

A capable process

Some Alternatives for Incapable Processes

When a process is in control* and it is assessed that the process is not capable of producing to specification limits, several alternatives can be considered. Listed here (not in order of selection) are a few of those alternatives.

1. Repair existing equipment.
2. Modify the existing process.
3. Select a better process.
4. Use 100% inspection (this is not advisable and should be considered only as a last resort).
5. Revise (loosen) drawing tolerances where possible.
6. Purchase new equipment.
7. Assemble selectively (this also is not recommended but is applicable in certain special cases).
8. Reconsider the make/buy decision (if the new decision is buy, the source must have a capable process).

Numerous other alternatives can be considered. The goal is to achieve a capable process at minimum costs.

Capability Studies: Control Chart Method

Certain conditions must be met before you can study the capability of a process and trust the answers you get. These are described in the following steps.

Step 1.   Make sure that the process is in stable statistical control before attempting to study the capability.

---

* If only one point is outside the control limits, it may be that the process is in control but is an isolated cause.

FIGURE 13.37
Frequency distribution of
individuals.

```
                              x
                          x x x   Individual measurements
                        x x x x x      (not averages)
                      x x x x x x x
                    x x x x x x x x x
```

Step 2. Make sure that the *individual measurements* are normally distributed. This is the point where you do not use the averages on the chart (as they will always be normally distributed). This is often done by a frequency distribution, as shown in Figure 13.37.

Note: The distribution should closely approximate the normal bell-shaped curve. In practice, do not expect it to be perfectly normal.

Step 3. Construct a frequency distribution on the individuals and draw lines on the scale to show $\overline{X}$ and each three-sigma limit (see Figure 13.38).

Step 4. Draw the nominal dimension and tolerance lines on the same distribution and calculate $\hat{\sigma}$, which is $\overline{R}$ divided by d_2.

Step 5a. If the process is centered (as shown in Figure 13.38), a capability index (C_p) can be used.

$$C_p = \frac{\text{total special tolerance}}{6\hat{\sigma}'} = \frac{.008}{.006} = 1.33$$

a. For ± 4 sigma capability, a C_p index of 1.33 minimum is needed.
b. The C_p index also requires that the distribution be normal and the process be closely centered on the specification limits. The C_p is the best performance that can be expected from a process unless the process is improved.

Step 5b. When the process is not centered on specification limits (as shown in Figure 13.39), a C_{pk} index can be used.

FIGURE 13.38
Tolerance lines and three sigma
lines drawn on the distribution.

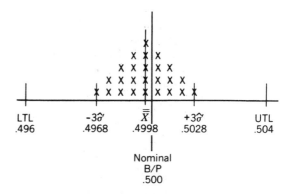

FIGURE 13.39
At ±3 (or six sigma) capability, there is no room for $\overline{X}$ to vary.

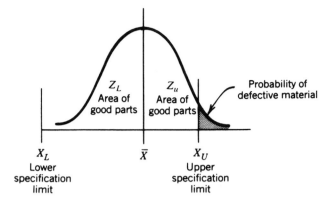

Using the C_{pk} Index for Bilateral Tolerances

The C_{pk} index is a worst-case capability index, which applies when processes are off-center of specification nominal, basic, or other ideal values between two specification limits. The C_{pk} is used in place of the C_p in these cases. Unlike the C_p discussed previously, the C_{pk} is applicable to specifications with two limits (e.g., limit tolerances, or bilateral tolerances). The C_{pk} is calculated as follows:

$$C_{pk} = \text{the lesser of } \frac{X_U - \overline{X}}{3\sigma} \text{ or } \frac{\overline{X} - X_L}{3\sigma} \text{ whichever is smaller}$$

where X_U is the upper specification limit
       X_L is the lower specification limit
       σ is the standard deviation
       $\overline{X}$ is the process average

Note: The estimated standard deviation ($\hat{\sigma}$) may be used with control charts.

The resultant C_{pk} value (like the C_p) should be 1.33 minimum. The C_{pk} is a measure of the performance of a process at the time the data were taken.

■ Example 13–16

The specification limits for the product are .500–.508 diameter. The process is normally distributed, the $\overline{X}$ value is .505, and the standard deviation (σ) is .0008 in. Calculate the C_{pk}.

Solution

$$C_{pk} = \text{the lesser of } \frac{X_U - \overline{X}}{3\sigma} \text{ or } \frac{\overline{X} - X_L}{3\sigma} \text{ whichever is smaller}$$

$$C_{pk} = \text{the lesser of } \frac{.508 - .505}{3(.0008)} \text{ or } \frac{.505 - .500}{3(.0008)}$$

$$\frac{.003}{.0024} \text{ or } \frac{.005}{.0024}$$

1.25 *Answer* or 2.08 *Answer*

The C_{pk} is the lesser of 1.25 or 2.08; therefore, it is 1.25 and the process does not meet the goal of 1.33 minimum C_{pk}.

Using the C_{pk} Index for Single Limit Specifications

The C_{pk} must be calculated carefully when dealing with single specification limits such as surface finish (e.g., AA32 max.), runout tolerances (e.g., .002 TIR max.), and other similar tolerancing methods which identify only one limit of concern. To calculate the C_{pk} properly for these tolerances, use the equation

$$C_{pk} \text{ (for single limits)} = \frac{\text{only limit} - \overline{X}}{3\sigma}$$

■ Example 13–17

A runout tolerance of .002 in. TIR is specified, the average of the process $(\overline{X})$ is .0006 in. runout, and the standard deviation is .0001 in. Calculate the C_{pk}.

Solution

$$C_{pk} \text{ (for single limits)} = \frac{\text{only limit} - \overline{X}}{3\sigma}$$

$$= \frac{.002 - .0006}{.0003} = \frac{.0014}{.0003}$$

$$= 4.67 \text{ Answer}$$

The C_{pk} of 4.67 is greater than the minimum goal of 1.33. The process is considered capable.

Limitations to Using the C_p and C_{pk}

These indexes provide reasonable ratios of process capability when they are properly used. Proper use of the C_p, of course, is to apply it to processes that are centered on specifications. Proper use of the C_{pk} is to apply it to bilateral limits when processes are off-center and to use the only specified limits when specifications are single sided. All of these indexes assume that the process is normally distributed (individual values, not averages). An understood assumption is that the process average $(\overline{X})$ is within specification boundaries.

Predicting the Percent Yield of a Process

Percent yield is the probability of good product expected from the process based on required specification tolerances and process variability. To compute the yield of a process, one must be able to define specific areas and percentages under the bell-shaped curve (refer to the table of areas under the normal curve (Table 13–10). This table helps calculate the area (or percentage) of the curve between $\overline{X}$ and a specification limit (based on the standard deviation of the process) by using Z values. The information necessary to compute a Z value is the average of the process ($\overline{X}$), the upper specification limit (for Z_U, the upper Z score), and the lower specification limit (for Z_L, the lower Z score). Once a Z value is computed (rounded to two decimal places) the area (percent) under the curve can be found using Table 13–10.

The table is used by finding the whole number and first decimal place of the Z value in the Z column, then finding the column that shows the second decimal place of the Z value. For example, to find the area under the curve for a Z value of 2 (or 2.00), first use 2.0 in the Z column, then use the second decimal place (0) to select the appropriate column (0 column) in this case. Converge these two columns to find the area under the curve (.47725 or 47.725%).

Note: The values found in the table are decimal fractions. To convert them to percentages, multiply by 100 (or move the decimal point two places to the right). The area found is the *percent yield* of good product expected to be between the process average ($\overline{X}$) and the specification limit.

For specifications with two specified limits, an upper Z value (using the upper tolerance limit) and a lower Z value (using the lower tolerance limit) must be obtained. It is important to note that for single specification limits, only one Z value is needed. The yield for the "other side" will be 49.999%. The following are equations for Z_U and Z_L:

$$Z_L = \frac{\overline{X} - X_L}{\sigma'}$$

$$Z_U = \frac{X_U - \overline{X}}{\sigma'}$$

where X_L is the lower spec. limit
X_U is the upper spec. limit
$\overline{X}$ is the mean
σ' is the standard deviation

■ Example 13–18

Calculate the percent yield (see Figure 13.40)

$$Z_L = \frac{\overline{X} - X_L}{\hat{\sigma}'}$$

$$= \frac{.502 - .496}{.001}$$

$$= \frac{.006}{.001}$$

$$= 6$$

TABLE 13–10
Areas Under the Normal Curve

Z	0	1	2	3	4	5	6	7	8	9
.0	.0000	.0040	.0080	.0120	.0160	.0199	.0239	.0279	.0319	.0359
.1	.0398	.0438	.0478	.0517	.0557	.0596	.0636	.0675	.0714	.0753
.2	.0793	.0832	.0871	.0910	.0948	.0987	.1026	.1064	.1103	.1141
.3	.1179	.1217	.1255	.1293	.1331	.1368	.1406	.1443	.1480	.1517
.4	.1554	.1591	.1628	.1664	.1700	.1736	.1772	.1808	.1844	.1879
.5	.1915	.1950	.1985	.2019	.2054	.2088	.2123	.2157	.2190	.2224
.6	.2257	.2291	.2324	.2357	.2389	.2422	.2454	.2486	.2518	.2549
.7	.2580	.2612	.2642	.2673	.2704	.2734	.2764	.2794	.2823	.2852
.8	.2881	.2910	.2939	.2967	.2995	.3023	.3051	.3078	.3106	.3133
.9	.3159	.3186	.3212	.3238	.3264	.3289	.3315	.3340	.3365	.3389
1.0	.3413	.3438	.3461	.3485	.3508	.3531	.3554	.3577	.3599	.3621
1.1	.3643	.3665	.3686	.3708	.3729	.3749	.3770	.3790	.3810	.3830
1.2	.3849	.3869	.3888	.3907	.3925	.3944	.3962	.3980	.3997	.4015
1.3	.4032	.4049	.4066	.4082	.4099	.4115	.4131	.4147	.4162	.4177
1.4	.4192	.4207	.4222	.4236	.4251	.4265	.4279	.4292	.4306	.4319
1.5	.4332	.4345	.4357	.4370	.4382	.4394	.4406	.4418	.4429	.4441
1.6	.4452	.4463	.4474	.4484	.4495	.4505	.4515	.4525	.4535	.4545
1.7	.4554	.4564	.4573	.4582	.4591	.4599	.4608	.4616	.4625	.4633
1.8	.4641	.4649	.4656	.4664	.4671	.4678	.4686	.4693	.4699	.4706
1.9	.4713	.4719	.4726	.4732	.4738	.4744	.4750	.4756	.4761	.4767
2.0	.4772	.4778	.4783	.4788	.4793	.4798	.4803	.4808	.4812	.4817
2.1	.4821	.4826	.4830	.4834	.4838	.4842	.4846	.4850	.4854	.4857
2.2	.4861	.4864	.4868	.4871	.4875	.4878	.4881	.4884	.4887	.4890
2.3	.4893	.4896	.4898	.4901	.4904	.4906	.4909	.4911	.4913	.4916
2.4	.4918	.4920	.4922	.4925	.4927	.4929	.4931	.4932	.4934	.4936
2.5	.4938	.4940	.4941	.4943	.4945	.4946	.4948	.4949	.4951	.4952
2.6	.4953	.4955	.4956	.4957	.4959	.4960	.4961	.4962	.4963	.4964
2.7	.4965	.4966	.4967	.4968	.4969	.4970	.4971	.4972	.4973	.4974
2.8	.4974	.4975	.4976	.4977	.4977	.4978	.4979	.4979	.4980	.4981
2.9	.4981	.4982	.4982	.4983	.4984	.4984	.4985	.4985	.4986	.4986
3.0	.4986500	.4990323	.4993128	.4995165	.4996630	.4997673	.4998409	.4998922	.4999276	.4999519
4.0	.4999683	.4999793	.4999866	.4999915	.4999940	.4999968	.4999979	.4999987	.4999992	.4999995
5.0	.4999997	.4999998	.4999999	.4999999	.5000000	.5000000	.5000000	.5000000	.5000000	.5000000

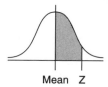

Mean   Z

$$Z = \frac{|X - \overline{X}|}{\sigma}$$

EXAMPLE: A "Z" value of 1.52 is .4357 or 43.57% of the area under the curve

NOTE: For Z values beyond 5.00 use 5.00.

463

FIGURE 13.40
Defective tail.

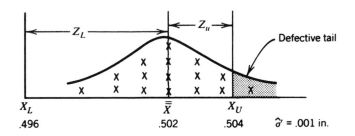

A value of 6 in the table will cause you to read the maximum value of .49999 or 49.999 percent.

$$Z_U = \frac{X_U - \overline{X}}{\hat{\sigma}'}$$

$$= \frac{.504 - .502}{.001}$$

$$= \frac{.002}{.001}$$

$$= 2$$

A value of 2 in the table will be .47725 or 47.725 percent.

Once you find both Z values in percentage (Z_L = 49.999 percent and Z_U = 47.725 percent) you simply add them to get the probability of good parts. These two values added together represent the percent of area under the normal curve that is expected to be acceptable. Therefore,

$$Z_L + Z_U = .97725 \text{ or } 97.725 \text{ percent}$$

This is the probability of good parts (percent yield) if nothing is changed. The probability of defective parts is 100 percent minus 97.725 percent = 2.275 percent in this case. This is the probability of "oversized" parts.

The Reason for ±4 Sigma Capability

All normal distributions have what is called the six-sigma (6σ) spread. From the center of the distribution ($\overline{X}$) there are three equal standard deviations to the right and three equal standard deviations to the left (see Figure 13.41).

If you have a normal distribution and you know what one standard deviation is equal to, for example, σ = .001 in., you can predict very accurately what the process will do from a given center ($\overline{X}$), which is typically the setup value. For example,

$$\hat{\sigma} = .001 \text{ in.}$$

$$\overline{X} = .500$$

FIGURE 13.41
Areas under the normal curve.

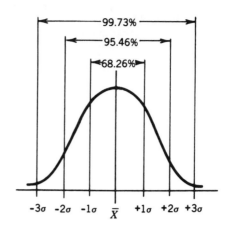

The process is in control. The distribution of the individuals is normal. If the process is set up on .500, the worst parts produced within plus or minus 3 sigma (or 99.73 percent of the time) will be as high as .503 and as low as .497 and most parts will be near .500 (see Figure 13.42).

You can see now that the *natural tolerance* of the process is equal to six sigma (6σ), which is 6 times .001 in. = .006 in.

> **Natural tolerance.** The ability of the process itself. This is what the process can do at best unless you reduce the variation (σ). This natural tolerance has nothing to do with the specification to tolerance (blueprint) other than the fact that you hope it is less than the drawing tolerance.

If the natural tolerance were exactly equal to the specification tolerance, there would be no room for setup error, or external variation (materials variation, for example). This is shown in Figure 13.43.

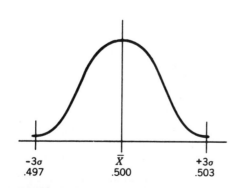

FIGURE 13.42
The six-sigma spread of a process.

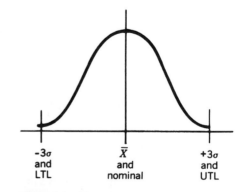

FIGURE 13.43
Six sigma equals the total tolerance.

FIGURE 13.44

± 4 (or 8 sigma) capability.

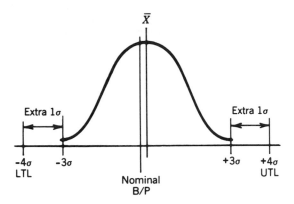

If the setup $(\overline{X})$ moves at all, the curve (or spread) moves with it. Therefore, any move in $\overline{X}$ here would cause defective parts to be produced.

To solve this problem, we include an extra standard deviation at each end of the curve to allow room for $\overline{X}'$ to move without the probability of making defective parts. This is shown in Figure 13.44.

Remember: The blueprint tolerance is equal to the upper limit minus the lower limit. The natural tolerance is equal to six sigma (6σ).

A basic point of view is to find $\hat{\sigma}$ (which is $\overline{R}/d_2$), then multiply $\hat{\sigma}$ by 8. If this value is equal to the total tolerance (and the process setup is closely centered on the blueprint nominal size), you have a process that is capable to plus or minus four sigma ($\pm 4\sigma$). Remember, the distribution must be normal.

The Potential Capability of a Process

The potential capability of a process can be studied by taking a relatively small sample of continuous products from the process, measuring them, and applying basic statistics to get an idea of the potential capability. The process potential study is a short-term study that does not give you the long-term accuracy of the control chart method, but does give you an idea of what the process can do.

Refer to Figure 13.46 for an example of a completed study.

Steps in a Process Potential Study

Step 1.   Identify the process that is to be studied and decide the purpose of the study.

Step 2.   Have the process set up ideally as close to the nominal specification size as possible. This is not a must but does simplify the study.

Step 3.   Make sure that the operator of the process understands what you are doing: You are not studying the person, just the process.

Step 4.   Have the operator let the process run the consecutive pieces (a minimum of 30 pieces, but 50 to 100 are recommended) without any adjustments.

Step 5. Have those pieces numbered in the sequence from which they came out of the process (in case further analysis is needed).

Step 6. Inspect the parts and record the sizes on the data sheet.

Step 7. Plot the data on a frequency distribution (check for normality). The process should be a normal distribution for the study to be accurate.

Step 8. Draw on the frequency distribution the nominal size and the specification limits.

Step 9. Calculate the average $(\overline{X})$ and the standard deviation (σ).

Step 10. Draw the average $(\overline{X})$ on the frequency distribution.

Step 11. Calculate the area under the normal curve (refer to the preceding pages on capability and to Table 13–10.)

Step 12. It is recommended for a short-term study like this that a capability within ± 4 sigma be demonstrated.

Step 13. Calculate the C_p and C_{pk} values.

Process potential studies (on a short-term basis) involve studying a small group of consecutively produced parts (usually 30, 50, and 100 pieces). However, to simplify the example given here, I choose to use only 10 pieces (not recommended for actual use) and a two place decimal for the tolerance.

■ **Example 13–19**

Machine number XX has been chosen to produce a shaft diameter with a tolerance of .50 $\pm$ .01 diameter. It is necessary to study machine XX to see if it is capable of producing this .02 total tolerance.

Step 1. Set up the machine to produce the shaft (preferably as close to .50 nominal dimension as possible on the setup).

Step 2. Once the machine is set up, run the entire sample (10 pieces here) without any adjustments. Let the machine run the parts.

Step 3. Inspect the sample (10 pieces) and record each measurement to at least one more decimal place accuracy.

Pieces recorded:	.503	.509	.507	.495	.498
	.506	.504	.492	.496	.506

Step 4. Calculate the standard deviation (using a statistical calculator or the method shown under standard deviation). Use $n - 1$ in the calculation, since it is a sample. The standard deviation of these numbers is .0059. Also calculate $\overline{X}$ (X-bar), which is .5016 in this problem.

Now we have computed the two main values necessary for finding the potential capability of the machine. It is time now to compare the blueprint tolerance of the shaft to the machine spread. Immediately, we can see that there is a problem because the six-sigma spread of the machine (which is $6 \times .0059 = .035$) is larger than the total tolerance of .02 that we want to produce. For a process (or machine) to be capable, the spread of the machine should be less than the total tolerance being produced.

Step 5. The next step involves calculating the C_p/C_{pk} or the use of the table of areas under the normal curve. This table will aid you in predicting the percentage yield (or the expected amount of good parts that this machine will produce according to the study without altering the machine). The table enables you to compare the central tendency and spread of the data to the tolerance limits on the blueprint (see Table 13–10).

Step 6. Draw a bell curve (see Figure 13.45) putting the blueprint tolerance limits at each end and, on that same curve, draw a line for the X-bar of the data (in its proper place). Also, note the standard deviation below the curve.

Step 7. Now you must calculate the area between the tolerance limits (X) and the central tendency ($\overline{X}$) using the standard deviation (σ). Since there are two tolerance limits, there are two areas you must calculate and add them together for the total capability. The formula for each area you need is the same:

$$Z = \frac{X = \overline{X}}{\sigma}$$

Z_1 is the area between the upper tolerance limit (.51) and X-bar (.5016).
Z_2 is the area between the lower tolerance limit (.49) and X-bar (.5016). So,

FIGURE 13.45
Capability study.

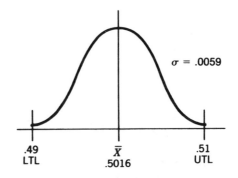

$\sigma = .0059$

.49
LTL

$\overline{X}$
.5016

.51
UTL

$$Z_1 = \frac{X - \overline{X}}{\sigma} \qquad\qquad Z_2 = \frac{X - \overline{X}}{\sigma}$$

$$= \frac{.51 - .5016}{.0059} \qquad\qquad = \frac{.49 - .5016}{.0059}$$

$$= \frac{.0084}{.0059} \qquad\qquad = \frac{-.0116}{.0059} \text{ (the minus sign means nothing here)}$$

$$= 1.42 \qquad\qquad = 1.97$$

Step 8. Now look at Table 13–10 to find that the area under the curve for a Z value of 1.42 is .4222 (or 42.22 percent) and the area for a Z value of 1.97 is .4756 (or 47.56 percent). Therefore, both areas added together are 42.22 + 47.56 percent = 89.78 percent.

For a process to be considered capable within $\pm 3\sigma$ limits, the total area must be 99.730 percent or higher. You can see that in this case machine XX has a capability of 89.78 percent and is close to but not quite capable of producing the shaft tolerance of 0.02 total. Usually, a capability of ± 4 sigma (99.994 percent) is preferred for a potential study. Again, see the table of areas under the normal curve (for the Z values above) in Table 13–10.

Process Capability Practice Problems and Solutions

1. You have measured the sample of parts and found the following:

$$\sigma = .0009 \qquad X\text{-bar} = .5015$$

If the blueprint limits are .505–.495, what is the % yield of the process?

$$\text{Yield} = Z_1 + Z_2$$

Solution

$$Z_1 = \frac{X_1 - \overline{X}}{\sigma} = \frac{.505 - .5015}{.0009} = 3.88$$

$$Z_2 = \frac{X_2 - \overline{X}}{\sigma} = \frac{.495 - .5015}{.0009} = 7.22$$

Referring to the table of areas under the normal curve (Table 13–10), we obtain

$$Z_1 = 3.88 = .5000 \text{ or } 50 \text{ percent}$$

$$Z_2 = 7.22 = .5000 \text{ or } 50 \text{ percent}$$

$Z_1 + Z_2 = 100$ percent (but you never say 100 percent; say 99.999 percent).

Note: You never say 100 percent in any case, because you are never 100 percent sure of anything.

2. Change the blueprint limits above to .503–.500 and predict the yield again.

Solution

$$Z_1 = .503 - .5015 \div .0009 = 1.67 = .4525 \text{ (in table)}$$

$$Z_2 = .500 - .5015 \div .0009 = 1.67 = .4525 \text{ (in table)}$$

$$Z_1 + Z_2 = .4525 + .4525 = .905 = 90.5 \text{ percent}$$

The yield is 90.5 percent here; therefore, the process is not capable, because for a process to be capable within ±3 sigma limits, the yield must be 99.730 percent or higher.

3. Now, change sigma above to .0001 and compute the yield.

$$Z_1 = .503 - .5015 \div .0001 = 15$$

$$Z_2 = .500 - .5015 \div .0001 = 15$$

Both of these Z values in the table are equal to .5000 or 50 percent, so when they are added, your yield will be 99.999 percent. The process is now capable.

4. Find the C_p and C_{pk} for Problem 1.

Solution

$$C_p = \frac{\text{Total tolerance}}{6\sigma} = \frac{.10}{.0054} = 1.85$$

$$C_{pk} = \text{Lesser of } \frac{X_U - \overline{X}}{3\sigma} \text{ or } \frac{\overline{X} - X_L}{3\sigma}$$

$$= \frac{.505 - .5015}{.0027} \text{ or } \frac{.5015 - .495}{.0027}$$

$$= 1.296 \text{ or } 2.41$$

$$C_{pk} = 1.296 \text{ (the lesser value)}$$

GAGE REPEATABILITY AND REPRODUCIBILITY STUDIES

Gage repeatability and reproducibility (gage R&R) studies are an important aspect of statistical process control (SPC). The purpose of SPC charts is to monitor and control the variation of the process. It is important at the start to study the measurement process and make sure it is capable of making accurate and repeatable measurements to avoid significant measurement error. The following gage R&R method is one approach (using ranges)

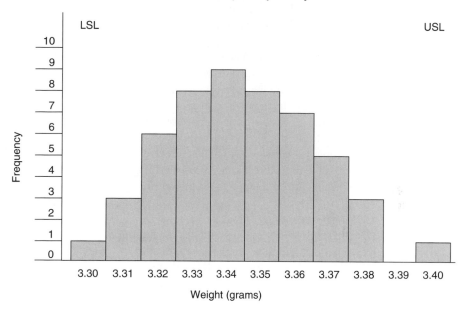

Histogram
Process Capability Study

Specification Limits: 3.30 – 3.40

Capability Study Results

Statistics	Results
Mean = 3.3443	C_p Index = .775
Standard Deviation (s) = .0215	C_{pk} Index = .69
6s = .129	% Yield = 97.55%
Z Upper = 2.59	Est. % Defective = 2.45%
Z Lower = 2.06	Parts Per Million (ppm) = 24,500

FIGURE 13.46

An example of a process capability study.

designed to measure the repeatability and reproducibility of the measurement process (the observer and the measuring instrument combined). Accurate and precise measurements help to avoid false signals on control charts.

Repeatability is the ability of one observer to obtain consistent results measuring the same part (or set of parts) using measuring equipment. **Reproducibility** is the overall ability of two or more observers to obtain consistent results repeatedly measuring the same part (or set of parts) using the same (or similar) measuring equipment. Gage R&R studies should be performed before the gage is used for data collection on control charts. Performing gage R&R studies at this time will prevent problems that are associated with taking data that are in error (such as false signals on control charts or false measures of process capability).

Gage R&R studies should be planned or they may fail due to fundamental measurement problems that will have to be corrected in any case. It is better to find these problems and correct them before taking data or studying the measurement process. Complete the following elements of planning before performing the study:

- Select the proper instrument.
- Make sure the measurement method is appropriate (e.g., datums contacted correctly).
- Follow the 10 percent rule of discrimination.
- Look for obvious training/skill problems with observers.
- Make sure all observers use the same gage (or same type of gage).
- Make sure all measuring equipment is calibrated.

Gage R&R Studies—Range Method

The range method is typically used when performing gage R&R studies. This method involves calculating the averages and range of observers' trials (making more than one measurement on each of a small sample of parts).

For example, suppose two observers are being studied using an outside micrometer to measure the outside diameter of a part. The part diameter dimension and tolerance is .047 ± .003 in. (a total tolerance of .006 in). It has been decided that the observers will measure the sample of 5 parts two times (trials) each. The measurements of each observer are shown in Figure 13.47.

The following steps can be used to perform a gage R&R study. Refer to Figure 13.47 for the data and Figure 13.48 for the results.

Step 1.   Obtain small samples (from 5 to 10 parts are recommended) from the process and number the samples 1 . . . n. These parts should represent the parts that are going to be measured for controlling the process. In this case, 5 parts have been selected for the study.

Step 2.   Decide on the number of trials (how many times each part will be measured). Two or three trials are recommended. In this example, two trials have been selected. Have each observer measure all n parts two trials (times) each and record the measurements.

Note: Devise a method for taking the data so that observers do not see their own previous measurements or any other observer's measurements. The purpose is to avoid bias.

Obs.	A					B					Statistics
Sample	Trial 1	Trial 2	Trial 3	Ave.	Range	Trial 1	Trial 2	Trial 3	Ave.	Range	$\bar{X}_A = \bar{X}_B =$
1	.0473	.0472				.0472	.0469				$\bar{X}_{DIFF} =$
2	.0468	.0471				.0471	.0469				$\bar{R}_A = \bar{R}_B =$
3	.0472	.0470				.0470	.0471				$\bar{\bar{R}} =$
4	.0471	.0473				.0470	.0471				
5	.0468	.0470				.0467	.0469				$UCL_R =$

Equipment Variation (EV)

$$EV = \bar{\bar{R}} \cdot K_1$$

Equipment Variation %

$$EV\% = \frac{EV}{Total\ tolerance} \times 100$$

Trials	2	3
K1	4.56	3.05

Appraiser Variation (AV)

$$AV = \sqrt{[\bar{X}_{DIFF} \cdot K_2]^2 - \frac{(EV)^2}{n \cdot r}}$$

$(n = parts,\ r = trials)$

Appraiser Variation %

$$AV\% = \frac{AV}{Total\ tolerance} \times 100$$

Observers	2	3
K2	3.65	2.70

Total R&R

$$R\&R = \sqrt{(EV)^2 + (AV)^2}$$

R&R%

$$R\&R\% = \frac{R\&R}{Total\ tolerance} \times 100$$

FIGURE 13.47

Gage R&R study—worksheet.

Obs.	A					B					Statistics
Sample	Trial 1	Trial 2	Trial 3	Ave.	Range	Trial 1	Trial 2	Trial 3	Ave.	Range	
1	.0473	.0472	---	.04725	.0001	.0472	.0469	---	.04705	.0003	$\overline{X}_A = .04708$ $\overline{X}_B = .04699$
2	.0468	.0471	---	.04695	.0003	.0471	.0469	---	.0470	.0002	$\overline{X}_{DIFF} = .00009$
3	.0472	.0470	---	.0471	.0002	.0470	.0471	---	.04705	.0001	$\overline{R}_A = .0002$ $\overline{R}_B = .00018$
4	.0471	.0473	---	.0472	.0002	.0470	.0471	---	.04705	.0001	$\overline{\overline{R}} = .00019$
5	.0468	.0470	---	.0469	.0002	.0467	.0469	---	.0468	.0002	$UCL_R = .00062$

Equipment Variation (EV)

$$EV = \overline{\overline{R}} \cdot K_1$$
$$= .00019 \cdot 4.56$$
$$= .00087$$

Equipment Variation %

$$EV\% = \frac{EV}{Total\ tolerance} \times 100$$
$$= \frac{.00087}{.006} \times 100 = 14.5\%$$

Trials	2	3
K1	4.56	3.05

Appraiser Variation (AV)

$$AV = \sqrt{[\overline{X}_{DIFF} \cdot K_2]^2 - \frac{(EV)^2}{n \cdot r}}$$
$$= \sqrt{[.00009 \cdot 3.65]^2 - \frac{.00087^2}{5 \cdot 2}}$$
$$= .00018$$
$$(n = parts,\ r = trials)$$

Appraiser Variation %

$$AV\% = \frac{AV}{Total\ tolerance} \times 100$$
$$= \frac{.00018}{.006} \times 100 = 3.0\%$$

Observers	2	3
K2	3.65	2.70

Total R&R

$$R\&R = \sqrt{(EV)^2 + (AV)^2}$$
$$= \sqrt{(.00087)^2 + (.00018)^2}$$
$$= .00089$$

R&R%

$$R\&R\% = \frac{R\&R}{Total\ tolerance} \times 100$$
$$= \frac{.00089}{.006} \times 100 = 14.83\%$$

FIGURE 13.48
Completed gage R&R study.

Step 3. For each observer, calculate the average $(\overline{X})$ of each trial and record it on the worksheet.

Step 4. For each observer, find the range (R) of each trial and record it on the worksheet.

Step 5. Calculate the average range $(\overline{R})$ for each observer using the ranges for all of that observer's trials. In this case, the average range for Observer A is .0002 in. and the average range for Observer B is .00018 in.

Step 6. Calculate the grand range $(\overline{\overline{R}})$ by averaging the $\overline{R}$'s of the observers. In this case, $\overline{\overline{R}}$ = .00019 in.

Step 7. Find the grand average of each observer's averages and enter the grand average for each observer in the space marked $\overline{X}_A$, $\overline{X}_B$, etc. In this case $\overline{X}_A$ = .04708 and $\overline{X}_B$ = .04699.

Step 8. Calculate the difference between the maximum observer average value and the minimum observer average value (called $\overline{X}_{DIFF}$). In this case, $\overline{X}_{DIFF}$ = .00009 in.

Step 9. Calculate the upper control limit for the ranges by multiplying the overall average range $(\overline{\overline{R}})$ by the appropriate D_4 value (refer to the table of factors for control charts, Table 13–1). The D_4 factor depends on the number of trials. If any of the observers' ranges exceed the upper control limit for ranges, either repeat those readings from that observer or discard the values, go back to Step 3, and recalculate. In this case, the UCL of ranges equals $\overline{\overline{R}}$ times D_4 or .00019 in. times 3.267 or .00062 in. There are no observer ranges that exceed the UCL of .00062 in., so the study can be continued.

Step 10. Calculate the equipment variation (EV) using the average range and the K_1 factor. The K_1 factor also depends on the number of trials.

$$EV = \overline{R}_0 \times K_1 = .00019 \times 4.56 = .00087$$

Step 11. Calculate the percent of tolerance consumed (EV%) by equipment variation:

$$EV\% = \frac{EV}{Total\ tolerance} \times 100 = \frac{.00087}{.006} \times 100 = 14.5\%$$

Step 12. Calculate the appraiser variation (AV).

$$AV = \sqrt{[\overline{X}_{DIFF} \cdot K_2]^2 \frac{(EV)^2}{n \cdot r}}$$

$$= \sqrt{[.00009 \cdot 3.65]^2 \frac{(.00087)^2}{5 \cdot 2}}$$

$$= .00018$$

where *n* is the number of parts

r is the number of trials

K_2 depends on the number of observers being studied

Step 13. Calculate the percent of tolerance consumed by appraiser variation (AV%):

$$AV\% = \frac{AV}{Total\ tolerance} \times 100 = \frac{.00018}{.006} \times 100 = 3.0\%$$

Step 14. Calculate the repeatability and reproducibility (R&R) of the measurements:

$$R\&R = \sqrt{(EV)^2 + (AV)^2} = \sqrt{.00087^2 + .00018^2} = .00089$$

Note: These calculations are based on predicting 5.15 sigma (or 99% of the area under the normal curve).

Step 15. Calculate the percent of tolerance consumed by the measurement process (R&R%):

$$R\&R\% = \frac{R\&R}{Total\ tolerance} \times 100$$

$$= \frac{.00089}{.006} \times 100$$

$$= 14.83\%$$

The total R&R% would be ideal if it were 10 percent or less, but each company must decide how much measurement process variation is acceptable based on internal or customer requirements. Typically 10 percent or less is considered very good R&R results, greater than 10 percent to 25 percent is considered marginally capable, and more than 25 percent is considered unacceptable.

Parts per Million (ppm)

Process quality levels that are demanded by consumers today are so high in terms of yield (or so low in terms of defectives) that percent defective, in many cases, is no longer a valid measure of quality. A relatively new measure of quality (or one might say "un-quality") is **parts per million (ppm).** Parts per million is a very straightforward calculation. It is the percent defective times 10,000. The assumption here is that a certain percent defective (per hundred parts) such as .065% may, in traditional quality terms, appear to be a good quality level from a process. In fact, in traditional quality terms, .065% was considered to be very good. But the .065% defective, if one million parts were made, translates to 650 ppm. In another example, if the yield of a process is 99.994% good product, then the estimated percent defective from this process is .006%. In this case, the ppm is .006% times 10,000 or 60 ppm.

Summary

For practice in calculating process capability indexes, the estimated percent yield, the estimated percent defective, and the ppm of a process, refer to Table 13–11. Table 13–11 also includes answers to the practice problems presented.

MISCELLANEOUS TOPICS

When Defectives Are Found in a Subgroup

In cases where some defectives are found in a subgroup on a control chart, the action to be taken is as follows.

Step 1. Identify the defectives found and separate them.

Step 2. Screen all products made since the last time the chart was in control.

Step 3. Take action on the process.

Assembly Control with SPC

In those cases where an assembly dimension must be controlled, you must look at the building blocks of that dimension.

■ **Example 13–20**

Parts A, B, and C are mounted together to form an assembly. The overall length of the assembly must be controlled.

Solution
Control the individual dimensions of the three parts that cause the overall length to vary.

___

■ **Example 13–21**

An important dimension that is created on the assembly line must be controlled.

Solution
Control that dimension at the assembly line.

___

Raw Material Control

For raw materials, such as casting and forgings, the control still may fall back on the "building blocks" approach. Control of the input and processing requirements is sometimes necessary.

For example, the control of castings may be done on the parameters that make the casting, such as temperature and alloy content, or attribute charts may be used to control the end result with respect to defects.

TABLE 13–11

Practice Problems and Solutions for Process Capability

Areas Under The Normal Curve and Capability Indexes

Process and specifications	"Z" scores and solution method	% yield	% def. (ppm)	C_p	C_{pk}
$\sigma = 3$ LSL 22 $\overline{X}$ 27 USL 34	$Z_{Upper} = \dfrac{USL - \overline{X}}{\sigma} = 2.33$ $Z_{Lower} = \dfrac{\overline{X} - LSL}{\sigma} = 1.67$ Add the areas	94.26%	$\dfrac{5.74\%}{(57,400)}$	0.67	0.56
$\sigma = 2$ $\overline{X}$ 16 USL 63 Max.	$Z_{Upper} = \dfrac{USL - \overline{X}}{\sigma} = 23.5$ Use 5.0 Other area is obvious. Add the areas	99.9997%	$\dfrac{.0003\%}{(3)}$	N/A	7.83
$\sigma = .0015$ LSL .493 $\overline{X}$ USL .501 .503	$Z_{Upper} = \dfrac{USL - \overline{X}}{\sigma} = 1.33$ $Z_{Lower} = \dfrac{\overline{X} - LSL}{\sigma} = 5.33$ Add the areas	90.82%	$\dfrac{9.18\%}{(91,800)}$	1.11	0.44
$\sigma = .003$ .502 .508 LSL USL $\overline{X}$ .501	$Z_{Upper} = \dfrac{USL - \overline{X}}{\sigma} = 2.33$ $Z_{Lower} = \dfrac{\overline{X} - LSL}{\sigma} = 0.33$ Subtract lower area from upper area	36.08%	$\dfrac{63.92\%}{(639,200)}$	0.33	−0.11
$\sigma = 1.2$ LSL USL 28 31 $\overline{X}$ 29	$Z_{Upper} = \dfrac{USL - \overline{X}}{\sigma} = 1.67$ $Z_{Lower} = \dfrac{\overline{X} - LSL}{\sigma} = 0.83$ Add the areas	74.92%	$\dfrac{25.08\%}{(250,800)}$	0.42	0.28

All of the above assumes a normal distribution.

In other cases, such as mixtures, a control chart may be used to control the amount of additives to that mixture to produce a favorable end result.

Sequential Control

One must always realize that *input material variation* at any stage of the process can cause considerable problems. At times it becomes necessary to back up and control the variation at a previous operation to produce a favorable end result.

For example, suppose Operation 10 produces a dimension that is located as a datum in several future operations. This dimension, if it varies widely, can cause the future operations on the part to vary as well.

If this dimension is identified as such, it can be controlled so that it will not cause problems down the line. Future operations cannot be brought into control unless the previous operation is in control.

REVIEW QUESTIONS

1. Nonrandom variation is due to _____ causes.
   a. inherent
   b. common
   c. special
   d. built-in
2. The ability of a process to hold a constant level of variation is called process _____.
   a. capability
   b. methods
   c. potential
   d. control
3. Which of the following graphical methods describes the variation of any process?
   a. cause and effect chart
   b. frequency distribution
   c. histogram
   d. none of the above
   e. b or c are correct
4. Calculate the mean of the following data: 2, 0, 4, 5, −1.
   a. 2.5
   b. 2
   c. 2.8
   d. 2.75
5. Calculate the range of the data in question 4.
   a. 6
   b. 4
   c. 3
   d. 7

6. Calculate the sample standard deviation of the data in question 4.
   a. 2.7
   b. 2.5
   c. −2.5
   d. 2.0

7. Find the mode of the following data: 5, 6, 7, 6, 5, 7, 8, 4, 5, 5, 7.
   a. 6
   b. 7
   c. 5
   d. 4

8. Find the median of the data in question 7.
   a. 6
   b. 5
   c. 7
   d. 8

9. If the sample standard deviation of a process is 9, then what is the sample variance?
   a. 27
   b. 81
   c. 33
   d. none of the above

10. What is the upper control limit of an $\overline{X}$ chart if $n = 3$, $\overline{\overline{X}} = .500$, and $\overline{R} = .002$?
    a. .503
    b. .502
    c. .499
    d. .505

11. What is the upper control limit of a range chart where $\overline{R} = .004$ and $n = 5$?
    a. .0085
    b. .007
    c. .0065
    d. none of the above

12. Calculate the upper control limit of a p chart if $\overline{p} = .05$ and $n = 30$.
    a. .09
    b. .08
    c. .14
    d. .169

13. Calculate the upper control limit of a c chart if $\overline{c} = 20$.
    a. 32
    b. 33.4
    c. 35
    d. 27

14. Calculate the lower control limit of a u chart if $\overline{u} = 12$ and $n = 30$.
    a. 0
    b. 1.10
    c. 10.1
    d. 12

15. Using the following information, calculate the C_p index.

    Sigma $= .001$ Upper spec. limit $= .758$ Lower spec. limit $= .750$

    a. 1.33
    b. .75
    c. 1.00
    d. 3.00

16. Using the information in question 15, if $\overline{X}$ was .755, what is the C_{pk}?
    a. 1.33
    b. 1.67
    c. .95
    d. 1.00

17. A process variable has a single maximum limit of 100. The average $(\overline{X})$ of the process is 32 and the standard deviation is 10. What is the C_{pk}?
    a. 1.07
    b. 2.27
    c. 1.33
    d. none of the above

18. What is the area under the normal curve between $\overline{X}$ plus and minus one standard deviation?
    a. 99.45%
    b. 99.73%
    c. 68.26%
    d. 99.994%

19. The average score on an exam taken by 30 students is 70 and the standard deviation of those scores is 5. If the minimum passing grade for the test is 70, *how many* students will pass? (Round answer to the nearest whole number.)
    a. 15
    b. 20
    c. 12
    d. 13

20. If the Z value is .60, what is the area under the normal curve?
    a. .2206
    b. .2304
    c. .1300
    d. .2257

21. A process has a mean of .501 and a standard deviation of .001. The upper specification limit is .505 and the lower specification limit is .495. What is the upper Z score?
    a. 3.00
    b. 2.50
    c. 4.00
    d. 5.00

22. If the Z score for a given area is 2.5, how many standard deviations is this from the mean?
    a. 2.5
    b. 1.75

c. 3

d. 4

23. Find the upper control limit of a sigma chart where $\bar{\sigma} = .0025$ and $n = 5$.

    a. .0027

    b. .0034

    c. .0052

    d. .006

24. What is the upper control limit for the sigma chart in question 23 if $n = 10$?

    a. .0052

    b. .0043

    c. .0027

    d. .0035

Use the following data to answer all remaining questions. Round all answers to two decimal places.

$$7 \quad 5 \quad 4 \quad 8 \quad -2 \quad 8 \quad 0$$

25. Find the mean of the data.
26. Find the sample standard deviation of the data.
27. Find the population standard deviation of the data.
28. Find the mode of the data.
29. Find the median of the data.
30. Find the sample variance of the data.
31. Find the population variance of the data.
32. If the specification limits are 4 to 14, what is the C_p (using the sample standard deviation)?
33. If the specification limits are 4 to 14, what is the C_{pk} (using the sample standard deviation)?
34. Does this process have the capability of producing a high yield of good product (yes or no)?
35. What is the range of the data?

CHAPTER

14 *Shop Mathematics*

Inspectors, operators, machinists, and engineers all use shop-related math at one time or another. Sometimes understanding shop math can be the difference between getting the job done or not. Shop math is encountered in a variety of problems on the shop floor, from tapers to threads, drill sizes, tolerances, and hole patterns. The list of uses of shop math related to the mechanical industry is very long.

In the following pages we outline the necessary shop mathematics for most of the problems encountered on the shop floor. In these pages we discuss

Fractions	Triangles
Equations	Percentages
Circles	Volumes
Decimals	Angles
Perimeters	Areas
Ratios	Tapers

These subjects encompass a good basis for shop math as used by inspectors, operators, machinists, and engineers. There is much more to shop math than these subjects, but most of them can be applied to solve a variety of problems.

ADDING AND SUBTRACTING FRACTIONS

The Least Common Denominator

When adding or subtracting fractions it is necessary to be able to find the *least common denominator* (LCD). The LCD is the smallest denominator in the group of fractions to be added or subtracted in which all of the denominators will evenly divide.

■ **Example 14–1**

$$\frac{3}{8} + \frac{5}{16} + \frac{1}{4}$$

The LCD in this problem is 16 because it is the least number that is common to the other denominators (4 and 8). The only step now is to change $\frac{3}{8}$ and $\frac{1}{4}$ to sixteenths, then add all three fractions with a common denominator of 16.

483

When working with fractions, before attempting to add or subtract them they must be "like" fractions (having the same denominator).

Finding the LCD.   There are methods for finding the LCD in any given problem. One of them is simply inspection.

■ **Example 14–2**

$$\frac{1}{4} + \frac{1}{2}$$

You know that $\frac{1}{2}$ is $\frac{2}{4}$, so

$$\frac{1}{4} + \frac{2}{4} = \frac{3}{4}$$

Another method (not always effective) is calculation.

■ **Example 14–3**

$$\frac{1}{3} + \frac{1}{4} + \frac{3}{8}$$

See if the largest denominator (8) can be evenly divided by the other ones (3 and 4). No, it cannot.

Now multiply the largest (8) times the smallest (3) and see if 24 is common. In this case, 24 is the common denominator.

Converting the Fractions to the LCD Found.   Once an LCD is found, convert the fraction to it in three steps.

Step 1.   Divide the denominator into the LCD.

Step 2.   Multiply this number times the numerator.

Step 3.   Place the answer in step 2 over the LCD for an equivalent fraction.

■ **Example 14–4**

Convert $\dfrac{3}{16}$ into an LCD of 64.

Step 1. $16\overline{)64}$ with quotient 4

Step 2. $3 \times 4 = 12$

Step 3.   *Answer:* $\frac{12}{64}$

■   **Example 14–5**

$$\frac{3}{8} + \frac{5}{16} + \frac{3}{64} + \frac{1}{3}$$

Solution
LCD is 192 (found simply by multiplying the lowest denominator, 3, by the highest denominator, 64).

$$\text{Conversion: } \frac{3}{8} \text{ to 192nds is } 192 \div 8 \text{ then} \times 3 = \frac{72}{192}$$

$$\frac{5}{16} \text{ to 192nds is } 192 \div 16 \text{ then} \times 5 = \frac{60}{192}$$

$$\frac{3}{64} \text{ to 192nds is } 192 \div 64 \text{ then} \times 3 = \frac{9}{192}$$

$$\frac{1}{3} \text{ to 192nds is } 192 \div 3 \text{ then} \times 1 = \frac{64}{192}$$

Next, add the numerators: $72 + 60 + 9 + 64 = 205$.

Now, you have an answer stated as an *improper fraction,* $\frac{205}{192}$. An improper fraction is one where the numerator is larger than the denominator. To convert this to the final answer (which will be a mixed number), divide the denominator, 192, into the numerator, 205, and put the remainder over the denominator.

$$\frac{205}{192} = 1\frac{13}{192} \quad Answer$$

Reducing Fractions

Sometimes you will add or subtract fractions and get an answer that is not in proper form (such as $\frac{2}{4}$). The fraction $\frac{2}{4}$ is not in proper form because it can be reduced to $\frac{1}{2}$. Reducing some fractions can get complicated unless you do it in a step-by-step manner.

■   **Example 14–6**

Reduce $\frac{20}{160}$ to lowest terms.

Solution

By inspection, find the largest number that will divide evenly into both 20 and 160. In this case it is 20. Now divide into both 20 and 160.

$$20 \div 20 = 1$$

$$160 \div 20 = 8$$

The fraction $\dfrac{20}{160}$ reduces to $\dfrac{1}{8}$.

Another way is to start dividing the number 2, 3, or 4 into both numerator and denominator, and continue to do this until the fraction is in lowest terms.

■   **Example 14–7**

$$\frac{16}{64} = \frac{8}{32} = \frac{4}{16} = \frac{2}{8} = \frac{1}{4}$$

With more practice, some of these steps can be omitted.

Adding and Subtracting Fractions Using Cross-Multiplication

Another way to add or subtract "unlike" fractions (where an LCD is necessary) is cross-multiplication. By cross-multiplying the two fractions and then reducing the "new fraction" to lowest terms, the answer is easily found. Other methods of finding the LCD are by inspection or deduction.

■   **Example 14–8**

$$\frac{3}{8} + \frac{1}{4} = ?$$

The answer here is $\dfrac{5}{8}$, since

$$\frac{1}{4} = \frac{2}{8} \qquad \left(\text{and } \frac{3}{8} + \frac{2}{8} = \frac{5}{8}\right)$$

To cross-multiply, simply follow these steps. Before long, with practice, you can perform these steps quickly in your head.

Step 1.   Set up the problem as shown.

$$\frac{3}{8} + \frac{1}{4} = \frac{\text{(new fraction)}}{}$$

Step 2. Multiply the denominator (8) of the first fraction times the numerator (1) of the second fraction and put it in the numerator of the "new fraction."

$$\frac{3}{8} + \frac{1}{4} = \frac{8}{}$$

Step 3. Multiply the numerator (3) of the first fraction times the denominator (4) of the second fraction and put it in the numerator of the new fraction.

$$\frac{3}{8} + \frac{1}{4} = \frac{8 \qquad\qquad 12}{}$$

Step 4. Multiply both denominators (8) and (4) together and use that answer for the denominator of the new fraction.

$$\frac{3}{8} + \frac{1}{4} = \frac{8 \qquad\qquad 12}{32}$$

Step 5. Now, since you are adding, put a plus sign between the two numbers in the numerator. (If you were subtracting, this would be a minus sign, and you would simply subtract the smallest number from the largest number there.)

$$\frac{3}{8} + \frac{1}{4} = \frac{8 \quad + \quad 12}{32}$$

Step 6. Do the math in the new fraction.

$$\frac{3}{8} + \frac{1}{4} = \frac{20}{32}$$

Step 7. Reduce the new fraction (if necessary) by finding the largest single number that will divide evenly into both the numerator and the denominator (the number 4).

$$\frac{20}{32} = \frac{5}{8} \; Answer$$

Some More Difficult Examples

1. $\dfrac{5}{16} + \dfrac{3}{64} = \dfrac{48 + 320}{1024} = \dfrac{368}{1024} = \dfrac{23}{64} \; Answer$

2. $\dfrac{3}{8} - \dfrac{13}{64} = \dfrac{192 - 104}{512} = \dfrac{88}{512} = \dfrac{11}{64} \; Answer$

MULTIPLYING AND DIVIDING FRACTIONS

Multiplying Fractions

Multiplying fractions is a three-step procedure: (1) multiply the numerators, (2) multiply the denominators, and (3) reduce the resulting fraction to lowest terms (if necessary).

■ **Example 14–9**

$$\frac{6}{32} \times \frac{1}{2} =$$

Step 1.   Multiply the numerators.

$$\frac{6}{32} \times \frac{1}{1} = \frac{6}{}$$

Step 2.   Multiply the denominators.

$$\frac{6}{32} \times \frac{1}{2} = \frac{6}{64}$$

Step 3.   Reduce the answer (if necessary) to lowest terms.

$$\frac{6}{64} = \frac{3}{32} \text{ (this answer is in lowest terms)}.$$

Dividing Fractions

When dividing fractions, the main thing to remember is to invert (or flip) the divisor and then multiply.

■ **Example 14–10**

$$\frac{\dfrac{3}{4}}{\dfrac{3}{8}} = \frac{3}{4} \times \frac{8}{3} = \frac{24}{12} = 2 \; Answer$$

Note: The answer is often an improper fraction that needs to be changed back into a proper fraction.

WORKING WITH DECIMALS

Decimals are widely used in industry. Some examples of these applications are in

- Drawing and part tolerances (English or metric)
- Inspection and machining formulas
- Business percentages

There are a few things about decimals that need to be covered:

- Reading decimals
- Converting fractions to decimals
- Adding and subtracting
- Multiplying and dividing
- Percentages

Reading Decimal Places

All decimal fractions are represented by numbers written to the right of a decimal point (.), and in some cases, this quantity of numbers could be infinite. Whole numbers are to the left of the decimal point. For the purpose of this book we will only go as far as six decimal places.

0. A B C D E F (Letters are used to identify the places of a decimal.)

0 place. The place for a whole number such as 1, 2, and so on.

A place. The tenths place. *Example:* The fraction $\frac{1}{10}$ is shown as 0.1 (called one tenth).

B place. The hundredths place. *Example:* The fraction $\frac{48}{100}$ is shown as 0.48 (called forty-eight hundredths).

C place. The thousandths place. *Example:* The fraction $\frac{1}{1000}$ is shown as 0.001 (called one-one thousandth).

D place. The ten-thousandths place. *Example:* The fraction $\frac{14}{10,000}$ is shown as 0.0014 (called fourteen ten-thousandths).

E place. The one-hundred thousandths place. *Example:* The fraction $\frac{15}{100,000}$ is shown as 0.00015 (called fifteen one-hundred-thousandths).

F place. The "millionths" place. *Example:* The fraction $\frac{2}{1,000,000}$ is shown as 0.000002 (called two millionths).

Here are some examples of reading decimals:

0.001 cm means one thousandth of a centimeter.

0.0015 in. means fifteen ten-thousandths of an inch.

0.2 mm means two tenths of a millimeter.

Converting a Fraction into a Decimal

This is done simply by dividing the numerator by the denominator of the fraction.

■ **Example 14–11**

Fraction   Decimal

$$\frac{5 \ \text{Numerator}}{8 \ \text{Denominator}} = 8\overline{)5} = 0.625$$

Adding and Subtracting Decimals

Rule: Keep all decimal points *in line*.

■ **Example 14–12**

Add 1.52 + 2.3 + 6.434.

Solution

$$
\begin{array}{r}
1.52 \\
2.3 \\
+ \quad 6.434 \\
\hline
10.254 \quad \textit{Answer}
\end{array}
$$

■ **Example 14–13**

Subtract 3.84 from 5.393.

Solution

$$
\begin{array}{r}
5.393 \\
- \quad 3.84 \\
\hline
1.553 \ \textit{Answer}
\end{array}
$$

Multiplying Decimals

Rule: You do not need to line up decimal points, but you must have the total number of decimal places in the answer as you have in the two numbers being combined.

■ **Example 14–14**

Multiply 4.536 × 3.12.

Solution

$$
\begin{array}{r}
4.536 \\
\times \quad 3.12 \\
\hline
9072 \\
4536 \\
13608 \\
\hline
1415232
\end{array}
$$

4.536   (there are three decimal places here)

× 3.12   (there are two decimal places here)

1415232   (since there are a total of five decimal places used above, the answer must also have five decimal places)

Answer: 14.152   (three places after rounding off)

Dividing Decimals

Rule: Make the divisor a whole number by moving the decimal point as many places to the right as necessary, then move the decimal point in the dividend the same amount of places. Next, bring the decimal point straight up in the answer.

■ Example 14–15

Divide 3.0 by 0.24:

$$
\text{divisor}\overline{)\,\text{dividend}} \quad \overset{\text{quotient}}{} \quad \textit{Answer}
$$

Solution

First, set up the problem.

$0.24\overline{)3.0}$. Next, make the divisor a whole number: $24\overline{)3.0}$ (had to move two places). Next, move the dividend point the same number of places: $24\overline{)300}$. Then the decimal point in the dividend comes straight up into the quotient.

$$
24\overline{)300.}
$$

Now, divide the numbers.

$$
\begin{array}{r}
12.5 \;\textit{Answer} \\
24\overline{)300.}
\end{array}
$$

Converting Decimals to Percentages

A decimal can be converted into a percentage simply by multiplying it by 100 and placing the % sign after the value. (An easier method is simply moving the decimal point exactly two places to the right.)

■ **Example 14–16**

0.12 (convert it to percent)

Solution

$$100 \times 0.12 = 12 \text{ percent}$$

or

0.12 (move decimal two places to the right) = 12 percent

If one knows and follows the rules for decimals, they are very easy to work with. A good beginning in the study of the metric system, in fact, is a thorough knowledge of working with decimals.

Converting Percentages to Decimals

A percentage can be converted into a decimal by dividing the percentage by 100 or by moving the decimal point two places to the left and removing the % sign. Keep in mind that whenever a whole number is written, it can be viewed as having a decimal point immediately after the whole number.

■ **Example 14–17**

When writing a whole number such as 10, the decimal point after the 10 is understood (such as 10.). If the whole number is stated as a percentage (e.g., 10%), it means 10.%.
    To change the 10.% to a decimal fraction, divide the value by 100.

$$10.\% \div 100 = 0.10$$

or move the decimal point two places to the left.

$$10.\% = 0.10 \text{ decimal fraction}$$

ADDING AND SUBTRACTING ANGLES

There are occasions where addition and subtraction of angles that are in degrees (°), minutes ('), and seconds (") become necessary to make a job easier, or simply to get the job done. Performing these two operations on angles of this kind is very simple, once you know the rules.

$$\text{Conversion:} \quad 1' = 60 \text{ seconds (")}$$
$$1° = 60 \text{ minutes (')}$$

The key conversion number is 60 for either one.

Addition of Angles

This is simply a matter of working with the problem from right to left and remembering the conversions.

■ **Example 14–18**

Add:

$$3° \; 12' \; 40''$$
$$+ \; 2° \; 51' \; 30''$$

Solution

First: Add the seconds: $40'' + 30'' = 70''$. Since there are 60 seconds in 1 minute, you can leave 10 seconds in the seconds column and carry over 1 minute to the minutes column.

$$1'$$
$$3° \; 12' \; 40''$$
$$+ \; 2° \; 51' \; 30''$$
$$10''$$

Next, add the minutes: $1' + 12' + 51' = 64'$. Since there are 60 minutes in 1 degree, leave $4'$ in the minutes column and carry over 1 degree to the degree column.

$$1° \quad 1'$$
$$3° \; 12' \; 40''$$
$$+ \; 2° \; 51' \; 30''$$
$$4' \; 10''$$

Last step: Add the degrees: $1° + 3 ° + 2° = 6°$.

$$1° \quad 1'$$
$$3° \; 12' \; 40''$$
$$+ \; 2° \; 51' \; 30''$$
$$6° \quad 4' \; 10''$$

Answer: 6° 04′ 10″

Subtraction of Angles

In subtraction, you still work from right to left in the same manner, but must often "borrow" from the next area to solve the problem.

■ **Example 14–19**

Subtract:

$$12° \; 13' \; 10''$$
$$- \; 9° \; 35' \; 40''$$

Solution

First: You can see that you cannot subtract 40″ from 10″, so you borrow 1 minute (which is 60 seconds). Now you have 70″ minus 40″ = 30″.

$$
\begin{array}{r}
12'\ 70'' \\
12°\ \cancel{13}'\ \cancel{10}'' \\
-\ \ 9°\ 35'\ 40'' \\
\hline
30''
\end{array}
$$

Next: You see that you cannot subtract 35′ from 12′, so you must borrow 1 degree (which is 60 minutes). Now you have 72′ minus 35′ = 37′.

$$
\begin{array}{r}
72' \\
11°\ \cancel{12}'\ 70'' \\
\cancel{12}°\ \cancel{13}'\ \cancel{10}'' \\
-\ \ 9°\ 35'\ 40'' \\
\hline
37'\ 30''
\end{array}
$$

Last step: Subtract 9 degrees from the remaining 11 degrees.

$$
\begin{array}{r}
72' \\
11°\ \cancel{12}'\ 70'' \\
\cancel{12}°\ \cancel{13}'\ \cancel{10}'' \\
-\ \ 9°\ 35'\ 40'' \\
\hline
2°\ 37'\ 30''
\end{array}
$$

Answer: 2° 37′ 30″

WORKING WITH EQUATIONS

Simple equations are frequently used in production and inspection to make or inspect products. Knowing how to work with these equations correctly is often a prerequisite to getting the job done.

Simple equations are defined as equations with only one unknown (or one variable). Examples of these equations are seen every day. For example,

$$\text{Sine}_{30°} = \frac{1}{2} \qquad C = \pi(12 \text{ in.}) \qquad A = .785(10^2)$$

Now consider an even simpler equation:

$$X = \frac{20}{2}$$

Note: In *all* equations, everything on one side of the "equals" sign is always equal to what is on the other side. You can see here that X (unknown value) equals 20 divided by 2.

All single-variable equations are equations that have only one unknown quantity.

■ **Example 14–20**

$$50(2) = \frac{2(X)}{4}$$

Note: ( ) means multiply, and the line means divide.

As you see here, the left side is simply numbers: (50 times 2), or 100. The right side cannot be worked out directly because you don't know what X is. The simplest way to find the number X is to get it all alone on one side of the equals sign. Then you can solve the other side, and, (since they are always equal) you will know what X is.

Important Rule: You can move any number from one side of the equals sign to the other side by changing its *sign* to the opposite sign. For example, if a number is multiplied ($\times$) on one side, move it and divide ($\div$) with it. Or, if a number is added ($+$) on one side, move it and subtract ($-$) it on the other side. The object is to get X all alone on one side of the equals ($=$) sign, and then do the math on the other side, to solve the problem.

Solution of Example 14–20

$$50(2) = \frac{2(X)}{4}$$

First, move the 4 and multiply: $50(2)(4) = 2(X)$.

Remember: When you move a number, never write it down again in the same area from which you moved it.

Next, move the multiplied 2 and divide with it on the other side:

$$\frac{50(2)(4)}{2} = X$$

Now you have X all by itself. Finally, you simply do the math.

50 times 2 is 100

100 times 4 is 400

400 divided by 2 is 200

Now you have $200 = X$, and since it is an equation, X is 200.

Checking Your Answer

If you want to prove that your answer of $X = 200$ is correct, you simply try 200 in place of X in the original problem.

$$\text{Original problem: } 50(2) = \frac{2(X)}{4}$$

$$\text{Now replace } X \text{ with your answer: } 50(2) = \frac{2(200)}{4}$$

Now do the math on both sides, and if the numbers on both sides are the same, your answer is correct.

$$50(2) = 100 \qquad 2(200) = 400 \text{ divided by } 4 = 100$$

$$100 = 100 \qquad \text{Your answer is correct.}$$

Here is another example:

$$X = \frac{10}{20} \qquad \begin{array}{l} 10 \text{ divided by 20 is 0.5; therefore,} \\ X = 0.5. \end{array}$$

Constants

In many formulas there are constants. These are simply letters or figures that always mean one particular number. For example, π(pi) is always 3.1416 (rounded to four decimal places). Anywhere you see this sign, just replace it with the number 3.1416. *Note:* The actual value of π (to nine places) is 3.141592654.

$$\text{Area of a circle} = 0.785 \, (D^2)$$

D means the diameter of the circle. 0.785 is a constant.

So, knowing the diameter of a circle, and using the constant number 0.785, you can find the surface area of that circle.

THE CIRCLE

Some situations arise during production and inspection that require the solution of parts of the circle. Those who have a working knowledge of circles can solve many of these problems. The circle has several different parts. Some of these are shown in Figure 14.1.

Angular Relationships

All circles have 360 degrees, as shown in Figure 14.2. Circles may also be divided into four "quadrants" of 90 degrees each, as shown in Figure 14.3. It is also a good idea to picture

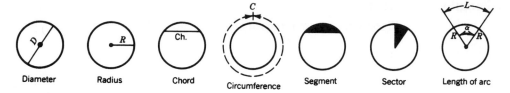

FIGURE 14.1

Parts of a circle.

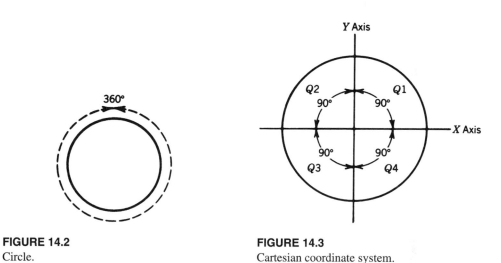

FIGURE 14.2

Circle.

FIGURE 14.3

Cartesian coordinate system.

the circle on the *X–Y* coordinate axes (called the *Cartesian coordinate system*) as shown in Figure 14.3.

All angular relationships start at quadrant 1 on the *X* axis and go counterclockwise from there. There are also inscribed angles (Figure 14.4) and central angles (Figure 14.5). A central angle forms the sector.

FIGURE 14.4

Inscribed angle.

FIGURE 14.5

Central angle.

Lengths

Examples of lengths related to a circle are

1. Length of circumference (or the distance around the circle)
2. Diameter (the distance from one point on the circle, going through the center to its opposite point)
3. Radius (exactly one-half of the diameter)
4. Length of chord (a straight line drawn from one point to another that does not go through the center)
5. Length of arc (the distance around the circumference of a circle between two intersecting points of an included angle) (see Figure 14.6).

FIGURE 14.6
Length of an arc.

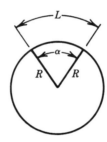

Areas

Surface area is always in square inches, square feet, square yards, and so on. There are three main surface area problems shown in Figure 14.7.

FIGURE 14.7
Areas of circle parts.

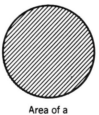

Area of a circle   Area of a segment   Area of a sector

Solving for Lengths and Areas

Abbreviations used in formulas:

D = diameter
L = length of arc
α = angle
Ch. = chord
A = area
R = radius

ϕ = one-half of an included angle (see Figure 14.10)
C = circumference
π = 3.1416 (constant called pi)
[ ], ( ) = multiply

Some Formulas Related to Circles

Note: Symbols or numbers next to each other must be multiplied with or without (   ), as shown in these formulas in Figures 14.8 to 14.11.

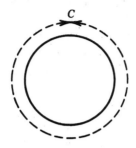

FIGURE 14.8
Circumference of a circle.

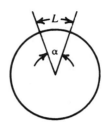

FIGURE 14.9
Length of an arc.

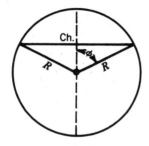

FIGURE 14.10
Length of a chord.

FIGURE 14.11
Area of a circle.

1. $C = (\pi)D$ or $C = 2(\pi)(R)$

2. $L = 0.017453(R)(\alpha)$ or $L = \pi(D)\dfrac{\alpha^{\circ}}{360^{\circ}}$

3. $Ch. = 2(R \sin \phi)$

4. $A = \dfrac{\pi(D^2)}{4}$ or $A = \pi(R^2)$

There are many uses for the formulas above and others that apply to the circle. One example is the bolt circle shown in Figure 14.12.

FIGURE 14.12
Bolt (hole) circle.

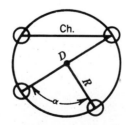

Circle Practice Problems and Solutions

1. Find the circumference of a circle that has a 3-in. diameter.

Solution
The formula is $C = (\pi)D$.
Answer: $C = 3.1416 \times 3 = 9.4248$ in.

2. Find the circumference of a circle with a 1.5-in. radius.

Solution
The formula is $C = 2(\pi)R$.
Answer: $C = 9.4248$ in.

   Note: This is the same answer as above because a 1.5-in. radius is equal to a 3-in. diameter.

3. Find the length of an arc produced by a 30-degree included angle on a circle with a diameter of 6 in.

Solution
The formula is

$$L = \pi(D) \left(\frac{\text{included angle}}{360 \text{ degrees}} \right)$$

$$= 3.1416 \times 6 \times \frac{30}{360}$$

$$= 3.1416 \times 6 \times 0.0833$$

Answer: $L = 1.57$ in.

4. Find the area of a circle that has a 10-in. diameter.

Solution
The formula is

$$A = \frac{\pi(D^2)}{4}$$

$$= \frac{3.1416 \times 10^2}{4}$$

$$= \frac{314.16}{4}$$

Answer: $A = 78.54$ sq in.

5. Find the length of a chord where the included angle is 30 degrees and the radius of the circle is 3 in.

Solution
The formula is

$$\text{Ch.} = 2[R(\sin\phi)]$$
$$= 2[3(0.26)]$$
$$= 2 \times 0.78$$

Note: The sine (sin) function of any angle can be found with a scientific calculator by entering the angle and pressing the "sin" button.
Answer: Ch. = 1.55 in.

6. If I begin at the starting point of all angles in the Cartesian coordinate system and revolve counterclockwise by 185 degrees, what quadrant would I be in?

Solution
See Figure 14.3. All angles begin on the right end of the X axis, and each quadrant has 90 degrees; I would be 5 degrees within the third quadrant.
Answer: Quadrant 3.

7. If I have a circle with a circumference of 6.2832 in., what is the diameter?

Solution
The basic formula is

$$C = \pi(D)$$

The formula, changed to solve for diameter, is

$$D = \frac{C}{\pi}$$

changed to

$$D = \frac{6.2832}{3.1416}$$

Answer: 2 in. diameter.

8. What is the radius of the circle in problem 7?

Solution

$$\frac{2'' \text{ diameter}}{2}$$

Answer: 1 in. radius.

PERIMETERS

The perimeter is simply the distance around any object. Perimeter is a length that is expressed in linear units such as inches, feet, yards, and meters. The only exception to the word *perimeter* when discussing the distance around something is a circle. The distance around a circle is called the *circumference.*

For the purpose of this chapter, we discuss only the perimeters of the four basic shapes: square, rectangle, triangle, and circle.

Perimeter of a Square

■ **Example 14–21**

Find the perimeter of the square in Figure 14.13. The formula is $P = 4s$ (Figure 14.14).

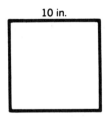

10 in.

FIGURE 14.13

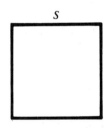

s

FIGURE 14.14
Square.

$$P = 4s$$
$$= 4(10)$$

Answer: $P = 40$ in.

Perimeter of a Rectangle

The formula is $P = 2S_1 + 2S_2$ (Figure 14.15).

FIGURE 14.15
Rectangle.

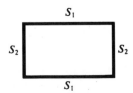

S_1

S_2 S_2

S_1

■ **Example 14–22**

Find the perimeter of the rectangle in Figure 14.16.

FIGURE 14.16

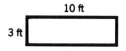

10 ft

3 ft

$$P = 2S_1 + 2S_2$$
$$= 2(3) + 2(10)$$
$$= 6 + 20$$

Answer: $P = 26$ ft.

Perimeter of a Right* Triangle

The formula is $P = H + B + A$ (see Figures 14.17 and 14.18).

■ **Example 14–23**

Find the perimeter of the right triangle.

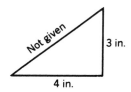

Not given

3 in.

4 in.

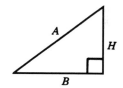

A

H

B

FIGURE 14.17

FIGURE 14.18
Right triangle.

$$P = H + B + \sqrt{B^2 + H^2}$$
$$= 3 + 4 + \sqrt{4^2 + 3^2}$$
$$= 3 + 4 + \sqrt{16 + 9}$$
$$= 3 + 4 + 5$$

Answer: $P = 12$ in.

_____

*   One angle is a right angle (90 degrees).

Using the Pythagorean theorem ($A^2 = B^2 + H^2$) and inserting values for the two given sides, the third (unknown) side may be calculated.

Circumference of a Circle

The formula is $C = \pi(D)$, where π is 3.1416 (Figure 14.19).

■ **Example 14–24**

Find the circumference of the circle in Figure 14.20.

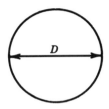

FIGURE 14.19
Circle.

FIGURE 14.20

$$C = \pi(D)$$
$$= 3.1416(3)$$

Answer: $C = 9.4248$ in.

AREAS

In some industrial applications, the surface area of an object must be calculated. The surface area has no thickness; it is simply the area of a surface, such as the top of a table or the floor of a building. In this chapter, only the four basic geometrical shapes will be discussed, but they can lead you to the solution of other shapes.

When you are concerned with the area of something, your answer is in *square* units such as square inches (sq in.) and square feet (sq ft). For example, the bottom surface of a box that is 1 foot long and 1 foot wide is equal to 1 square foot.

Area of a Square

■ **Example 14–25**

Find the area of the square in Figure 14.21. The formula is $A = S^2$ (Figure 14.22).

$$A = S^2$$
$$A = 10^2$$

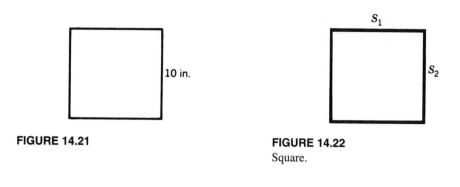

FIGURE 14.21   **FIGURE 14.22**
                                                   Square.

Answer: $A = 100$ sq in.

Area of a Rectangle

To find the area of a rectangle, such as shown in Figure 14.23, the formula is $A = L \times W$.

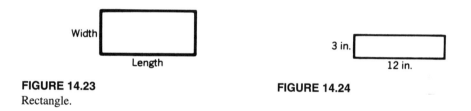

FIGURE 14.23   **FIGURE 14.24**
Rectangle.

■ **Example 14–26**

Find the area of the rectangle shown in Figure 14.24.

$$A = L \times W$$
$$= 12 \text{ in.} \times 3 \text{ in.}$$

Answer: $A = 36$ sq in.

Area of a Right Triangle

Note: A right triangle is one that contains one angle of 90 degrees.

■ **Example 14–27**

Find the area of a right triangle (Figure 14.25). The formula is $A = \frac{1}{2}(B)(H)$ (Figure 14.26).

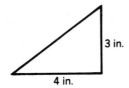

3 in.

4 in.

FIGURE 14.25

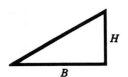

H

B

FIGURE 14.26
Right triangle.

$$A = \frac{1}{2}(B)(H)$$

$$= \frac{1}{2}(4)(3)$$

$$= \frac{1}{2}(12)$$

Answer: $A = 6$ sq in.

Area of a Circle

One formula is $A = \pi R^2$, where π is 3.1416 (Figure 14.27a).

FIGURE 14.27
Circle: (*a*) radius; (*b*) diameter.

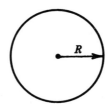
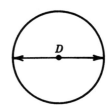

R

D

■ **Example 14–28**

Find the area of the circle.

$$R = 4 \text{ in.}$$

$$A = \pi(R^2)$$

$$= 3.1416 \times 4^2$$

$$= 3.1416 \times 16$$

Answer: $A = 50.2656$ sq in.

The alternate formula is $A = \dfrac{\pi(D^2)}{4}$ (Figure 14.27b).

■ **Example 14–29**

Find the area of the circle.

$$R = 4 \text{ in. or } D = 8 \text{ in.}$$

$$A = \frac{\pi(D^2)}{4}$$

$$= \frac{3.1416 \times 8^2}{4}$$

$$= \frac{3.1416 \times 64}{4}$$

$$= \frac{201.0624}{4}$$

Answer: $A = 50.2656$ sq in.

VOLUMES

Volume is a *cubic* measure. It is simply the area times height (or thickness) expressed in cubic units such as cubic inches (cu in.) and cubic feet (cu ft). Volume is the total space within a solid geometrical shape.

The following examples are only of simple geometrical shapes, but are the foundation for the more complex examples found in industry.

Volume of a Cube

The formula is volume = $L \times W \times H$ (Figure 14.28).

■ **Example 14–30**

Find the volume of the cube in Figure 14.29.

$$V = L \times W \times H$$

$$= 10 \text{ in.} \times 10 \text{ in.} \times 10 \text{ in.}$$

Answer: $V = 1000$ cu in.

Volume of a Rectangular Solid

The formula is volume = $L \times W \times H$ (Figure 14.30).

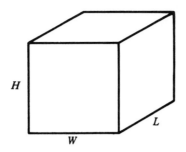

FIGURE 14.28
Cube.

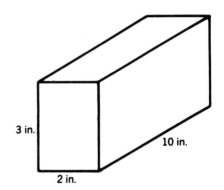

FIGURE 14.29

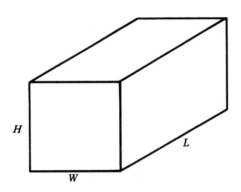

FIGURE 14.30
Rectangular solid.

FIGURE 14.31

■ **Example 14–31**

Find the volume of the rectangular solid in Figure 14.31.

$$V = L \times W \times TH$$
$$= 10 \text{ in.} \times 2 \text{ in.} \times 3 \text{ in.}$$

Answer: $V = 60$ cu in.

Volume of a Right Triangular Solid

■ **Example 14–32**

Find the volume of the right triangular solid in Figure 14.32. The formula is volume $= \frac{1}{2}$ $(B \times H)(Th)$ (Figure 14.33).

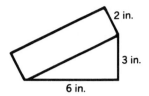

FIGURE 14.32

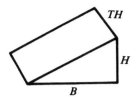

FIGURE 14.33
Right triangle solid.

$$V = \frac{1}{2}(B \times H)(Th)$$

$$= \frac{1}{2}(6 \times 3)(2)$$

$$= \frac{1}{2}(18)(2)$$

$$= \frac{1}{2}(36)$$

Answer: $V = 18$ cu in.

Volume of a Cylinder

The formula (if the radius is known) is: volume $= \pi(R^2)(L)$, where π is 3.1416 (Figure 14.34).

■ **Example 14–33**

Find the volume of the cylinder in Figure 14.35.

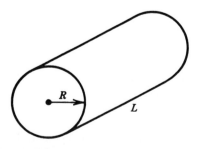

FIGURE 14.34
Cylinder.

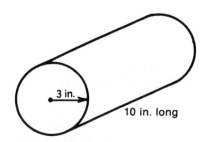

FIGURE 14.35

$$V = \pi (R^2)(L)$$
$$= 3.1416\,(3^2)(10)$$
$$= 3.1416\,(9)(10)$$
$$= 3.1416\,(90)$$

Answer: $V = 282.744$ cu. in.

Another formula (if the diameter is known) is: volume $= \dfrac{\pi(D^2)}{4}(L)$.

Volume of a Cone

The formula is $V = \frac{1}{3}\pi r^2 h$ (Figure 14.36).

FIGURE 14.36
Cone.

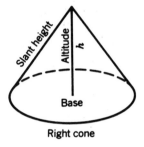

Right cone

■ **Example 14–34**

Find the volume of a cone where the radius is 5 in. and the height is 6 in.

$$V = \frac{1}{3}\pi\, r^2 h$$

Answer: $V = 157.1$ cu. in. $= \frac{1}{3}\cdot 3{,}1416 \cdot 5^2 \cdot 6$

SOLVING THE RIGHT TRIANGLE FOR SIDES ONLY: THE PYTHAGOREAN THEOREM

In many manufacturing and inspection applications, it becomes necessary to be able to find the length of one side of a right triangle when you know the length of the other two sides. It is always an asset to have the ability to see a right triangle within certain geometrical shapes. Then, once you see the triangle, you can solve it for an unknown side by using the

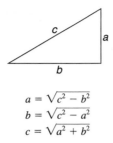

$$a = \sqrt{c^2 - b^2}$$
$$b = \sqrt{c^2 - a^2}$$
$$c = \sqrt{a^2 + b^2}$$

FIGURE 14.37

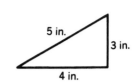

FIGURE 14.38
A 3–4–5 triangle.

Pythagorean theorem. The Pythagorean theorem is simply stated by Figure 14.37 and the formula $c^2 = a^2 + b^2$.

To avoid the lengthy solving of the one formula each time to calculate a side, it is better simply to write out each of the three possible formulas and have them handy to use when needed. These are shown in Figure 14.37. Now you can quickly solve for the unknown side (*a, b,* or *c*).

Remember: The three formulas work only when you call the sides *a, b,* and *c,* as shown in the figure.

An example of how to use the formulas is best described by the 3-4-5 triangle, shown in Figure 14.38. This triangle says that side $c = 5$ in., side $a = 3$ in., and side $b = 4$ in.

For practice, let's say that we had the same triangle but did not know the length of the longest side (the 5-in. side). So when looking at the three choices of formulas, we see that the side we are looking for is called side *c.* Now use the formula for side *c,* which is

$$c = \sqrt{a^2 + b^2}$$

The steps in using the formula are as follows:

Step 1.   Enter the sides you know into the formula.

$$c = \sqrt{3^2 + 4^2}$$

Step 2.   Square the numbers. (Squaring is simply multiplying the number times itself.)

$$c = \sqrt{9 + 16}$$

Step 3.   Add the two numbers together.

$$c = \sqrt{25}$$

Step 4. Find the square root. Square roots are any number, when multiplied by itself, results in the number being considered.

$$\sqrt{25} = 5$$

Answer: $c = 5$ in.

As you can see, the answer is 5, as shown in the original triangle. Be sure to follow the steps above at all times.

Practice Problems and Answers (Figure 14.39)

FIGURE 14.39
Right triangles with one unknown side.

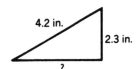

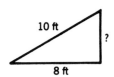

$$b = \sqrt{4.2^2 - 2.3^2}$$

$$= \sqrt{12.35}$$

$$= 3.5$$

Answer: 3.5 in.

$$a = \sqrt{10^2 - 8^2}$$

$$= \sqrt{36}$$

$$= 6$$

Answer: 6 ft.

Examples of Seeing the Triangle in a Problem

If you know the length of A and B (in Figure 14.40), you could solve the triangle for side C, which is the radius of the bolt circle, then double it to find the bolt circle diameter. Or find the corner-to-corner (or diagonal) distance of a rectangle if sides a and b are given (Figure 14.41).

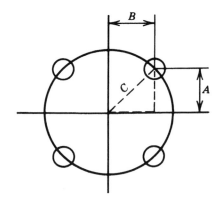

FIGURE 14.40
Finding the diameter of a bolt circle.

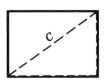

FIGURE 14.41
Finding the diagonal distance, C, of a rectangle.

Diagonal of a Square

The Pythagorean theorem can also be used to calculate the diagonal (or corner-to-corner distance) of a square. In this case sides *a* and *b* are the same, and the diagonal is equal to side *c*.

■ **Example 14–35**

A 5-in. square (Figure 14.42).

$$\text{Diagonal } (c) = \sqrt{a^2 + b^2}$$

$$= \sqrt{25 + 25}$$

$$= \sqrt{50}$$

$$= 7.07 \text{ in. } Answer$$

FIGURE 14.42
Diagonal of a square.

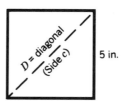

5 in.

SOLVING THE RIGHT TRIANGLE

The right triangle is used in many ways to solve many problems in the production and inspection of machined parts and castings. Knowing how to solve the right triangle for unknown sides and angles is an asset to everyone involved.

A right triangle has three sides and three angles as shown in Figure 14.43. The three angles are *A*, *B*, and *C*. The three sides are *a*, *b*, and *c*. Angle *C* is always 90 degrees. The reason you need to solve for the rest of the angles and sides is that in most cases, you are already given part of the information and you need to know the rest.

FIGURE 14.43
The three sides and three angles of a
right triangle.

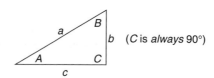

Angles:

$$A = 180° - (B + C)$$
$$B = 180° - (A + C)$$
$$C = 180° - (A + B)$$

Sides:

$$a = \sqrt{b^2 + c^2}$$
$$b = \sqrt{a^2 - c^2}$$
$$c = \sqrt{a^2 - b^2}$$

The steps to solving the right triangle are simple once you practice with them. There are six trigonometry functions used to solve these triangles. These functions are simply formulas to use and should not be regarded as too challenging. The functions are

sine   consine   tangent   cotangent   secant   cosecant

(or abbreviated):

sin   cos   tan   cot   sec   csc

The formulas are

$$\sin = \frac{\text{opp.}}{\text{hyp.}} \qquad \cos = \frac{\text{adj.}}{\text{hyp.}} \qquad \tan = \frac{\text{opp.}}{\text{adj.}}$$

$$\cot = \frac{\text{adj.}}{\text{opp.}} \qquad \sec = \frac{\text{hyp.}}{\text{adj.}} \qquad \csc = \frac{\text{hyp.}}{\text{opp.}}$$

Usually, the only ones used are the first three (the other three are reciprocals of the first three). Read the formulas as follows:

$$\sin_A = \frac{\text{opp.}}{\text{hyp.}}$$

This says that the sine function is equal to the length of the opposite side, divided by the length of the hypotenuse side.

There are three sides to a right triangle: hypotenuse, adjacent, and opposite. The hypotenuse is always the *longest* side of any right triangle.

One thing to remember is that the adjacent side and the opposite side change, depending on which of the angles you are dealing with. To establish which is which, remember that the opposite side is the side directly *opposite* the angle you are working with at the time, and the adjacent side is always the side next to the angle you are working with at the time.

The drawings in Figure 14.44 show how to name the sides of the triangle. For example, if you are working with angle *a* above, the opposite side is as shown, but if you are working with angle *b* above, the other side is the opposite side. *It is important to understand this.*

The next step (after knowing the formulas, and how to name the sides) is to decide which formula to use for the problem. A good rule is to take what you know about a problem, plus what you want to find, and look for the only formula that has both.

FIGURE 14.44
Naming the opposite sides of a right triangle.

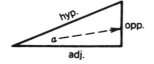

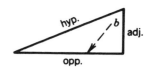

■ **Example 14–36**

The right triangle in Figure 14.45 has a 5-in. hypotenuse and a 1-in. opposite side. You can see in the formulas that the only formula having these things in it is the sine formula. This is the one you need to solve the problem. You will also need a table (such as Table 14–1) or calculator to convert the sine function of 0.2 into the angle because the formula only gives you the sine function of the angle, which is simply a number related to the angle.

FIGURE 14.45
A right triangle with hyp. = 5 in.
and opp. = 1 in.

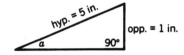

This sine 0.2 can be changed into an angle by using a calculator or a Table of Trigonometric Functions.

$$\sin = \frac{\text{opp.}}{\text{hyp.}} = \frac{1}{5} = 0.2$$

Right Triangle Practice Problems and Solutions

Below are some practice problems with step-by-step solutions to assist you in learning to solve the right triangle.

■ **Example 14–37**

Solve the triangle in Figure 14.46 for the length of side A.

FIGURE 14.46
Right triangle with unknown sides A
and B.

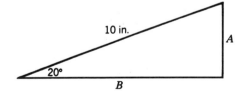

Solution

   Step 1.   Name the sides you are working with. They are the 10 inch (hyp.) and the A (opp.) side.

   Step 2.   Since the sides are hyp. and opp. use the formula

$$\sin = \frac{\text{opp.}}{\text{hyp.}}$$

TABLE 14–1

Basic Trigonometric Functions

Function⇒ Angle ⇓	Sin ⇓	Cos ⇓	Tan ⇓	Cot ⇓	
0	.0000	1.0000	.0000	----	90
1	.0175	.9998	.0175	57.2900	89
2	.0349	.9994	.0349	28.6363	88
3	.0523	.9986	.0524	19.0811	87
4	.0698	.9976	.0699	14.3007	86
5	.0872	.9962	.0875	11.4301	85
6	.1045	.9945	.1051	9.5144	84
7	.1219	.9925	.1228	8.1443	83
8	.1392	.9903	.1405	7.1154	82
9	.1564	.9877	.1584	6.3138	81
10	.1736	.9848	.1763	5.6713	80
11	.1908	.9816	.1944	5.1446	79
12	.2079	.9781	.2126	4.7046	78
13	.2250	.9744	.2309	4.3315	77
14	.2419	.9703	.2493	4.0108	76
15	.2588	.9659	.2679	3.7321	75
16	.2756	.9613	.2867	3.4874	74
17	.2924	.9563	.3057	3.2709	73
18	.3090	.9511	.3249	3.0777	72
19	.3256	.9455	.3443	2.9042	71
20	.3420	.9397	.3640	2.7475	70
21	.3584	.9336	.3839	2.6051	69
22	.3746	.9272	.4040	2.4751	68
23	.3907	.9205	.4245	2.3559	67
24	.4067	.9135	.4452	2.2460	66
25	.4226	.9063	.4663	2.1445	65
26	.4384	.8988	.4877	2.0503	64
27	.4540	.8910	.5095	1.9626	63
28	.4695	.8829	.5317	1.8807	62
29	.4848	.8746	.5543	1.8040	61
30	.5000	.8660	.5774	1.7321	60
31	.5150	.8572	.6009	1.6643	59
32	.5299	.8480	.6249	1.6003	58
33	.5446	.8387	.6494	1.5399	57
34	.5592	.8290	.6745	1.4826	56
35	.5736	.8192	.7002	1.4281	55
36	.5878	.8090	.7265	1.3764	54
37	.6018	.7986	.7536	1.3270	53
38	.6157	.7880	.7813	1.2799	52
39	.6293	.7771	.8098	1.2349	51
40	.6428	.7660	.8391	1.1918	50
41	.6561	.7547	.8693	1.1504	49
42	.6691	.7431	.9004	1.1106	48
43	.6820	.7314	.9325	1.0724	47
44	.6947	.7193	.9657	1.0355	46
45	.7071	.7071	1.0000	1.0000	45
	⇑ Cos	⇑ Sin	⇑ Cot	⇑ Tan	⇑ Angle ⇐Function

Examples: $Sin_{30} = 0.5000$, $Sin_{52} = .7880$, $Tan_{41} = .8693$

516

Step 3.   Change the formula to solve for the opposite side *A*.

$$\sin(\text{hyp.}) = \text{opp.}$$

Step 4.   Enter the known values into the formula.

$$\sin 20°(10) = \text{opp.}$$

Step 5.   Find the sine of 20 degrees and enter that number in the proper place.

$$0.3420(10) = \text{opp.}$$

Step 6.   Work the simple math for the answer.

$$0.3420(10) = 3.420 \text{ in.}$$

Answer: Side *A* = 3.420 in.

■   **Example 14–38**

Solve the triangle in Figure 14.46 for the length of side *B*.

Solution

   Step 1.   Name the sides you are working with. They are the 10 inch (hyp.) and the *B* (adj.) side.

   Step 2.   Since the sides are hyp. and adj., use the only formula with them in it.

$$\cos = \frac{\text{adj.}}{\text{hyp.}}$$

   Step 3.   Change the formula to solve for the adj. side.

$$\cos(\text{hyp.}) = \text{adj.}$$

   Step 4.   Enter the known values where they belong in the formula.

$$\cos 20°(10) = \text{adj.}$$

   Step 5.   Find the cosine of 20 degrees and enter it into the formula. (The cosine of 20 degrees is 0.9397.)

   Step 6. $0.9397(10) = 9.397$ in.

   Answer: Side *B* = 9.397 in.

■ **Example 14–39**

Find the angle α in Figure 14.47.

FIGURE 14.47
Right triangle with unknown
angle α.

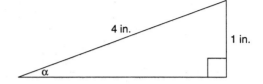

Solution

Step 1. Name the sides you are working with. They are the 4-in. hypotenuse and the 1-in. opposite side.

Step 2. Choose the only formula (of the basic three) that has both of these sides in it.

$$\text{Formula: } \sin = \frac{\text{opp.}}{\text{hyp.}}$$

Step 3. The formula above says that the sine function of the angle is equal to the opposite side length divided by the hypotenuse side length. *This formula solves only for the sine function of the angle, not the angle itself.*

Step 4. Enter all known values in their proper place.

$$\sin_? = \frac{\text{opp.}}{\text{hyp.}} = \frac{1}{4}$$

Step 5. Do the simple math.

$$\sin_\alpha = \frac{1}{4} = 0.25$$

Step 6. Convert the sine function of 0.25 to degrees. If $\sin_\alpha = 0.25$, then the angle is $= 14.48$ degrees (rounded off).

Answer: 14.48 degrees.

■ **Example 14–40**

What is angle α in Figure 14.48?

FIGURE 14.48
Right triangle with unknown
angle α.

Solution

The sum of all angles in a triangle is 180 degrees; therefore, the unknown angle must be

$$180° - 20° - 90° = 70°$$

Answer: 70 degrees.

■ **Example 14–41**

Solve the triangle in Figure 14.49 for side *B* using the Pythagorean theorem.

FIGURE 14.49
Right triangle with unknown side *B*.

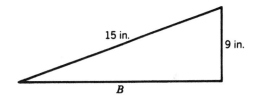

Solution

$$B = \sqrt{15^2 - 9^2} = 12 \text{ in.}$$

Answer: 12 in.

■ **Example 14–42**

A ladder is 8 feet long (Figure 14.50) and is leaning against a building. The point where the top of the ladder is touching is 7 feet high. How far is the base of the ladder from the base of the building?

Step 1. Draw the problem so that there is a triangle (as in Figure 14.50).

Step 2. Solve this problem with the Pythagorean theorem.

Step 3. $? = \sqrt{8^2 - 7^2}$.

FIGURE 14.50
Right triangle formed by a ladder
leaning against a building.

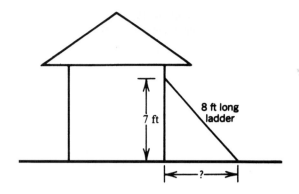

Step 4.   ? = 3.873 ft (rounded off).

Answer: 3.873 ft.

Triangle Practice Problems and Solutions

Note: The total of the internal angles of a triangle always equals 180 degrees (Figure 14.51).

1. Find angle *B*.

Solution

$$180 - (90 + 15) = 75°$$

Answer: 75 degrees.

2. Find the opposite side.

Solution

$$\sin \text{ angle } A = \frac{O}{H}$$

$$O = \sin \times H$$
$$= 0.199 \times 5 \text{ in.}$$
$$= 0.995 \text{ in.}$$

Answer: 0.995 in.

FIGURE 14.51
(*a*) Right triangle with unknown
angle *B*. (*b*) Right triangle with
unknown opposite side.

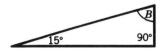

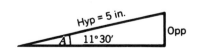

3. Find the sine of 11° 0′.

Answer: 0.19081.

4. Find the tangent of 78° 30′.

Answer: 4.9151.

MEMORIZING THE TRIGONOMETRY FORMULAS FOR THE RIGHT TRIANGLE

The trigonometry formulas discussed earlier are necessary in solving problems in manufacturing and inspection. Some people choose to memorize the formulas as they become comfortable with solving a right triangle. This is not easily done without some kind of system.

The system discussed here is very effective for accurately memorizing the formulas. It is simply a short cut to memorizing the formulas without having to look them up each time. If used correctly, it can be very helpful.

$$\text{sine} = \frac{\text{opposite}}{\text{hypotenuse}} \qquad \text{cosine} = \frac{\text{adjacent}}{\text{hypotenuse}} \qquad \text{tangent} = \frac{\text{opposite}}{\text{adjacent}}$$

$$\text{cotangent} = \frac{\text{adjacent}}{\text{opposite}} \qquad \text{secant} = \frac{\text{hypotenuse}}{\text{adjacent}} \qquad \text{cosecant} = \frac{\text{hypotenuse}}{\text{opposite}}$$

Next are five steps to memorizing the formulas.

Step 1. Memorize these letters *in order.* O   A   O   A   H   H

Step 2. Next, underline them. O   A   O   A   H   H

Step 3. Then write them backward under the first set.

O	A	O	A	H	H
H	H	A	O	A	O

Step 4. Now draw vertical lines separating each set of letters.

O	A	O	A	H	H
H	H	A	O	A	O

Step 5. Now (in the proper order as shown) put the abbreviated names of the functions over the sets of letters. The proper order is sin, cos, tan, cot, sec, csc:

sin	cos	tan	cot	sec	csc
O	A	O	A	H	H
H	H	A	O	A	O

You can see that by remembering *OAOAHH* and the order of the functions, you can effectively memorize the six trigonometry functions of the right triangle for quick use in the shop or anywhere.

SOLVING OBLIQUE TRIANGLES

Triangles that have no 90-degree angles are called *oblique triangles*. It is necessary to know three things to solve oblique angles. When three parts are known (e.g., two sides and an angle), the formulas can be used to solve for the other parts.

There are still three sides and three angles (the same as the right triangle). The formulas that follow match the picture in Figure 14.52 of an oblique triangle. When solving a problem you simply take the parts you know and match them to this sample triangle; then find what you want to know from the formula.

To Find This	Knowing This	Use This Formula
A	B and C	$180° - (B + C)$
B	A and C	$180° - (A + C)$
C	A and B	$180° - (A + B)$
b	a, A, B	$b = \dfrac{a \times \sin B}{\sin A}$
c	a, A, C	$c = \dfrac{a \times \sin C}{\sin A}$
area of	a, b, C	$\text{area} = \dfrac{a \times b \sin C}{2}$
c	a, b, C	$c = \sqrt{a^2 + b^2 - (2ab \times \cos C)}$
$\cos A$	a, b, c	$\cos A = \dfrac{b^2 \times c^2 - a^2}{2bc}$
$\sin B$	a, A, b	$\sin B = \dfrac{b \times \sin A}{a}$
$\tan A$	a, b, C	$\tan A = \dfrac{a \times \sin C}{b - (a \times \cos C)}$

Note above that the abbreviations sin, cos, and tan are used. In formulas where they are used, remember that they mean the sine, cosine, and tangent function of the angle. For example, the sin 30 = 0.5 (or 0.5 is the sine function of 30 degrees).

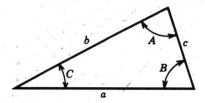

FIGURE 14.52
Oblique triangle.

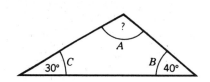

FIGURE 14.53
Oblique triangle with unknown angle A.

Oblique Triangle Practice Problems and Solutions

For the following problems, refer to the previous table of formulas to find an unknown side or angle.

1. Find the unknown angle in Figure 14.53.

Solution
The unknown angle is angle A. The formula is

$$A = 180 - (B + C)$$
$$= 180 - (40 + 30)$$
$$= 180 - 70$$

Answer: $A = 110$ degrees.

2. Find the unknown side in Figure 14.54.

Solution
The unknown side is the side c in the reference triangle. The formula is

$$c = \frac{a \times \sin C}{\sin A}$$

$$= \frac{3 \times 0.5}{0.9397}$$

$$= \frac{1.5}{0.9397}$$

Answer: $c = 1.596$ in.

FIGURE 14.54
Oblique triangle with unknown
side c.

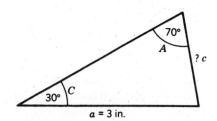

a = 3 in.

FIGURE 14.55
Finding the area of an oblique
triangle.

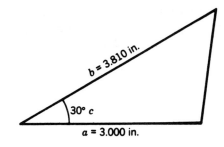

$b = 3.810$ in.

30° c

$a = 3.000$ in.

3. Find the area of the triangle in Figure 14.55.

Solution
The formula is

$$\text{area} = \frac{a \times b \times \sin C}{2}$$

$$= \frac{3 \times 3.810 \times 0.5}{2}$$

$$= \frac{5.715}{2}$$

Answer: Area = 2.8575 sq in.

4. Find the unknown angle in Figure 14.56.

Solution
Sides *a, b,* and *c* are known. The unknown angle is angle *A*. The formula is

$$\cos A = \frac{b^2 + c^2 - a^2}{2bc}$$

$$= \frac{8^2 + 9^2 - 7^2}{2 \times 8 \times 9}$$

$$= \frac{64 + 81 - 49}{144}$$

$$= \frac{96}{144}$$

Answer: cos *A* = 0.66667.
   Now, if the cosine of angle *A* is 0.66667, then convert the cosine of *A* to angle *A* using a calculator, or the trigonometry tables.
   Answer: arccos 0.66667 = 48.2 degrees.

FIGURE 14.56
Triangle with unknown angle A.

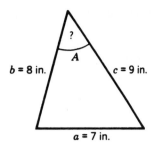

$b = 8$ in. $c = 9$ in.

$a = 7$ in.

Miscellaneous Geometric Problems

1. Area of a trapezoid (Figure 14.57).

Example	**Equation**

FIGURE 14.57
Trapezoid.

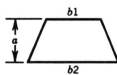

$$\text{Area} = \frac{1}{2}\, a(b_1 + b_2)$$

2. Area and volume of a sphere (Figure 14.58).

Example	**Equation**

FIGURE 14.58
Sphere.

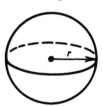

$$\text{Area} = 4\pi r^2$$

$$\text{Volume} = ;d4,3,\pi r^3$$

3. Diameter of a circle inscribed in an equilateral triangle (Figure 14.59).

Example	**Equation**

FIGURE 14.59
Equilateral
triangle with an
inscribed circle.

$$D = S(0.57735)$$

4. Side of an equilateral triangle inscribed inside a circle (Figure 14.60).

Example   **Equation**

FIGURE 14.60
Equilateral
triangle with a
circumscribed
circle.

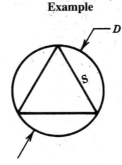

$S = D(0.866)$

TAPERS

A variety of applications on engineering drawings require tapers. Taper on a single surface can be explained by referring to it as the *rise versus the run* (see Figure 14.61).

Surface Taper

For 1 degree of taper on a single surface, the rise will be approximately 0.0175 in. per 1 in. of run. The computation necessary for this is

$$\text{tangent}_\alpha \times \text{adjacent side} = \text{opposite side}$$

where "tangent" is the tangent function of the angle
α is the angle
"adjacent side" is the side nearest the angle
"opposite side" is the side opposite the angle

Diametral Taper

With respect to a diameter (or thickness) where two surfaces are involved, there are three parts to consider:

1. Large diameter (or thickness)
2. Small diameter (or thickness)
3. Length between diameters (or thicknesses)

FIGURE 14.61
Triangular representation of a
tapered surface. Angle $\alpha = 1°$.

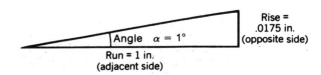

The taper is usually referred to as *taper per inch* (TPI) or *taper per foot* (TPF). In many cases, the inspector or machinist is given only parts of the information for a taper and must calculate the remaining parts.

■ **Example 14–43: Three Parts Are Given, But Not the Taper**

To find the taper in Figure 14.62, subtract the small diameter from the large diameter and divide the answer by the length.

$$\frac{\text{large diameter} - \text{small diameter}}{\text{length}} = \text{taper (inch per inch)}$$

$$\frac{1.000 \text{ in.} - 0.500 \text{ in.}}{4.000 \text{ in.}} = 0.125 \text{ in. per in.}$$

Tapers

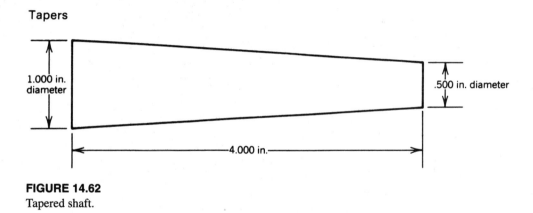

FIGURE 14.62
Tapered shaft.

■ **Example 14–44: The Amount of Taper is Given, But One of the Diameters Is Not Given**

To find the taper in Figure 14.63, find diameter (*d*) by first multiplying the taper per inch (TPI) times the length (4.000 in.), then subtract that amount from the large diameter.

$$
\begin{aligned}
\text{Small diameter } (d) &= \text{large diameter} - (\text{TPI} \times \text{length}) \\
&= 1.000 \text{ in.} - (0.125 \text{ in.} \times 4.000 \text{ in.}) \\
&= 1.000 \text{ in.} - 0.500 \text{ in.} \\
&= 0.500 \text{ in.}
\end{aligned}
$$

If the missing diameter were the large diameter, the equation would be

$$\text{large diameter } (D) = \text{small diameter} + (\text{TPI} \times \text{length})$$

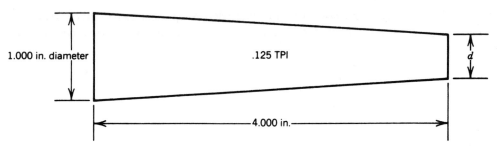

FIGURE 14.63
Tapered shaft.

Converting TPI to TPF

Taper per inch (TPI) can be converted to taper per foot (TPF), or vice versa, using the following equations:

$$\frac{TPF}{12} = TPI$$

$$TPI \times 12 = TPF$$

Converting Taper into the Corresponding Angle

The angle (per side) can be found by dividing the taper per foot (TPF) by 24, then finding the angle whose tangent function corresponds to that quotient.

■ **Example 14–45**

Find the angle in Figure 14.64.

FIGURE 14.64
1.5 in. taper per foot.

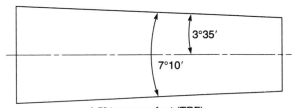

1.5" taper per foot (TPF)

$$Tangent\ angle = \frac{TPF}{24}$$

$$= \frac{1.5}{24}$$

$$= 0.0625$$

The angle can then be found by referring to the table of natural trigonometric functions in Table 14–1. The angle is 3° 35'. To find the included angle, multiply the taper angle by 2.

Using Gage Spheres to Inspect an Angle

At times, there are shop problems that involve producing or inspecting tapered features such as a tapered inside diameter shown in Figure 14.65. The unknown variable in these cases could be the angle, size, or the location of the feature. The included angle of the tapered hole can be inspected using two precision gage spheres as shown in the figure. The following equation is used in conjunction with the two height measurements shown. Note: The equation solves for one-half of angle A.

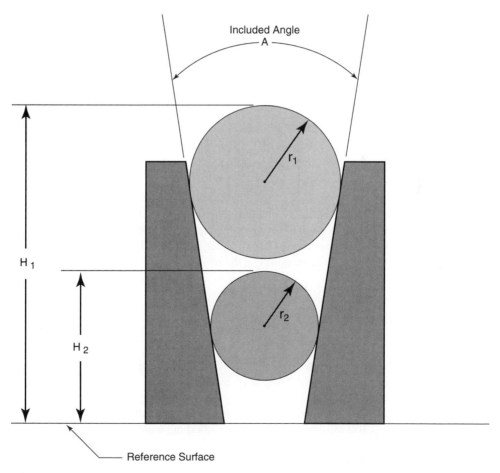

FIGURE 14.65
Inspecting an angle of a tapered hole using precision gage spheres.

$$Sin\frac{A}{2} = \frac{r_1 - r_2}{(H_1 - r_1) - (H_2 - r_2)}$$

where: r_1 = the radius of ball #1
r_2 = the radius of ball #2
H_1 = the height measurement to the top of ball #1
H_2 = the height measurement to the top of ball #2

■ **Example 14–46**

A tapered feature of a part has been produced and the unknown variable of interest is angle A. Two precision balls at exactly 0.750″ and 0.375″ diameter (within gage tolerances) have been selected in order to inspect the angle. With each ball inserted into the cone, the height measurements H_1 and H_2 are measured. Given the following results, find angle A.

$$r_1 = 0.375 \qquad r_2 = 0.1875 \qquad H_1 = 1.500″ \qquad H_2 = 0.500″$$

$$Sin\tfrac{A}{2} = \frac{r_1 - r_2}{(H_1 - r_1) - (H_2 - r_2)}$$

$$Sin\tfrac{A}{2} = \frac{.375 - .1875}{(1.500 - .375) - (0.500 - .1875)}$$

$$Sin\tfrac{A}{2} = \frac{.1875}{.8125}$$

$$Sin\tfrac{A}{2} = .23076923$$

Therefore, the sine of one-half the angle is 0.23076923. Using a calculator, to find one-half of the angle is 13.34236375°. Multiplying one-half the angle times 2, we find that the included angle is 26.6847275°. When converted from a decimal-based angle to degrees (°), minutes (′), and seconds (″), the resulting angle is **26° 41′ 5″**.

Practice Problems and Solutions

1. If the rise were 0.035 in. for a run of 1 in., what would be the corresponding angle?

   Solution
   There is approximately 0.0175 in. rise per inch per degree; therefore, 0.035 in. is twice that amount, or 2 degrees.

2. If a diameter were tapered and the large diameter were 0.500 in., the small diameter were 0.100 in., and the length were 1.000 in., what would the taper be?

   Solution

   $$\frac{0.500 - 0.100 \text{ in.}}{1.000 \text{ in.}} = 0.400 \text{ in. per in.}$$

3. A shaft that is 4.000 in. long is tapered. The small diameter is 1.000 in. and the TPI is 0.125 in. What is the large diameter?

Solution

The larger diameter (D) = the small diameter + (TPI × length).

$$D = 1.000 \text{ in.} + (0.125 \times 4.000 \text{ in.})$$
$$= 1.000 \text{ in.} + 0.500 \text{ in.}$$
$$= 1.500 \text{ in.}$$

4. If there were a shaft that had a taper per inch (TPI) of 0.100 in., what would be the corresponding TPF?

Solution

$$\text{TPI} \times 12 = \text{TPF}$$
$$0.100 \text{ in.} \times 12 = \text{TPF}$$
$$\text{TPF} = 1.200 \text{ in.}$$

5. If a shaft had a taper per foot (TPF) of 0.144 in., what is the corresponding taper per inch (TPI)?

Solution

$$\text{TPI} = \frac{\text{TPF}}{12}$$
$$= \frac{0.144 \text{ in.}}{12}$$
$$= 0.012 \text{ in.}$$

UNITS OF MEASUREMENT

Metrology spells out different units of measurement for different characteristics (e.g., length, force, weight, speed, distance, etc.). There are three different systems of measurement; English, metric, and SI (*Système international d'unités*). The English system for measurement has different bases (e.g., 12 inches per foot, 3 feet per yard, etc.). The metric and SI systems are both base 10 systems (i.e., the measurement units and conversions are based on factors of 10). The SI system was developed as a modification of the original metric system as technology advanced over the years. Table 14–2 is a comparison of some measurement units.

For further information on the units of measurement, refer to an appropriate math textbook.

TABLE 14–2

Examples of Units of Measurement

Measurement	English system	Original metric system	SI
Length	foot	meter	meter
Time	second	second	second
Force	pound	kilogram	newton
Mass	slug	kilogram-second^2/meter	kilogram
Temperature	Fahrenheit	Celsius	kelvin

METRIC OR ENGLISH: A COMPARISON

In comparing the English and metric systems, several factors are involved. The conversion from English to metric units is a subject that has been under study for a long time. The reason for the length of time is the impact the conversion would have on U.S. industry. One major consideration is the large amount of money that would have to be spent for drawings, tooling, training, and so on.

The bottom line is to answer the question: Is it something we should do, or not?

The English System: Conversion

In the English system, conversions (such as feet to miles, yards to inches, gallons to ounces, etc.) usually cause confusion because nothing is straightforward. For example, there are many different constants.

1 mile = 5280 feet = 1760 yards

1 yard = 3 feet = 36 inches

1 foot = 12 inches

1 inch can be divided into a wide variety of fractions and decimal fractions (usually, halves, fourths, eighths, sixteenths, thirty-seconds, and sixty-fourths)

All of these constants (5280 or 1760, 3 or 36, 12, etc.) cause confusion in conversion. For example, there are many calculations involved in converting 1 mile to inches.

1 mile = 5280 feet = 5280 × 12 inches/feet = 63,360 inches

or 1 mile = 1760 yards = 1760 × 36 inches/yard = 63,360 inches

Our system (English) is convenient simply because we were trained to use it in childhood and have been using it ever since. Many people do not like the thought of changing over to metric units because they do not understand the system. The metric system is much more logical than the English system and much easier to use.

The Metric System: Conversion

The metric system is easier to use than the English system because it uses a base of 10. For example, any number multiplied by 10 simply moves the decimal point one place to the right.

$$6.0 \times 10 = 60$$

Any number divided by 10 simply moves the decimal point one place to the left.

$$5.7 \div 10 = 0.57$$

When converting a metric measurement to another, the decimal point is simply moved to the right or left. The following example shows how easy conversion is in the metric system.

0	0	0	0	1.	0	0	0	0
↑	↑	↑	↑	↑	↑	↑	↑	↑
megameter	kilometer	hectometer	dekameter	meter	decimeter	centimeter	millimeter	micron
1,000,000 meters	1000 meters	100 meters	10 meters	1 meter	0.1 meter	0.01 meter	0.001 meter	0.000001 meter

You can see above that 1 centimeter equals 10 millimeters, 1 meter equals 100 centimeters, 1 dekameter equals 10 meters, and so on. All of these conversions have just one multiplier to use in converting from one unit to another, for example, from meters to centimeters.

Remember the terms:

mega   means 1,000,000

kilo   means 1000

hecto   means 100

deka   means 10

deci   means $\dfrac{1}{10}$

centi   means $\dfrac{1}{100}$

milli   means $\dfrac{1}{1000}$

micron   means $\dfrac{1}{1,000,000}$

These terms (and the conversion) are used in all measurements, whether they be grams, liters, or meters. Further study of the metric system reveals how easy the system is to use. See Tables A–2 and A–3 in the Appendix for metric-to-English conversion tables.

Metric Conversion Examples

1. Convert 5 meters to centimeters.

   Solution
   1 meter = 100 centimeters, so 5 meters = 5 × 100 = 500 centimeters.

2. Convert 0.5 millimeters to centimeters.

   Solution
   1 millimeter = 0.1 centimeter, so 0.5 millimeters = 0.5 × 0.1 = 0.5 centimeters.

3. Convert 20 dekameters to meters.

   Solution
   There are 10 meters in a dekameter, so 20 dekameters = 20 × 10 = 200 meters.

4. Convert 2 kilometers to meters.

   Solution
   1 kilometer equals 1,000 meters, so 2 kilometers = 2 × 1,000 = 2,000 meters.

RATIOS

Most people use ratios every day without realizing it. When working with fractions, you are actually working with ratios. For example, the fraction $\frac{1}{2}$ is simply a ratio of 1 to 2, written as 1 : 2.

Ratios are simply a statement that compares two numbers. When a pie is cut into eight equal pieces and you get one of those pieces, your piece is a ratio of 1 : 8 (because there are a total of eight pieces, and you got one of them).

Another example is the sprocket on a bicycle. The size of the rear sprocket related to the size of the front sprocket is important. If the front sprocket turns in a ratio of 4 : 1 compared to the back sprocket, the rider will have to pedal faster. If the front sprocket were a ratio of 2 : 1, the rider will pedal slower but hold the same speed. Why, you ask?

A 4:1 ratio between the front and rear sprocket means that the front sprocket must make four full turns to turn the rear sprocket just one full turn. Therefore, the rider pedals faster. A 2 : 1 ratio means that the front sprocket has to turn only two full turns for the rear sprocket to make a full turn. Therefore, the rider can pedal slower than a rider on the same type of bike with a 4 : 1 ratio.

Ratios are used throughout industry in several applications, including machines and measuring equipment. Regardless of use, if you understand ratios, you will be in a better position to work with them.

TEMPERATURE CONVERSION

At times it becomes necessary to convert temperature from one scale to another. Today, the main scales to be considered are Fahrenheit and Celsius (sometimes called centigrade).

Converting from Fahrenheit to Celsius, or from Celsius to Fahrenheit, is very simple. Formulas can be used for each conversion, or tables are available to read the conversions directly. (See Table A–4.)

The formulas are

$$T_c = \frac{5}{9}(T_f - 32)$$ T_c stands for temperature in Celsius

$$T_f = \frac{9}{5}(T_c) + 32$$ T_f stands for temperature in Fahrenheit

■ Example 14–47

Convert 68 degrees Fahrenheit to Celsius. The formula is $T_c = \frac{5}{9}(T_f - 32)$.

Solution

$$T_c = \frac{5}{9}(68 - 32)$$

$$= \frac{5}{9}(36)$$

$$= \frac{180}{9}$$

Answer: T_c = 20 degrees Celsius (20°C).

■ Example 14–48

Convert 38 degrees Celsius to Fahrenheit. The formula is $T_f = \frac{9}{5}(T_c) + 32$.

Solution

$$T_f = \frac{9}{5}(38) + 32$$

$$= 68.4 + 32$$

Answer: T_f = 100.4 degrees Fahrenheit (100.4°F).

The ability to convert temperatures (specifically Celsius and Fahrenheit) will assist you in many ways. An example is in temperature measurement. If the specification calls for degrees Celsius and you have equipment that measures Fahrenheit, you can measure it and convert your value to Celsius.

THE REAL NUMBER LINE

In many mathematical situations an understanding of the real number line can be an asset to the person trying to solve the problem. Most frequent are those problems that contain positive and negative numbers. The real number line is helpful in gaining an understanding of positive and negative numbers and looking at them in terms of distance. The line itself is very simple. It begins with a value of 0 in the middle and shows that all numbers to the right are plus and all numbers to the left are minus, as shown in Figure 14.66.

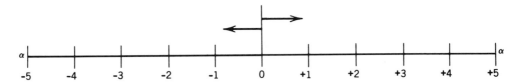

FIGURE 14.66
Real number line.

If one simply looks at the real number line in terms of distance, one can clearly see that the distance between $+3$ and -2 is 5 since the distance is 5 divisions, regardless of whether the values are positive or negative. Everyone understands that the distance from 0 to $+5$ is 5, and the distance from 0 to -5 is also 5. When working with positive and negative numbers, remember the real number line and distance and they will not be such a problem.

Practice Problems

1. What is the distance from -10 to $+2$?
2. What is the distance from $+2$ to $+8$?
3. What is the distance from -3 to -12?
4. What is the distance from 0 to $+7$?

Solutions to Practice Problems

1. 12
2. 6
3. 9
4. 7

THE CARTESIAN COORDINATE SYSTEM

An understanding of the Cartesian coordinate system is essential to solving some geometrical relationships and many inspection problems. The system begins with two axes, X and Y. The X axis is horizontal and the Y axis is vertical. These axes separate the space into four *quadrants*. The quadrants begin in the upper right space called quadrant 1 and proceed counterclockwise to quadrant 4. An example of the system is shown in Figure 14.67.

FIGURE 14.67
Cartesian coordinate system.

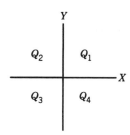

Plotting Points

When a scale is placed on the X and Y axes with certain divisions, points can be plotted. From the center of the system where the axes converge, values are assigned a positive or negative number, depending on which direction they are plotted. For example, on the X axis, a value to the right is positive and a value to the left is negative. On the Y axis, a value upward is positive and a value downward is negative.

Now the points plotted in any given quadrant can be assigned positive or negative signs. Points in quadrant 1 are positive, since it takes an X value to the right and a Y value upward. Points in quadrant 2 have a positive and negative sign. Points in quadrant 3 are both negative, and points in quadrant 4 are positive and negative. Examples of this relationship are shown in Figure 14.68.

When plotting points on the system the first value spoken is the X value. For example, referring to the point at $(+2, +3)$ means an X value of $+2$ and a Y of $+3$. The point for these values would fall in the first quadrant, two units to the right on the X axis and three units upward on the Y axis.

Graphs

When graphs (such as a line graph) are used, they usually appear in the first quadrant of the system. In this application, the horizontal X axis is usually referred to as the *abscissa* and the vertical Y axis is called the *ordinate*. In this relationship, all values plotted on the graph are plus unless a specific scale is assigned to the ordinate to dictate otherwise.

FIGURE 14.68
Points plotted on the X axis.

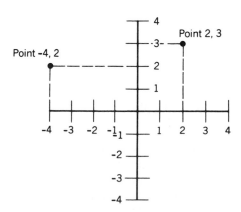

FIGURE 14.69
Angular relationship on the
Cartesian coordinate system.

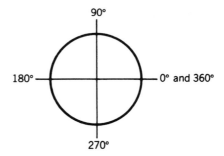

Angles

Angles on the Cartesian coordinate system begin on the first quadrant X axis at 0 degrees and rotate counterclockwise. This angular relationship is shown in Figure 14.69. The relationship is clearer when a circle is superimposed on the system (or vice versa).

Practice Problems

1. One point is plotted. The X value is $+3$ and the Y value is $+5$. In which quadrant will the plotted point for these values fall?
2. Another point is plotted. The X value is -2 and the Y value is $+3$. In which quadrant will the point be plotted?
3. Which quadrant of the system contains the line for a 75-degree angle?
4. One point is plotted at $X = +3$ and $Y = +3$. When you plot this point and draw a line from the center of the system through this point, what will the angle be?

Note: Draw the problem before answering.

Solutions to Practice Problems

1. Quadrant 1 (since both values are plus)
2. Quadrant 2
3. Quadrant 1 (since it contains everything from 0 to 90 degrees)
4. 45 degrees (since the line goes through the middle of quadrant 1 and quadrant 1 contains everything from 0 to 90 degrees)

SCIENTIFIC NOTATION

Some mathematical problems involve very large (or very small) numbers. Many calculators and software programs handle these large or small numbers by expressing them in *scientific notation*. Scientific notation is a method of expressing a very large or very small number in a small space by using a positive (or negative) exponent to tell how many places (right or left) to move the decimal point.

■ **Example 14–49: Large Number**

A value of 1,250,000 can be expressed in scientific notation. The scientific notation method of expressing 1,250,000 is 1.25^{+06} (or move the decimal point 6 places to the right).

■ **Example 14–50: Small Number**

A value of 0.000009 can be expressed in scientific notation. The scientific notation method of expressing 0.000009 is 9^{-06} (or move the decimal point six places to the left).

Practice Problems

1. Convert 2.54^{-03} to the actual value.
2. Convert $.75^{+05}$ to the actual value.

Solution to Practice Problems

1. 0.00254
2. 75,000

REVIEW QUESTIONS

1. Add the following fractions: $\frac{3}{8} + \frac{6}{16}$.
2. Reduce $\frac{12}{32}$ to lowest terms.
3. Multiply the following fractions: $\frac{5}{8} \times \frac{3}{4}$.
4. Write the decimal forty-three hundredths inch.
5. Multiply the following decimals: 0.502×0.003 (round the answer to four places).
6. Divide the following decimals: $0.502 \div 0.2$.
7. If $646 = X \div 12$, what is X?
8. Find the circumference of a 5-in.-diameter circle (round off to one decimal place).
9. Find the area of a circle that has a 7-ft. diameter (round off to one decimal place).
10. What is the perimeter of a 12-in. square?
11. What is the area of the above 12-in. square?
12. Find the volume of a cylinder with a 5-in. radius and a height of 10 in.
13. The hypotenuse of a right triangle is 10 in. long, and the shortest side is 6 in. long. What is the length of the other side?
14. Find the diagonal distance of a 0.010-in. square (round the answer to three decimal places).
15. The sum of two angles in a right triangle is 140°. What is the other angle?
16. Find the largest square that will fit inside a 0.010-in.- diameter circle (round the answer to three decimal places).
17. Convert 98°F to Celsius (round the answer to one decimal place).
18. Refer to question 12. What is the volume of the cylinder expressed in cubic centimeters (cm^3)?
19. If two angles of a right triangle are 90 degrees and 40 degrees, what is the third angle?

20. Find the volume of a cone. The radius of the cone is 3 in. and the height of the cone is 5 in.
21. Which of the following units is the SI unit of measure for temperature?
    a. Fahrenheit
    b. Celsius
    c. kelvin
    d. joule
22. Which of the following is a common unit of measurement in the English, metric, and SI systems?
    a. Celsius
    b. second
    c. Fahrenheit
    d. meter
23. Find the diagonal distance of a rectangle. The length of the rectangle is 12 in. and the width is 5 in.
24. Convert 68 degrees Fahrenheit to Celsius.
25. Convert 12 liters to U.S. gallons.
26. Convert 12 dekameters to meters.
27. Convert 12 in. to centimeters.
28. Which of the following units is common to the metric and the SI systems?
    a. kelvin
    b. Celsius
    c. meter
    d. kilogram
29. A box is 24 in. long, 12 in. wide, and 8 in. high. What is the volume of the box expressed in cubic centimeters (cm^3)?
30. If a bicycle front sprocket has a 10 in. diameter and the rear sprocket has a 4 in. diameter, what is the front-to-rear ratio of the diameters?
31. What is the value 12,300,000 expressed in scientific notation?
32. Express the value 0.00000037 in scientific notation.

GLOSSARY A
Quality Terms and Definitions

Acceptable quality level. The maximum percentage or proportion of variant units in a lot or batch that, for the purposes of acceptance sampling, can be considered satisfactory as a process average. The term *variant unit* should be replaced by more specific terms, such as *nonconforming unit* or *defective unit,* where appropriate. (2)*

Acceptance number. The maximum number of defects or defective units in the sample that will permit acceptance of the inspection lot or batch. (1)*

Accuracy (of measurement). An unbiased true value. The difference between the average of several measurements and the true value.

Addendum. The distance from the pitch diameter to the crest on an external thread (for example).

Allowance. The intended dimensional difference between mating parts (whether it be clearance or interference).

Alloy. A substance composed of two or more metals, or metals and nonmetals, combined for various reasons.

Aluminum. A lightweight, shiny, nonmagnetic metal that resists oxidation.

Attribute. A characteristic or property that is appraised in terms of whether it does or does not exist (e.g., Go or No Go) with respect to a given requirement. (1)

Attribute gages. Gages that measure on a Go–NoGo basis. An example is a plug gage for a hole.

Average outgoing quality (AOQ). The expected quality of outgoing product following the use of an acceptance sampling plan for a given value of incoming product quality. (2)

Average outgoing quality limit (AOQL). For a given acceptance sampling plan, the maximum AOQ over all possible levels of incoming quality. (2)

---

* Ref (1): Department of Defense, Mil-Std-109B, U.S. Government Printing Office. Ref. (2): ASQC, *Glossary and Tables for Statistical Quality Control;* reprinted by permission.

Batch. A definite quantity of some product or material produced under conditions that are considered uniform. *Note:* A batch is usually smaller than a lot or population. (2)

Bevel. An angled edge other than a right angle.

Bias in measurement. This bias occurs when one uses a micrometer to measure a part and tends to squeeze the thimble tighter to get the desired reading, not the *actual* reading.

Bolt circle (pattern). A collection of holes on a circular centerline related to a common center.

Bore. Making a hole using a lathe, boring bar, drill press, and so on, or the size of the hole (i.e., the bore).

Brass. An alloy, essentially made of copper and zinc.

Bronze. An alloy made mostly of copper and some tin.

Buff. Polishing, sometimes with a wheel made of fabric with abrasives added.

Burnish. To rub a material with a tool for compacting or smoothing or for turning an edge.

Calibration. A comparison of two instruments or measuring devices—one of which is a standard of known accuracy traceable to national standards—to detect, correlate, report, or eliminate by adjustment any discrepancy in accuracy of the instrument or measuring device being compared with the standard. (1)

Calibration interval. A specified amount of time between calibrations during which the accuracy of gages (or test equipment) is considered valid.

Case harden. Hardening the outer surface of steel by heating, then quenching. *See also* Quench.

Center drill. To drill holes in the ends of a part that is to be mounted on centers.

Characteristic. A property that helps to differentiate between items of a given sample or population. *Note:* The differentiation may be either quantitative (by variables) or qualitative (by attributes). (2)

Chill. For example, hardening the outer surface of a casting by quick cooling.

Coin. Forming a part, or a surface of a part, by stamping using a mold or die.

Cold rolling (steel). Steel that is rolled while it is cold, producing smooth, accurate stock.

Consumer's risk (beta risk), B. The probability of accepting a bad lot.

Counterbore. A larger diameter drilled on the same centerline of an existing hole.

Countersink. A bit or drill for making a countersink or a funnel-shaped enlargement at the outer end of a drilled hole. It is generally used to seat the head of a conical head screw, or make it easier to start a bolt.

Critical defect. This classification of defect is one that experience or judgment indicates is likely to cause unsafe conditions for those who use, maintain, or depend on the product; or a defect likely to prevent performance of the function of a major end-item.

Dedendum. The distance from the pitch diameter of an external thread to the root of the thread (for example).

Defect. A departure of a quality characteristic from its intended level or state that occurs with a severity sufficient to cause an associated product or service not to satisfy intended normal, or reasonably foreseeable, usage requirements. (2)

Defective (defective unit). A unit of product or service containing at least one defect, or having several imperfections that in combination cause the unit not to satisfy intended normal, or reasonably foreseeable, usage requirements. (2)

Defects per hundred units. The number of defects per hundred units of any given quantity of product is the number of defects contained therein divided by the total number of units of product, the quotient multiplied by 100 (one or more defects being possible in any unit of product). Expressed as an equation.

$$\text{defects per 100 units} = \frac{\text{number of defects} \times 100}{\text{number of units}} \quad (1)$$

Deviation. Written authorization, granted prior to the manufacture of an item, to depart from a particular performance or design requirement of a contract, specification, or referenced document for a specific number of units or specific period of time. (1)

Deviation (measurement sense). The difference between a measurement or quasi-measurement and its stated value or intended level. (2)

Die. A tool made of hard metal, used to cut or form a required shape in sheet metal, forgings, and so on.

Differential measurement. The use of a device that transforms actual movement into a known value, a dial indicating gage, for example.

Direct measurement. Where the standard is directly applied to the part, and a reading can be taken. A steel rule to measure length is an example.

Discrimination. The direct distance between two lines on a scale.

Discrimination rule. Never attempt to "read between the lines"; use an instrument with the proper discrimination for the measurement.

Draft. Tapered shapes in parts to allow them to be easily taken out of a mold or die.

FAO. Term meaning "to finish all over."

Fit. The degree of tightness (or looseness) between mating parts.

Fixture. A device for holding the workpiece.

Graduations. Accurate divisions on a scale or dial face.

Harden. Heating metal to a specified temperature and then quenching (cooling) it in oil or water, or other cooling material.

Heat treating. Changing the properties of metals by heating, then cooling them.

Inspection. The process of measuring, examining, testing, gaging, or otherwise comparing the unit with the applicable requirements. *Note:* The term *requirements* sometimes is used broadly to include standards of good workmanship. (2)

Inspection by attributes. Inspection whereby either the unit of product or characteristics thereof are classified simply as defective or nondefective, or the number of defects in the unit of product is counted, with respect to a given requirement. (1)

Inspection by variables. Inspection wherein certain quality characteristics of sample are evaluated with respect to a continuous numerical scale and expressed as precise points along this scale. Variable inspection records the degree of conformance or nonconformance of the unit with specified requirements for the quality characteristics involved. (1)

Inspection level. A feature of a sampling scheme relating the size of the sample to that of the lot. *Note:* Selection of an *inspection level* may be based on simplicity and cost of a unit of product, inspection cost, destructiveness of inspection, or quality consistency between lots. In some of the more widely used sampling systems, lot or batch size is grouped into convenient sets, and the sample size appropriate to each set is furnished for a designated *inspection level.* (2)

Inspection lot. A collection of similar units, or a specific quantity of similar material, offered for inspection and subject to a decision with respect to acceptance.

Inspection record. Recorded data concerning the results of inspection action.

Inspection, reduced. A feature of a sampling scheme permitting smaller sample sizes than are used in normal inspection. *Reduced inspection* is used in some sampling schemes when experience with the level of submitted quality is sufficiently good and other stated conditions apply. (2)

Inspection, tightened. A feature of a sampling scheme using stricter acceptance criteria than those used in normal inspection. *Tightened inspection* is used in some sampling schemes as a protective measure to increase the probability of rejecting lots when experience shows the level of submitted quality has deteriorated significantly. (2)

Interchangeable. Parts made to dimensions so that they will fit and function when interchanged among the same upper-level part number assemblies.

Knurl. A particular pattern of small ridges or beads on a metal surface to aid in gripping (e.g., a handle).

Lap. To produce a fine surface finish by sliding the lapping material (with or without abrasive powder) over the surface.

Lot. A definite quantity of a product or material accumulated under conditions that are considered uniform for sampling purposes. (2)

Lot size (N). The number of units in the lot. (2)

Major defect. This is a defect other than critical that may cause the product to fail, cause poor performance or shortened life, or prevent interchangeability.

Measured surface. That surface of a measuring tool that is movable and from which the measurement is taken: for example, the spindle of a micrometer.

Measurement error. Something that can be determined and is simply the difference between the measured value and the actual value.

Measurement pressure. Should be positive but not excessive; the most important factor (in most cases) is that the pressure used on the workpiece be the same pressure as that used when the tool was calibrated.

Measurement standard. A standard of measurement that is a true value and is recognized by all as a basis for comparison.

Measuring and test equipment. All devices used to measure, gage, test, inspect, diagnose, or otherwise examine materials, supplies, and equipment to determine compliance with technical requirements. (1)

Metrology. The science of measurement.

Minor defect. A defect that is not likely to reduce materially the usability of the unit. Example: A scratch on the side of a dishwasher.

Nonconforming unit. A unit of product or service containing at least one nonconformity. (2)

Nonconformity. A departure of a quality characteristic from its intended level or state that occurs with a severity sufficient to cause an associated product or service not to meet a specification requirement. (2)

Normal inspection. Inspection, under a sampling plan, that is used when there is no evidence that the quality of the product being submitted is better or poorer than the specified quality level. (1)

One hundred percent (100%) inspection. Inspection in which specified characteristics of each unit of product are examined or tested to determine conformance with requirements. (1)

Operating characteristic curve (OC curve). (1) For isolated or unique lots or a lot from an isolated sequence: A curve showing, for a given sampling plan, the probability of accepting a lot as a function of the lot quality (type A). (2) For a continuous stream of lots: A curve showing, for a given sampling plan, the probability of accepting a lot as a function of the process average (type B). (3) For continuous sampling plans: A curve showing the proportion of submitted product over the long run accepted during the sampling phases of the plan as a function of the product quality. (4) For special plans: A curve showing, for a given sampling plan, the probability of continuing to permit the process to continue without adjustment as a function of the process quality. (2)

Original inspection. The first inspection of a lot as distinguished from the inspection of a lot that has been resubmitted after previous nonacceptance. (2)

Parallax error. The apparent shifting of an object caused by shifting of the observer. An example is the act of viewing an indicator dial face from an angle, when it should be viewed directly.

Percent defective. The percent defective of any given quantity of units of product is 100 times the number of defective units of product contained therein divided by the total number of units of product, that is,

$$\text{percent defective} = \frac{\text{number of defectives} \times 100}{\text{number of units inspected}} \quad (1)$$

Precision. The closeness of agreement between randomly selected individual measurements or test results. (2)

Probability of acceptance (Pa). The probability that a lot will be accepted under a given sampling plan. (2)

Process average. The average percent of defective or average number of defects per hundred units of product submitted by the supplier for original inspection. (1)

Producer's risk (alpha risk), α. The probability of rejecting a good lot.

Quality. The totality of features and characteristics of a product or service that bear on its ability to satisfy given needs. (2)

Quality assurance. All those planned or systematic actions necessary to provide adequate confidence that a product or service will satisfy given needs. (2)

Quality control. The operational techniques and the activities that sustain a quality of product or service that will satisfy given needs; also, the use of such techniques and activities. (2)

Quality management. The totality of functions involved in the determination and achievement of quality. (2)

Quench. To cool a heated piece of metal suddenly by immersion in water, oil, or other coolants.

Random sampling. The process of selecting units for a sample of size n in such a manner that all combinations of n units under consideration have an equal or ascertainable chance of being selected as the sample. (2)

Reduced inspection. Inspection under a sampling plan using the same quality level as for normal inspection but requiring a smaller sample. (1)

Reference surface. The surface of a measuring tool that is fixed: for example, a micrometer anvil.

Rejection number. The minimum number of variants or variant units in the sample that will cause the lot or batch to be designated as not acceptable. (2)

Reliability. The probability that an item will perform its intended function for a specified interval under stated conditions. (1)

Resubmitted lot. A lot that previously has been designated as not acceptable and that is submitted again for acceptance inspection after having been further tested, sorted, reprocessed, and so on. (2)

Sample. A group of units, portion of material, or observations taken from a larger collection of units, quantity of material, or observations that serves to provide information that may be used as a basis for making a decision concerning the larger quantity. (2)

Sample percent defective. This percent is found by dividing the number of defectives found in the sample by the sample size and then multiplying by 100, or in a formula.

$$\frac{D}{n} \times 100$$

Sample size (n). One or more units selected at *random* from a lot without regard for their quality.

Sampling, double. Sampling inspection in which the inspection of the first sample of size n_1 leads to a decision to accept a lot, not to accept it, or to take a second sample of size n_2. (2)

Sampling, multiple. Sampling inspection in which, after each sample is inspected, the decision is made to accept a lot, not to accept it, or to take another sample to reach the decision. There may be a prescribed maximum number of samples, after which a decision to accept or not to accept the lot must be reached. (2)

Sampling plan. A statement of the sample size or sizes to be used and the associated acceptance and rejection criteria. (1)

Sampling, sequential. Sampling inspection in which, after each unit is inspected, the decision is made to accept the lot, not to accept it, or to inspect another unit. (2)

Sampling, single. Sampling inspection in which the decision to accept or not to accept a lot is based on the inspection of a single sample of size *n*. (2)

Screening inspection. Inspection in which each item of product is inspected for designated characteristics and all defective items are removed. (1)

Secondary reference standards. Standards used to perform standards/equipment calibration. They are a lower (second) level standard, usually compared calibrated to primary standards.

Specification limits. Limits that define the conformance boundaries for an individual unit of a manufacturing or service operation. (2)

Taper. Gradual decrease of diameter, thickness, or width in an object.

Testing. A means of determining the capability of an item to meet specified requirements by subjecting the item to a set of physical, chemical, environmental, or operating actions and conditions. (2)

Tightened inspection. Switching to tightened inspection is usually done when quality levels are observed to be getting worse.

Tolerance. The amount of permissible variation in a dimension.

Unit. A quantity of product, material, or service forming a cohesive entity on which a measurement or observation may be made. (2)

Universe. A group of populations, often reflecting different characteristics of the items or material under consideration. (2)

Variable gage. Gages that are capable of measuring the actual size of a part. A dial-indicating gage is an example.

Variables, method of. Measurement of quality by the method of variables consists of measuring and recording the numerical magnitude of a quality characteristic for each of the units in the group under consideration. This involves reference to a continuous scale of some kind. (2)

Working standards. Standards used to perform equipment calibration. They are a lower (third) level standard, usually compared calibrated to secondary standards.

Various Manufacturing Processes

Many different kinds of manufacturing processes are used to manufacture products for industry. This brief glossary is intended to cover a few of the most widely used processes in manufacturing companies today. Some processes are still manual (operator-controlled); others are controlled by computers (e.g., numerically controlled). Some processes are fully automated.

The quality technician should have a basic understanding of manufacturing processes so that elements of quality (such as process control, problem solving, corrective action, inspection, etc.) can be performed more successfully.

MACHINING PROCESSES (MATERIAL REMOVAL)

Broaching. Produces a variety of possible characteristics (e.g., round holes, other shaped holes, etc.). The cutting tool, which has multiple progressive cutting edges, is pushed or pulled through the part.

Chemical machining (CHM). Producing desired characteristics through removal of metal by controlled chemical attack or etching.

Counterboring. A process used to enlarge an existing hole to a specified depth. *Also see* Spotfacing.

Countersinking. A conical cutting tool produces a tapered enlargement at the opening of a hole. The purpose of a countersink can be for accepting conical bolt heads, ease of assembly, or producing a machining center.

Drilling. A rotary end-cutting tool (drill bit) produces a round hole (through or blind hole).

Electrical discharge machining (EDM). Producing holes, slots, and other characteristics in electroconductive materials through melting (or vaporization) by high-frequency electrical sparks.

Etching. Removal of material, for example, using acids.

549

Grinding. Metal removed from the workpiece using an abrasive grinding wheel.

Honing. Stock removed by the bonded abrasive grains of a honing stone or "stick." Used for finishing a characteristic to a fine surface finish and dimensional accuracy.

Lapping. A low-speed, low-pressure abrasive operation that produces extreme dimensional accuracy and a fine surface finish.

Milling. A machining process that removes metal with a rotating multiple-toothed cutter; each tooth in the cutter removes a small amount of material. There are a variety of milling processes and characteristics that can be milled.

Planing. A process in which metal is removed from surfaces in horizontal, vertical, or angular planes. The workpiece is moved against one or more single-point cutting tools.

Reaming. A rotary tool (reamer) that is often used to take a light cut on a hole (already produced by drilling). The reamer improves the dimensional accuracy and surface finish of the hole.

Shaping. A process in which metal is removed from surfaces in horizontal, vertical, or angular planes. A single-point cutting tool is moved against the fixed workpiece.

Spotfacing. A process, similar to counterboring, that produces a shallow flat seat at the opening of a hole to accept the head of a bolt.

Tapping (threads). Cutting internal threads via a cylindrical or conical thread-cutting tool.

Thread grinding. Producing internal or external screw threads by contact between a rotating workpiece and a rotating grinding wheel. The grinding wheel has been shaped to the desired thread form.

Thread milling. Cutting internal or external screw threads using a milling cutter in a thread mill.

Trepanning. A hole, groove, disk, cylinder, or tube is produced (usually by a single-point cutting tool) by being cut out of solid stock. For example, in trepanning, the cutting tool is surrounded by solid stock on both sides, hence a hole can be cut into the stock by cutting out a disk from the stock.

Turning. Machining external surfaces of revolution (e.g., diameters) where the workpiece is rotating and the cutting tool traverses along the axis of the workpiece. For internal surfaces, the term is "boring."

MACHINING PROCESSES (NO MATERIAL REMOVAL)

Roller burnishing. Pressure-rolling the surface of the characteristic by moving metal to improve dimensional accuracy and surface finish and to work harden a surface.

Thread rolling. A cold forming process that produces threads by rolling the impression of hardened steel dies into the surface of the workpiece.

METAL JOINING PROCESSES

Brazing. Used to join metals by heating them above 800 degrees F but below the melting point of the metals being joined. Clean surfaces are very important for successful brazing.

Welding. Localized bonding of metals by heating to suitable temperatures with (or without) the use of pressure and with (or without) the use of filler metal.

CASTING PROCESSES

Die casting. Molten metal is taken from a basin and forced under pressure into a metal mold (or die).

Mold Casting. Hot (fluid) metal is poured into molds and around cores without external pressure.

SHEET METAL PROCESSES

Bending. Forming materials via the use of a press break and various dies. The primary characteristic produced is a bend angle.

Drawing. Metal is pulled (or drawn) through suitable tools (e.g., dies) to form specific shapes or to reduce the size of the shape.

Shearing. Shearing operations include blanking, piercing, shaving, broaching, trimming, and slitting.

HEAT TREATING PROCESSES

Annealing. Heating and cooling applied typically for the purpose of inducing softening (or removing stresses).

Carburizing. Carbon induced into a solid iron-based alloy (followed by quenching) to harden materials.

Case hardening. Using heat treatment to change the composition of the outer layer of an iron-based alloy followed by appropriate thermal treatment.

Stress relieving. Reducing internal residual stresses in a metal object through heating the object to a suitable temperature and holding it at that temperature for a specific time.

Tempering. Reheating hardened (or normalized) steel to a temperature that is below its transformation temperature range, then performing any desired rate of cooling.

GLOSSARY C
Terms Used in Reliability

Reliability is the probability that an item will perform its intended function for a specified period of time under stated environmental conditions. Reliability goes beyond quality. It refers to the life of the product, not just whether the product is within (or not within) specifications.

RELIABILITY TERMS, DEFINITIONS, AND KEY TOPICS

Accelerated test. A test that exceeds the normal stresses of regular use of a product to shorten the time necessary to witness the response of the product to test conditions.

Bathtub curve. The life cycle of products, with respect to the probability of failures over time, results in a bathtub-shaped curve, as shown in Figure C.1. In the early phase of a product (called the *infant mortality* or *burn-in* phase), the probability of failures is expected to be high. In the *normal useful life* or *catastrophic* phase, the probability of failures is expected to be low. In the *wear out* phase, the probability of failures is expected to be high again.

Derating. The act of using materials, components, etc., in a design that are rated much higher in performance than necessary for that design. An example is using a component part that has 30,000 psi tensile strength when the stresses of the application only require 5,000 psi.

Failure rate. The reciprocal of the mean time between failures (see MTBF), expressed as lambda (λ). Failure rate is also expressed as the number of failures divided by the total operating hours.

$$\lambda = \frac{1}{\text{MTBF}} \qquad \lambda = \frac{\text{\# of failures}}{\text{Total operating hours}}$$

Failure rate is expressed in different ways, such as failures per hour, percent failures per 1,000 hours, or failures per 10^6 hours.

Limit design. In a limit design the worst possible stresses have been allowed in the design.

FIGURE C.1
Bathtub curve.

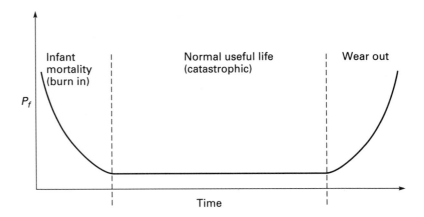

Mean time between failures (MTBF). The average time between failures of repairable equipment (usually expressed in terms of operating hours). The MTBF is the reciprocal of the failure rate.

$$\text{MTBF} = \frac{1}{\lambda}$$

For example, the failure rate (λ) of a unit is .0005 hours. The MTBF is:

$$\text{MTBF} = \frac{1}{\lambda} = \frac{1}{.0005} = 2{,}000 \text{ hours}$$

Parallel system reliability. Figure C.2 shows an example of a parallel system. The reliability of the system is calculated per the example shown in the figure: as one minus the product of the unreliabilities of the components. Parallel system reliability can be much better than the reliability of the components because of the redundant circuits. Note that it is also a more expensive design.

FIGURE C.2
Parallel system reliability.

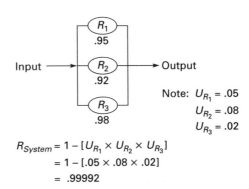

$$R_{System} = 1 - [U_{R_1} \times U_{R_2} \times U_{R_3}]$$
$$= 1 - [.05 \times .08 \times .02]$$
$$= .99992$$

Probability of failure (P_F). The probability that the unit will fail. The probability of failure is $1 - P_S$ (the probability of success).

Probability of success (P_S). The probability that the unit will not fail. The probability of success is $1 - P_F$.

Redundancy. Redundancy means that a unit has backup systems that will take over if a component fails. An example is a parallel system. If one electrical circuit fails, another circuit will take over to prevent system failure.

Series system reliability. Figure C.3 shows an example of a series system and the method of calculating the reliability of the system. The system reliability in a series connection is the product of the reliability of each component. The system reliability, then, is generally worse than the worst component.

FIGURE C.3
Series system reliability.

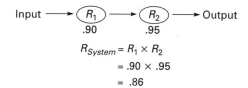

$$R_{System} = R_1 \times R_2$$
$$= .90 \times .95$$
$$= .86$$

Simulation. Establishing an exact replica of certain conditions in which the product will be used.

PRODUCT AVAILABILITY

Availability. The probability that equipment will not be down due to required maintenance actions. There are two aspects of availability: operational availability (A_0) and inherent availability (A_i).

Downtime. The time period that equipment is inoperative because of maintenance activities on that equipment. Downtime is the amount of time it takes to diagnose the problem plus the amount of time required to correct the problem.

$$\text{Downtime} = \text{Diagnosis} + \text{Repair}$$

Operational availability (A_0). The probability that equipment (used under stated conditions in an actual support environment) will operate satisfactorily at any time.

$$A_0 = \frac{\text{MTBM}}{\text{MTBM} + \text{MDT}}$$

where MTBM is the mean time between maintenance actions. *Note:* The MTBM becomes the MTBF when preventive maintenance downtime is zero (or not considered).

MDT is the sum of the mean preventive and corrective action time intervals. For example, if MTBM = 90 days and MDT = 60 days,

$$A_o = \frac{90}{90 + 60} = \frac{90}{150} = .6$$

Inherent availability (A_i). The probability that equipment (used under stated conditions in an ideal support environment) will operate satisfactorily at any given time. *Note:* Scheduled or preventive maintenance downtime is excluded.

$$A_i = \frac{MTBF}{MTBF + MTTR}$$

where MTBF is the mean time between failures; MTTR is the mean time to repair. For example, if MTBF = 3,000 hours and MTTR = 1.5 hours,

$$A_i = \frac{3,000}{3,000 + 1.5} = .9995$$

Overview of Nondestructive Evaluation (NDE) Methods

Nondestructive evaluation (NDE) was formerly called nondestructive testing (NDT). NDE is an inspection method for identifying flaws (or discontinuities) at or below the surface of the product. It is *not a test*. NDE is nondestructive inspection (e.g., does not destroy the part or make it unfit for use), and the specific method is typically identified by NDE specialists in the quality assurance department. Here we provide an overview of various popular NDE methods for inspection.

Destructive tests such as tensile strength tests and spectrographic tests, are often replaced by specific nondestructive tests. For example, hardness tests (such as the Brinell or Rockwell tests covered in Chapter 11) are nondestructive tests that can be used to determine tensile strength or ultimate strength requirements since there is a correlation between hardness and these variables. Many nondestructive tests are used in industry, more than can be covered here.

MAGNETIC PARTICLE INSPECTION

Magnetic particle inspection is a method for finding surface and certain subsurface flaws in ferromagnetic materials. The part must be ferromagnetic (able to be magnetized) or this method will not work. Magnetic particles are drawn into flaws by a magnetic field; then the outline of the flaw can be seen under black light.

LIQUID PENETRANT INSPECTION

Liquid penetrant inspection is used for finding surface flaws (e.g., surface cracks, laps, porosity, laminations, etc.) in solid nonporous metals. The liquid penetrants seep into flaws by capillary action. The method is used on both ferrous and nonferrous materials. For ferrous materials, liquid penetrant can be more sensitive than the magnetic particle inspection method.

ULTRASONIC INSPECTION

In ultrasonic inspection, beams of high-frequency sound waves (between 1 and 25 MHz) are introduced into the material being inspected to detect surface and subsurface flaws in the material. Cracks, laminations, and shrinkage cavities are some of the flaws that are detectable using ultrasonic methods. Ultrasonic inspection is one of the most widely used NDE methods.

EDDY CURRENT INSPECTION

Eddy current inspection is used to identify many different physical, structural, or metallurgical conditions in electrically conductive metals. The metals can be ferrous or nonferrous. It is an electromagnetic induction technique. Eddy current inspection can be used to measure/identify:

1. Electrical conductivity, magnetic permeability, grain size, heat treatment condition, and physical dimensions
2. Seams, laps, cracks, voids, and inclusions
3. Dissimilar metals
4. Thickness of nonconductive coatings on a conductive metal, or nonmagnetic metal coatings on a magnetic metal

RADIOGRAPHIC INSPECTION

Radiographic inspection methods are based on differential absorption of penetrating radiation in the form of x-rays (or gamma rays). Although the method can detect certain sizes of surface flaws, the primary use for this method is to find subsurface flaws (such as gas pockets, foreign materials, etc.). Ultrasonic methods can also be used. The three main advantages of using the radiographic method are:

1. The ability to detect internal flaws
2. The ability to detect significant variations in composition
3. Permanent recording of raw inspection data

HARDNESS TESTING

Hardness testing (such as the Brinell or Rockwell tests covered in Chapter 11) is a nondestructive method even though both types of tests leave an indentation in the product. It is assumed that the tests will be performed in areas of the product where the indentation will not render it unfit for use.

Bibliography

1. Griffith, Gary. *Measuring and Gaging Geometric Tolerances.* Prentice Hall.
2. Arter, Dennis. *Quality Audits for Improved Performance,* 2nd ed. ASQC Quality Press.
3. Juran, Joseph. *Quality Control Handbook.* McGraw-Hill.
4. *ASM Metals Handbook,* Vol. 3.
5. *Standard Handbook for Mechanical Engineers,* 8th ed. McGraw-Hill.
6. *Machinery's Handbook,* 20th ed.
7. Griffith, Gary. *Statistical Process Control Methods for Long and Short Runs,* 2nd ed. ASQC Quality Press.
8. A.I.A.G. *Statistical Process Control Reference Manual.* Automotive Industry Action Group.

APPENDIX
Tables

TABLE A–1

Decimal Equivalents for Common Fractions

FRACTION	DECIMAL EQUIV.	FRACTION	DECIMAL EQUIV.	FRACTION	DECIMAL EQUIV.
1/2	.500	15/32	.46875	27/64	.421875
1/4	.250	17/32	.53125	29/64	.453125
3/4	.750	19/32	.59375	31/64	.484375
1/8	.125	21/32	.65625	33/64	.515625
3/8	.375	23/32	.71875	35/64	.546875
5/8	.625	25/32	.78125	37/64	.578125
7/8	.875	27/32	.84375	39/64	.609375
1/16	.0625	29/32	.90625	41/64	.640625
3/16	.1875	31/32	.96875	43/64	.671875
5/16	.3125	1/64	.015625	45/64	.703125
7/16	.4375	3/64	.046875	47/64	.734375
9/16	.5625	5/64	.078125	49/64	.765625
11/16	.6875	7/64	.109375	51/64	.796875
13/16	.8125	9/64	.140625	53/64	.828125
15/16	.9375	11/64	.171875	55/64	.859375
1/32	.03125	13/64	.203125	57/64	.890625
3/32	.09375	15/64	.234375	59/64	.921875
5/32	.15625	17/64	.265625	61/64	.953125
7/32	.21875	19/64	.296875	63/64	.984375
9/32	.28125	21/64	.328125	1	1.00000
11/32	.34375	23/64	.359375		
13/32	.40625	25/64	.390625		

TABLE A–2
Conversion Chart: Millimeters to Inches

1millimeter = .03937 inches

MM	INCHES	MM	INCHES	MM	INCHES	MM	INCHES
.01	.00039	.34	.01339	.67	.02638	1.0	.03937
.02	.00079	.35	.01378	.68	.02677	2.0	.07874
.03	.00118	.36	.01417	.69	.02717	3.0	.11811
.04	.00157	.37	.01457	.70	.02756	4.0	.15748
.05	.00197	.38	.01496	.71	.02795	5.0	.19685
.06	.00236	.39	.01535	.72	.02835	6.0	.23622
.07	.00276	.40	.01575	.73	.02874	7.0	.27599
.08	.00315	.41	.01614	.74	.02913	8.0	.31496
.09	.00354	.42	.01654	.75	.02953	9.0	.35433
.10	.00394	.43	.01693	.76	.02992	10.0	.39370
.11	.00433	.44	.01732	.77	.03032	11.0	.43307
.12	.00472	.45	.01772	.78	.03071	12.0	.47244
.13	.00512	.46	.01811	.79	.03110	13.0	.51181
.14	.00551	.47	.01850	.80	.03150	14.0	.55118
.15	.00591	.48	.01890	.81	.03189	15.0	.59055
.16	.00630	.49	.01929	.82	.03228	16.0	.62992
.17	.00669	.50	.01969	.83	.03268	17.0	.66929
.18	.00709	.51	.02008	.84	.03307	18.0	.70866
.19	.00748	.52	.02047	.85	.03346	19.0	.74803
.20	.00787	.53	.02087	.86	.03386	20.0	.78740
.21	.00827	.54	.02126	.87	.03425	21.0	.82677
.22	.00866	.55	.02165	.88	.03465	22.0	.86614
.23	.00906	.56	.02205	.89	.03504	23.0	.90551
.24	.00945	.57	.02244	.90	.03543	24.0	.94488
.25	.00984	.58	.02283	.91	.03583	25.0	.98425
.26	.01024	.59	.02323	.92	.03622	26.0	1.02362
.27	.01063	.60	.02362	.93	.03661	27.0	1.06299
.28	.01102	.61	.02402	.94	.03701	28.0	1.10236
.29	.01142	.62	.02441	.95	.03740	29.0	1.14173
.30	.01181	.63	.02480	.96	.03780	30.0	1.18110
.31	.01220	.64	.02520	.97	.03819	31.0	1.22047
.32	.01260	.65	.02559	.98	.03858	32.0	1.25984
.33	.01299	.66	.02598	.99	.03898	33.0	1.29921

TABLE A-3
English-to-SI Conversions

LENGTH MEASURES

1 millimeter (mm) = 0.03937 inches (in)
1 centimeter (cm) = 0.3937 inches (in)
1 meter (m) = 39.37008 inches (in)
 = 3.2808 feet (ft)
 = 1.0936 yards (yd)
1 kilometer (km) = 0.62137 miles (mi)
1 inch (in) = 25.4 millimeters (mm)
 = 2.54 centimeters (cm)
1 foot (ft) = 304.8 millimeters (mm)
 = 0.3048 meters (m)
1 yard (yd) = 0.9144 meters (m)
1 mile (mi) = 1.609 kilometers (km)

AREA MEASURES

1 square millimeter = 0.00155 square inch
1 square centimeter = 0.155 square inch
1 square meter = 10.764 square feet
 = 1.196 square yards
1 square kilometer = 0.3861 square miles
1 square inch = 645.2 square millimeters
 = 6.452 square centimeters
1 square foot = 9.2903 sq. centimeters
 = 0.0929 square meters
1 square yard = 0.836 square meters
1 square mile = 2.5899 square kilometers

VOLUME (CAPACITY) DRYMEASURES

1 cubic centimeter = 0.061 cubic inch
1 liter = 0.0353 cubic foot
 = 61.023 cubic inches
 = 0.908 quarts
1 cubic meter = 35.315 cubic feet
 = 1.308 cubic yards
1 cubic inch = 16.38706 cubic centimeters
1 cubic foot = 0.02832 cubic meters
 = 28.317 liters
1 cubic yard = 0.7646 cubic meters
1 quart = 1.101 liters

VOLUME (CAPACITY) LIQUID MEASURES

1 liter = 1.0567 quarts
 = 0.2642 gallons
1 cubic meter = 264.2 gallons
1 quart = 0.9463 liters
1 gallon = 3.785 liters
 = 0.1337 cubic feet
 = 231 cubic inches
1 cubic foot = 7.48 gallons

WEIGHT MEASURES

1 gram (g) = 15.432 grains
 = 0.03215 ounce troy
1 kilogram (kg) = 2.2046 pounds (lb)
1000 kilograms (kg) = 1.1023 tons (t)
 = 1 metric ton (t)
1 ounce troy = 31.103 grains
1 pound (lb) = 453.6 grams (g)
 = 0.4536 kilograms (kg)
1 grain = 0.0648 grams (g)
1 metric ton = 0.9842 tons

TABLE A–4
Temperature Conversions

Use this table to convert Fahrenheit degrees (F°) directly to Centigrade degrees (C°) and vice versa. It covers the range of temperatures used in most hardening, tempering and annealing operations.

Lower, higher and intermediate conversions can be made by substituting a known Fahrenheit (F°) or Centigrade (C°) temperature figure in either of the following formulas.

$$F° = \frac{C° \times 9}{5} + 32$$

$$C° = \frac{F° - 32}{9} \times 5$$

F°	C°	F°	C°	F°	C°	F°	C°	F°	C°
− 160	− 107	340	171	840	449	1340	727	1840	1004
− 140	− 96	360	182	860	460	1360	738	1860	1016
− 120	− 84	380	193	880	471	1380	749	1880	1027
− 100	− 73	400	204	900	482	1400	760	1900	1038
− 80	− 62	420	216	920	493	1420	771	1920	1049
− 60	− 51	440	227	940	504	1440	782	1940	1060
− 40	− 40	460	238	960	516	1460	793	1960	1071
− 20	− 29	480	249	980	527	1480	804	1980	1082
0	− 18	500	260	1000	538	1500	816	2000	1093
20	− 7	520	271	1020	549	1520	827	2020	1104
40	4	540	282	1040	560	1540	838	2040	1116
60	16	560	293	1060	571	1560	849	2060	1127
80	27	580	304	1080	582	1580	860	2080	1138
100	38	600	316	1100	593	1600	871	2100	1149
120	49	620	327	1120	604	1620	882	2120	1160
140	60	640	338	1140	616	1640	893	2140	1171
160	71	660	349	1160	627	1660	904	2160	1182
18ʋ	82	680	360	1180	638	1680	916	2180	1193
200	93	700	371	1200	649	1700	927	2200	1204
220	104	720	382	1220	660	1720	938	2220	1216
240	116	740	393	1240	671	1740	949	2240	1227
260	127	760	404	1260	682	1760	960	2260	1238
280	138	780	416	1280	693	1780	971	2280	1249
300	149	800	427	1300	704	1800	982	2300	1260
320	160	820	438	1320	716	1820	993	2320	1271

TABLE A–5

Depth of Single Threads (Using the Turns Method)

(TPI)								
9	.555	.656	.777	.888	1.000	N/A	N/A	N/A
10	.500	.600	.700	.800	.900	1.000	N/A	N/A
11	.454	.545	.636	.727	.818	.909	1.000	N/A
12	.416	.500	.583	.666	.750	.833	.916	1.000
13	.384	.461	.538	.615	.692	.769	.846	.923
14	.357	.428	.500	.571	.642	.714	.785	.856
16	.312	.375	.437	.500	.562	.625	.687	.750
18	.277	.333	.388	.444	.500	.555	.610	.666
20	.250	.300	.350	.400	.450	.500	.550	.600
24	.208	.250	.291	.332	.374	.416	.458	.499
28	.178	.214	.250	.285	.321	.357	.393	.428
32	.156	.187	.218	.250	.281	.312	.343	.374
36	.138	.166	.194	.222	.249	.277	.305	.332
40	.125	.150	.175	.200	.225	.250	.275	.300

	NUMBER OF TURNS							
THREADS PER INCH (TPI)	13	14	15	16	17	18	19	20
16	.812	.875	.937	1.000	N/A	N/A	N/A	N/A
18	.721	.777	.832	.888	.943	1.000	N/A	N/A
20	.650	.700	.750	.800	.850	.900	.950	1.000
24	.540	.582	.624	.666	.707	.749	.790	.932
28	.464	.500	.535	.571	.607	.643	.678	.714
32	.406	.437	.468	.499	.530	.562	.593	.624
36	.360	.388	.415	.443	.471	.499	.526	.554
40	.325	.350	.375	.400	.425	.450	.475	.500

EXAMPLE:

A 1/2-16 THREAD PLUG GAGE (ONCE FULLY INSERTED INTO THE TAPPED HOLE) IS REMOVED AND 8 FULL TURNS ARE COUNTED. THIS MEANS THAT THE DEPTH OF FULL THREADS IS .500".

Subject Index

Formulas Index